W9-DFQ-376

Biodiesel

TP
359
.B46
P34
2008

#22850322/

Biodiesel
Growing a New Energy Economy

SECOND EDITION

GREG PAHL

Foreword by Bill McKibben

Chelsea Green Publishing Company
White River Junction, Vermont

LIBRARY
WAUKESHA COUNTY TECHNICAL COLLEGE
800 MAIN STREET
WITHDRAWN
PEWAUKEE, WI 53072

Copyright © 2004, 2008 Greg Pahl. All rights reserved.
No part of this book may be transmitted or reproduced in any form by any
means without permission in writing from the publisher.

Developmental Editor: Ben Watson
Project Manager: Emily Foote
Copy Editor: Susan Barnett
Proofreader: Ellen Brownstein
Book Designer: Peter Holm

Printed in the United States of America
First printing, August, 2008
    10 9 8 7 6 5 4 3 2 1   08 09 10 11 12 13

Our Commitment to Green Publishing

Chelsea Green sees publishing as a tool for cultural change and ecological stewardship. We strive to align our book manufacturing practices with our editorial mission and to reduce the impact of our business enterprise on the environment. We print our books and catalogs on chlorine-free recycled paper, using soy-based inks whenever possible. This book may cost slightly more because we use recycled paper, and we hope you'll agree that it's worth it. Chelsea Green is a member of the Green Press Initiative (www.greenpressinitiative.org), a nonprofit coalition of publishers, manufacturers, and authors working to protect the world's endangered forests and conserve natural resources.

   Biodiesel, Second Edition, was printed on 50-lb. Natures Book Natural, a 30-percent postconsumer-waste recycled, FSC (Forest Stewardship Council)-certified paper supplied by Thomson-Shore.

Library of Congress Cataloging-in-Publication Data

Pahl, Greg.
   Biodiesel : growing a new energy economy / Greg Pahl ; foreword by Bill McKibben. -- 2nd ed.
        p. cm.
   Includes bibliographical references and index.
   ISBN 978-1-933392-96-7
 1. Biodiesel fuels. I. Title.

   TP359.B46P34 2008
   333.95'39--dc22

2008021200

Chelsea Green Publishing Company
Post Office Box 428
White River Junction, VT 05001
(800) 639-4099
www.chelseagreen.com

This book is dedicated to the many people around the world who have worked tirelessly to make biodiesel a reality.

"The use of vegetable oils for engine fuels may seem insignificant today, but such oils may become, in the course of time, as important as petroleum and the coal-tar products of the present time."

RUDOLF DIESEL, 1912

CONTENTS

Acknowledgments — xi
List of Illustrations and Tables — xiii
Preface to the Second Edition — xv
Foreword — xix
Introduction — xxi

Part One *Biodiesel Basics*
  1. Rudolf Diesel — 3
  2. Vegetable-Oil Revival — 15
  3. Biodiesel 101 — 33
  4. Biodiesel's Many Uses — 54

Part Two *Biodiesel around the World*
  5. Europe, the Global Leader — 77
  6. Other European Countries — 105
  7. Africa and Asia — 136
  8. Australia, New Zealand, and the Americas — 169

Part Three *Biodiesel in the United States*
  9. A Brief History — 195
  10. The Main Players — 214
  11. Biodiesel Politics — 240
  12. Recent Developments — 263

Part Four *Biodiesel in the Future*
  13. Looking Ahead — 289

Organizations and Online Resources — 321
Bibliography — 328
Glossary — 329
Notes — 335
Index — 359

ACKNOWLEDGMENTS

While I worked on this project, I met and spoke with some wonderful, enthusiastic people who are committed to helping the global community free itself from its dependency on fossil fuels.

I would like to acknowledge the many people who were so generous with their time and advice. This book would not have been possible without people like Dr. Charles Peterson, University of Idaho, Moscow, Idaho; Dr. Thomas Reed, Golden, Colorado; Professor Leon Schumacher, University of Missouri at Columbia; Professor Jon Van Gerpen, University of Idaho, Moscow, Idaho; Bill Ayres, Kansas City, Kansas; Bob King, president, and Daryl Reece, vice president, Pacific Biodiesel Inc., Kahului, Hawaii; Jerrel Branson, former president, Best BioFuels, LLC, Austin, Texas; Tom Leue, president, Homestead Inc., Williamsburg, Massachusetts; John Hurley, Dog River Alternative Fuels, Berlin, Vermont; Joel Glatz, vice president, Frontier Energy, Inc., South China, Maine; Gene Gebolys, president, World Energy Alternatives LLC, Chelsea, Massachusetts; Dennis Griffin, chairman, Griffin Industries, Cold Spring, Kentucky; Gary Haer, West Central Coop, Ralston, Iowa; Joe Loveshe, esters sales manager, Columbus Foods, Chicago, Illinois; Jeff Probst, president and CEO, Blue Sun Biodiesel, Denver, Colorado; Bob Clark, sales manager, biodiesel division, Imperial Western Products, Coachella, California; John Plaza, founder and president, and Brian Young, Imperium Renewables, Inc., Seattle, Washington; John Williams, Scoville Public Relations, Seattle, Washington; Tom Bryan, editor, *Biodiesel* magazine, Grand Forks, North Dakota; Kyle Althoff, BBI International, Lakewood, Colorado; Joe Jobe, CEO, Charles Hatcher, former regulatory director, Manning Feraci, vice president of federal affairs, Scott Hughes, director of governmental affairs, Jenna Higgins, director of communications, and Amber Pearson, communications specialist, the National Biodiesel Board, Jefferson City, Missouri; Neil Caskey, former special assistant to the CEO, American Soybean Association, St. Louis, Missouri; Bill Schuermann, American Soybean Association, St. Louis,

Missouri; Gerhard H. Knothe, National Center for Agricultural Utilization Research, Peoria, Illinois; Alan Weber, MARC-IV Consulting, Columbia, Missouri; John Van de Vaarst, deputy area director, Beltsville Agricultural Research Center, Beltsville, Maryland; Professor Andrew Foley, United States Coast Guard Academy, New London, Connecticut; Nicole Cousino, San Francisco, California; Sarah Lewison, San Francisco, California; Maria "Mark" Alovert, San Francisco, California; Michael Sturtz, The Crucible, Oakland, California; Professor Phanindra Wunnava, Middlebury College, Middlebury, Vermont; Kyoko Davis, Middlebury College, Middlebury, Vermont; Terry Mason, North Wolcott, Vermont.

I also want to thank Raffaello Garofalo, secretary general, and Christian Schaible, European Biodiesel Board, Brussels, Belgium; Werner Körbitz, chairman, Austrian Biofuels Institute, Graben, Austria; Manfred Wörgetter, deputy director, head of Research Agricultural Engineering, Federal Institute for Agricultural Engineering, Wieselburg, Austria; Dr. Martin Mittelbach, Institute of Organic Chemistry, University of Graz, Graz, Austria; Lourens du Plessis, Bio/Chemtek division, CSIR, Pretoria, South Africa; and Darryl Melrose, Biodiesel SA, Merrivale, South Africa.

I offer my sincere thanks to all the wonderful folks at Chelsea Green Publishing who helped bring this book to completion. In particular I want to thank Ben Watson and Marcy Brant, my editors, who helped guide me through the process. And I especially want to thank Margo Baldwin for her unwavering enthusiasm and support for this project.

I also want to thank the Austrian Biofuels Institute, the Union for the Promotion of Oil and Protein Plants (Germany), the European Biodiesel Board, and the National Biodiesel Board, who supplied the information for most of the charts, and anyone else I may have forgotten to mention. All of your contributions, both large and small, are greatly appreciated.

Last, but by no means least, I want to thank my wife, Joy, for her help in chasing down obscure facts, proofreading, making suggestions, and generally putting up with me while I was trying to meet deadlines.

LIST OF ILLUSTRATIONS AND TABLES

1. World Proved Oil Reserves by Geographic Region as of January 1, 2007 *U.S. Energy Information Administration* — xxiii
2. Rudolf Diesel's 1897 engine *Deutsches Museum* — 10
3. Raw Materials for Biodiesel *Author's estimate* — 39
4. Shares of World Biodiesel Production (2000–2006) *European Biodiesel Board* — 79
5. Biodiesel Production Capacity in Germany, 2000–2007 *Various sources* — 91
6. 2006 World Palm Oil Production *U.S. Department of Agriculture* — 163
7. U.S. Biodiesel Demand *National Biodiesel Board* — 215
8. Commercial Biodiesel Production Plants *National Biodiesel Board* — 218
9. Imperium Grays Harbor *Imperium Renewables* — 231
10. Piedmont Biofuels Co-op Members *Piedmont Biofuels Cooperative* — 236
11. "Die Moto" *Chip Chipman* — 282
12. U.S. Biodiesel Production Capacity, 1999–2008 *National Biodiesel Board* — 284

PREFACE TO THE SECOND EDITION

What a difference four years make. In 2004, when I was researching and writing the first edition of *Biodiesel: Growing a New Energy Economy*, the biodiesel industry was on a roll. In Europe, the global leader in biodiesel production, most countries were investing heavily in promoting and encouraging the industry with a broad range of tax and other benefits, biodiesel refinery construction was proceeding at a breakneck pace, profits were generally high, and the mood was ebullient. In the United States, the industry had finally begun to enter the mainstream of public awareness, plant construction was ramping up, federal and state vehicle fleets were beginning to use significant quantities of biodiesel blends, public biodiesel pumps were starting to spread across the nation, and the industry finally succeeded in getting its landmark blending tax credit passed by Congress. In other countries around the world, biodiesel was just beginning to show up on the radar screen, but a number of early initiatives were already taking place in Africa, Asia, Australia, and South America. And, in general, most environmental groups and nongovernmental organizations were supportive of the industry.

As I write this in early 2008, a very different picture is emerging. Beset by record high feedstock prices, enormous production overcapacity, vanishing tax breaks, less expensive imports, and declining public support, the European biodiesel industry is struggling. Despite years of concerted effort, the EU uses less than 2 percent of nonfossil fuels in its transportation sector and is on track to miss meeting its ambitious biofuels targets in the years ahead. In just one year, since January 2007, the price of some biodiesel feedstock crops has doubled, driving the cost of biodiesel up over 50 percent to about $5.50 per gallon. Although prices for petrodiesel have risen sharply too, by May 2008 it was selling for only around $4.39 a gallon. The prices for both of these fuels have continued to climb since then, but biodiesel has generally remained more expensive. This trend is at odds with the conventional wisdom that higher petroleum prices make biofuels more competitive. In the United States, federal and state support for biodiesel remains strong, with a new package of tax credits and other incentives recently passed into law. But the U.S. biodiesel industry has been struggling with the same high feedstock and overcapacity issues that their European competitors are facing. And around the world, many biodiesel refineries (even new ones) are operating at reduced output or have closed to try to minimize their losses. Some have gone out of business. Also, as food prices have increased around the globe, at least partly in response to strong demand for biofuel feedstocks, concerns about food versus fuel have multiplied, causing a number of environmental groups and NGOs to begin to turn against biofuels, or at least some biofuels.

As a result of this dramatic shift in the biodiesel landscape, it became obvious that the first edition of this book needed a complete revision so it would better reflect current reality. I have tried to present as balanced and accurate a picture of the biodiesel industry around the world as possible. That picture is definitely mixed and sometimes confusing, and it's changing so fast that it is hard to really keep it in proper focus. It's probably best to consider this new second edition to be a snapshot of the state of the industry as of early 2008, just as the first edition was a snapshot of the industry in 2004. At the moment, governmental and public support in Europe seems to be on the wane while support in the United States, Brazil, India, China, and other nations seems to be on the rise or at least holding steady. Many biodiesel industry observers view the current turmoil as a temporary situation, and believe that, once supply and demand come back into balance, the industry will be poised for further expansion. Others aren't so sure. What is increasingly clear, however, is that if the industry is going to live up to its full potential, it will have to rely on a new "second generation" of inedible feedstocks that don't compete directly with the food sector.

Finally, no responsible member of the industry has ever claimed that biodiesel will solve all of our energy problems. It won't. They also readily acknowledge that biodiesel is not perfect. Every fuel has its advantages and its disadvantages, and biodiesel is no exception. But since there are very few viable alternatives in the liquid transportation fuel sector aside from ethanol (which has its own set of problems and limitations), biodiesel is better than nothing. The era of cheap petroleum is over, and as supplies tighten in the years ahead, biodiesel will at least offer some energy security as we try to adjust to this new reality. Along the way, biodiesel needs to find its niche in the broad range of renewable-energy strategies that will be needed as we move deeper into the twenty-first century and an increasingly uncertain future.

FOREWORD

For a hundred years we've powered our lives the easiest possible way—by tapping into those pools of hydrocarbons left by the aeons. That energy is highly concentrated, easily portable. It was just waiting there, for us to come along and scoop up, and so we did.

Now we've begun to realize that petroleum can't be what we use to power our next century. Not only is the supply starting to dwindle, and hence getting more expensive to extract (both in terms of money and blood), but it's also become clear just what a high environmental price we've paid for the convenience. If nothing else, the news that our planet is likely to warm five degrees this century should be enough to set us looking for new paths.

Biodiesel is one of the most intriguing of those new possibilities. For ancient biology, compressed by the weight of time into petroleum, it substitutes present-day biology: crops of soybeans and rapeseed and maybe even algae, grown by present-day farmers, processed into a diesel fuel substitute that works just fine in modern Volkswagens and Mack trucks and school buses—even in the oil-burning furnace down in the basement. It is potentially a truly sweet solution, offering a new market for hard-pressed local farmers even as it begins to help solve some of our most pressing environmental problems. Greg Pahl's book, though it is impeccably careful and well-documented, nonetheless brims over with a justified excitement at the possibility of this homegrown energy.

It also manages to raise the right questions (and raise them early enough) so that we can perhaps build a structure for this developing industry that serves local farmers and processors instead of simply corporate agribusiness giants: since this project is largely dependent on public funding for a jump start, that is not too much to ask. The proper scale is a key question—clearly, though, it's somewhere between the guy in his garage brewing old fryer oil into fuel and the Cargills and Archer Daniels Midlands of the world simply adding energy to their portfolios. By tempering his enthusiasm with reality on these questions, Pahl does an enormous service to the future.

He's also realistic about an important fact: Biodiesel is not going to solve our energy and environment woes by itself. It might replace 10 or 20 percent of our current diesel fuel use. That's good, but it's not a silver bullet against global warming. There are no silver bullets—every solution, from new lightbulbs to windmills to solar rooftops to higher mileage standards to biodiesel is going to get us a few percentage points of the way to where we need to go. Energy of the future will be far more diffuse, and harder to gather, than the current concentrated pools of oil. It's crucial that we recognize that fact and its key implication—that every ounce of effort put into new fuel supplies must be matched by an equal attention to conservation, to learning to live elegantly with less. This is completely possible: Europe, whose efforts on biodiesel Pahl chronicles in comprehensive fashion, manages to use about half as much energy per capita overall—with no discernable harm to living standards (in fact just the opposite is more likely).

This book excited me enormously. I can imagine the day when the schoolbuses on the rural rounds in my county run on the oilseed crops that their passengers can see out the window; when the ferries across Lake Champlain give off that slight french-fry whiff as they ply the waters; when the dairy farmers who are going broke raising milk have something else to grow. And when we can hold our heads a little higher, realizing that we're taking new responsibility for the energy we use. Pahl is a visionary, but a visionary with his feet firmly planted in the soil. May his vision flower, and soon!

BILL McKIBBEN

INTRODUCTION

We are running out of oil. This is an undeniable fact. The only remaining question is not *if* but *when*.

However, pumping the global oil barrel dry is not the immediate problem. The more imminent danger is what happens when demand outstrips supply: dramatic price increases for oil, followed by exponential price increases. Recent record-high gasoline prices of well over $4.00 per gallon in the United States are only a hint of what is coming in the near future. Since 2000, oil prices have quadrupled, and we can expect more of the same in the years ahead. The main problem is that, even now, annual demand for oil is four times greater than the volume of new oil reserves discovered (which peaked back in the late 1960s and early 1970s). Moreover, many of the highly publicized "huge" new recent oil finds will add only a few *days'* supply to the global oil market, which currently consumes 86 million barrels a day. By 2025, that demand is expected to climb to around 121 million barrels a day, according to the U.S. Energy Information Administration. This is troubling, since a number of industry experts now say that global production will probably never exceed 100 million barrels a day due to future supply limitations. What's more, as demand continues to increase, production from most of the largest existing oil fields is declining at a rate of 5 percent or more annually, and world production of oil is expected to peak sometime in the near future. Sooner or later, the line on the chart for demand that's heading up will cross the line on the chart for supply that's coming down. Whenever that occurs, we will have reached the critical "tipping point."

The utter chaos this would cause in the global economy is almost too frightening to contemplate. But we'd better take a good, close look, because the date for this scenario will arrive long before we actually run out of oil. And this date is coming much sooner than most people realize. The experts, as always, are divided on when it will take place. One of the most sanguine views, promoted by the U.S. Department of Energy, maintains that oil production won't peak until 2037. Many observers, though, feel this estimate is far too optimistic, especially considering the huge increase in demand from countries like China, which overtook Japan as the world's second-largest oil consumer in 2003. Renowned petroleum geologist Colin Campbell estimates that global extraction of oil will peak around 2011. Geophysicist Kenneth Deffeyes says the date for maximum production was December 2005.

Although these predictions may sound alarmist, the massive 2004 accounting scandal involving oil giant Royal Dutch/Shell and the subsequent 22 percent (4.35 *billion* barrels) cut in the company's petroleum reserve estimates is viewed by some industry experts as just the tip of the iceberg of overinflated reserve figures. If these calculations are as misleading as some people suspect, then we are in for a very rough ride in the very near future. Regardless of who turns out to be right about the timing of oil's tipping point, most of today's middle-aged population probably will live to see the consequences. They may wish they hadn't. And as for the younger generations, well, they just may be out of luck.

Nearly two-thirds of the world's proven oil reserves are located in the 13 countries that make up the Organization of the Petroleum Exporting Countries (OPEC): Algeria, Angola, Ecuador, Indonesia, Iran, Iraq, Kuwait, Libya, Nigeria, Qatar, Saudi Arabia, the United Arab Emirates, and Venezuela. The fact that the majority of these nations are located in the increasingly unstable Middle East is not especially reassuring. And the fact that the former colonial powers of Europe, and more recently the United States, have been involved in numerous wars in the region in an ongoing attempt to protect the uninterrupted flow of oil is even more troubling and does not augur well for the future. Figure 1, a chart from the U.S. Energy Information Administration, shows where most of the world's remaining reserves are located.

Most of the global economy is so dependent on the price of oil that any substantial price increase means big trouble. The huge price increases for oil predicted for the time after we reach the tipping point will lead unquestionably to much higher food prices, as well as to disruptions in the increasingly globalized system of food production and distribution. As a result, millions of people, particularly in struggling Third World coun-

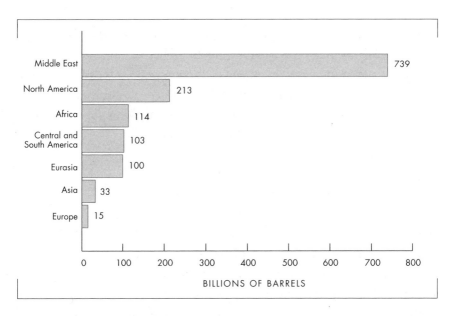

Figure 1. World Proved Oil Reserves by Geographic Region as of January 1, 2007 U.S. Energy Information Administration

tries, almost certainly will starve. Prices for most other goods also will rise dramatically. A severe global economic depression, massive unemployment, political instability, and more international conflict are almost certain to follow. It's conceivable that industrialized society as we know it could collapse—even before we are decimated by the consequences of global warming caused by the use of fossil fuels.

This second scenario, which environmental activists and scientists have been warning about for many years, is beginning to look increasingly likely. Eight of the warmest years in recorded weather history have taken place in the past 10 years, according to the United States Historical Climate Network. And it's not just environmentalists who are worried. In February 2004, a secret Pentagon study that the Bush administration had tried to suppress was leaked to the press. The study warned about the possible consequences of sudden climate changes caused by global warming and offered a terrifying picture of a global catastrophe costing millions of lives due to wars and natural disasters. The threat to worldwide stability posed by global warming far surpasses that of terrorism, according to the study. More recent studies and reports by scientists from around the world have simply confirmed the bad news about global warming. These scientists have added increasingly urgent calls for immediate action to try to halt—or at least slow down—some of the worst effects of global warming. Will the international economic meltdown be caused by oil prices or by global warming? Either way, we're in deep trouble.

RUNNING OUT OF TIME

Is there any hope of avoiding this terrible scenario? Perhaps, but time is rapidly running out. In fact, it may have already run out. If so, we won't know for sure until it's too late to do anything about it. Ominously, a number of nations that have been working tirelessly to get the United States to sign the 1997 Kyoto agreement to cut greenhouse gas emissions recently have begun to suggest that, in the absence of meaningful progress, the world should "prepare for the worst." Unfortunately, many politicians—especially in the United States—are still in denial and have long resisted doing anything about cutting greenhouse gas emissions, saying that it "would damage the U.S. economy" or "cost jobs." What these head-in-the-sand politicians fail to grasp is that if the worst-case scenarios about global warming come to pass, the catastrophic economic damage caused will be far worse than any possible costs of trying to meet the targets of the Kyoto accord. This kind of shortsighted stupidity imperils all of us, and it is particularly galling to the millions of concerned citizens around the world (and even in the United States) who rightly view the United States as the world's largest consumer of energy—and the biggest polluter. This country, with just 4.5 percent of the global population, consumes about 25 percent of the world's energy and releases roughly 25 percent of the global carbon dioxide emissions, according to the Energy Information Administration. Without active, constructive participation by the United States to help deal with these critical issues, the rest of the global community is effectively stymied. The recent increased public awareness in the U.S. about global warming is an encouraging sign, but that awareness must be translated into immediate action if it is going to do any good.

It's going to take a huge cooperative effort on the part of the entire global community to wean ourselves from our present addiction to fossil fuels in general, and petroleum in particular. Many alternative strategies that rely on various forms of renewable energy, including wind, solar, and geothermal, are gaining in popularity. But running most vehicles directly on these forms of renewable energy, given present technology, is not practical. "And even though technology allows for greater fuel efficiency than ever before, cars and other forms of transportation account for nearly 30 percent of world energy use and 95 percent of global oil consumption," according to the Worldwatch Institute.

Ninety-five percent of global oil is consumed for transportation! This statistic points right to the heart of the problem and emphasizes why we are so vulnerable. Some people suggest that compressed natural gas (CNG) could serve as a substitute for oil. But using CNG would, at the very least, require expensive retrofitting of vehicles. Unfortunately, natural gas, though cleaner-burning, is still a fossil fuel, and natural gas prices have been soaring while world reserves are shrinking almost as fast as those for oil. Hydrogen-powered fuel cells are widely viewed as the ultimate solution for the transportation sector. But hydrogen is produced by the electrolysis of water, and the electricity required to produce enough hydrogen to fuel all the cars in the United States would require four times the present capacity of the national grid. (Unfortunately, the present grid relies on nonrenewable energy sources for more than 90 percent of its capacity and is barely able to keep up with present demand.) What's more, there is no infrastructure for the production and delivery of the vast amounts of hydrogen that would be required. The transition to hydrogen is, at best, a long, long way off.

AVAILABLE NOW

In the meantime, there is one liquid fuel that is both renewable and can be used in a wide range of vehicles without any modifications to the engines. That fuel is biodiesel. For many years, farmers, environmentalists, and renewable-energy advocates in Europe and the United States have been promoting the use of biodiesel as an alternative to at least a portion of the petroleum-based diesel fuel market. But it wasn't until the attacks on the World Trade Center on September 11, 2001, that most Americans finally began to realize the implications of their overreliance on oil—especially Middle Eastern oil—and its heavy economic, political, social, and military costs. The U.S.-led military campaign in Afghanistan and the subsequent ill-advised invasion of Iraq, with its terrible and costly aftermath, have added urgency to the movement seeking to wean the United States from its almost total addiction to petroleum-based fuels.

While many strategies are currently being pursued to accomplish that end, biodiesel is one of the most intriguing and, until fairly recently, one of the least publicized in the United States.

In most of Europe, the general public is aware of biodiesel due to long-standing governmental support, but the sudden emergence of biodiesel from relative obscurity in the United States in the past few years has taken many Americans by surprise. While other renewable-energy strategies such as solar, wind, ethanol, and fuel cells have received most of the media attention, a group of Midwestern soybean farmers and other entrepreneurs have been quietly building biodiesel production capacity and infrastructure. At the same time, a number of federal and state agencies and independent organizations have been testing and evaluating biodiesel performance and setting up fuel production standards, laying the foundation for a new sustainable energy industry. Based on that firm foundation, the biodiesel industry is experiencing dramatic growth, both in production capacity and in the number of retail fuel outlets across the country. Despite that growth, however, many people still have only a vague idea of what biodiesel is, and fewer still understand that it can be used for more than fueling diesel-powered cars or pickup trucks.

What, exactly, is biodiesel, and why is it generating so much excitement? First, it's important to understand that even though *diesel* is part of its name, pure biodiesel does not contain petroleum diesel or fossil fuel of any sort. Biodiesel generally falls under the category of *biomass*, which refers to renewable organic matter such as energy crops, crop residues, wood, municipal and animal wastes, et cetera, that are used to produce energy. More specifically, *biofuels*, a subcategory of biomass, includes three energy-crop-derived liquid fuels: ethanol (usually referred to as grain alcohol), methanol (usually referred to as wood alcohol), and biodiesel. Technically a fatty acid alkyl ester, biodiesel can be easily made through a simple chemical process from virtually any vegetable oil, including (but not limited to) soy, corn, rapeseed (canola), cottonseed, peanut, sunflower, avocado, hemp, and mustard seed. But biodiesel also can be made from recycled cooking oil or animal fats. There have even been some promising experiments with the use of algae as a biodiesel feedstock.

And the process is so simple that biodiesel can be made by virtually anyone, although the chemicals required (usually lye and methanol) are hazardous and need to be handled with extreme caution.

Best of all, biodiesel feedstock sources are renewable and can be produced locally. Although fossil fuels were formed over millions of years (and are being rapidly depleted), biodiesel feedstocks can be created in just a few months. The source of the energy content in biodiesel is solar energy captured by feedstock plants during the process of photosynthesis, inspiring some to refer to the fuel as "liquid solar energy." And the plants grown to make more biodiesel naturally balance most of the carbon dioxide emissions created when the fuel is combusted, eliminating a major contributing factor to global warming. What's more, the resulting fuel is far less polluting than its petroleum-based counterpart; biodiesel produces lower quantities of cancer-causing particulate emissions, is more biodegradable than sugar, and is less toxic than table salt. And because it can be produced from domestic feedstocks, biodiesel reduces the need for foreign imports of oil, while simultaneously boosting the local economy. No wonder there is so much enthusiasm, especially in much of the agricultural community, about biodiesel: Farmers can literally grow their own fuel.

Although biodiesel may be a relative newcomer to the United States, in Europe it has enjoyed widespread acceptance as a vehicle fuel (as well as a heating fuel in some countries) due to deliberate government tax policies that, until very recently, have favored its use. In most European countries, diesel-powered cars represent a substantial proportion (40 to 50 percent) of the automobile fleet, and biodiesel powers many of them. The expansion of the European Union in May 2004 offered a good deal of additional potential for continued growth in biodiesel production and use in the so-called new accession nations. There is good potential for the industry in many other countries around the world as well.

WHAT'S INSIDE

This book describes biodiesel's dramatic growth and its potential to help pave the way for an eventual transition from fossil fuels to a wide range of renewable energy sources. It is divided into four parts. In Part One we begin this exploration with a look at biodiesel basics. We'll travel back in time to the nineteenth century to discover the roots of the device that has made the whole biodiesel movement possible—the diesel engine— and we'll learn about its tireless inventor, Rudolf Diesel, and his renewable-fuel vision that is only now being realized. Then we'll fast-forward to the 1970s to see how and why biodiesel was developed. We'll also go through the biodiesel production process and examine the many (sometimes quirky) raw materials from which biodiesel can be made, and we'll explore the fuel's environmental impact. Finally, we'll focus on the modern diesel engine and the many uses of biodiesel fuels.

In Part Two we'll travel to Europe, the leader in global biodiesel production, to see why Germany, France, and Italy combined produce nearly five times more biodiesel than the United States, and how they managed to gain such a decisive lead. We'll also visit the other European nations that have expanded their biodiesel industries and check out some of the more interesting developments there as well. Then we'll travel around the world to see other new biodiesel projects from India to Australia and Japan to Brazil.

In Part Three, our biodiesel odyssey finally arrives in the United States. We'll look at some early biofuels projects sponsored by Henry Ford and see how the fledgling biofuels industry was eliminated by Big Oil and other business pressures. Then we'll follow the revival of biofuels after the 1973 OPEC oil crisis and the subsequent development of the biodiesel movement in the United States. We'll also learn about some of the main players in today's biodiesel industry and take a look at the complex world of biodiesel politics. Then we'll hear from a number of high-profile celebrities about their use of biodiesel and also see what many different people all across the country are doing with this renewable biofuel today.

In Part Four we take stock of the current state of the industry and explore some of the key issues that need to be confronted if it is going to be successful. We'll also look at a number of concerns that some observers have raised about the dramatic growth of the industry. And, finally, we'll look into a crystal ball with some of the industry's key players to try to envision the outlines of where the biodiesel industry may be headed in the future.

Although biodiesel is not the single solution to all our energy problems, it can be part of the transition from our current near-total dependency on fossil fuels, while at the same time creating jobs, assisting farmers, reducing harmful emissions, and promoting greater energy security. Biodiesel, along with a wide range of other renewable energy strategies, coupled with dramatically increased energy efficiencies and conservation, should help us to meet our energy needs as we make that transition. However, in order to achieve that goal, we need to increase the pace of the shift to a new energy economy—today, immediately, while there's still time.

A few technical notes: The word *diesel* is used throughout this book but has different meanings in different contexts. In an attempt to clarify this, the word *diesel* (in lowercase) refers to the engine or the industry. The capitalized word *Diesel* refers to the engine's inventor, Rudolf Diesel, or his family. The word *petrodiesel* refers to the petroleum-based diesel fuel that has been used to run diesel engines since the early 1900s. And, finally, *biodiesel* refers to the renewable biofuel that is the main focus of this book.

Also, the tangle of various units of measure used in different countries around the world was a particular challenge. Metric and U.S. equivalents for these units of measure are provided here and there to help readers in most countries make sense of the statistics cited. But in an attempt to avoid overkill, not every single unit of measure has been converted. Unless otherwise noted, tons refer to metric tons and gallons refer to U.S. gallons.

PART ONE

Biodiesel
Basics

1

Rudolf Diesel

The death of Rudolf Diesel is shrouded in mystery. Diesel, the inventor of the engine that now bears his name, was sailing from Antwerp, Belgium, to Harwich, England, to attend the annual directors' meeting of the British Diesel Company and the groundbreaking ceremonies for a new diesel engine plant at Ipswich when he disappeared. On the evening of September 29, 1913, Diesel embarked on the steamship *Dresden*, accompanied by fellow director George Carels and Alfred Luckmann, the chief construction engineer of the diesel engine company.

After the ship departed from Antwerp, the three men had dinner, followed by a leisurely stroll on deck, where they talked for some time. "Dr. Diesel was in the very best spirits. The conversation was cheery and buoyant," Carels is quoted as saying in a September 30, 1913, *New York Times* article about the incident. Around 10 P.M., Diesel and his companions decided to turn in for the night, and they descended to their cabins. Diesel stopped briefly in his cabin, but then came down the corridor to shake Carels's hand and wish him good night. "I will see you tomorrow morning," Diesel said to Carels.

But the next morning, when Diesel failed to show up for breakfast, Carels and Luckmann went to his cabin to find him. They knocked on the door, and when there was no response, they looked inside. The cabin was empty. A quick search of the cabin revealed that Diesel's bed had not been slept in, although his nightshirt was neatly laid out and his watch had been left where he could see it from the bed. Nothing appeared to have been disturbed. The *Dresden's* crew was immediately notified and a quick search of the ship was conducted, but Diesel could not be found. The missing man's hat and overcoat, however, were discovered neatly folded beneath the afterdeck railing. After the ship arrived in Harwich, the German vice-consul was notified, and he instigated a thorough search of the entire ship—to no avail. Diesel had completely vanished. "As his landing ticket had not been given up, we felt certain that Dr. Diesel could not have landed, and, as he was not to be found on board, we could not think otherwise than that he disappeared overboard in the course of the night," Carels said. But how? And why? Was it an accident? Had he committed suicide? Or had he, perhaps, been murdered?

There did not appear to be any logical explanation for Diesel's disappearance. The sea had been relatively calm (for the English Channel) on the night in question, eliminating the possibility that Diesel had been accidentally swept overboard during a storm. The press had a field day. Headlines such as "Diesel Murdered by Agents from Big Oil Trusts" or "British Secret Service Eliminates Diesel" quickly appeared in European newspapers. In the United States, "Creator of the Diesel Engine Executed as a Traitor to Secure U-boat Secrets" and "Inventor Thrown into the Sea to Stop Sale of Patents to British Government" screamed from page one. It also was suggested by some that "embittered enemies" or "business competitors" of Diesel might have arranged his death. There was at least some credibility to the suspicions raised about the U-boat issue, since Diesel had already convinced the French Navy in 1904 to install his engine in a number of their submarines. And one of his less publicized reasons for traveling to England was to try to do the same for the British Navy.

On October 10, 1913, the Belgian pilot steamer *Coertsen* spotted a body floating in the North Sea. Despite stormy weather, a small lifeboat was lowered into the water and the body was taken on board long enough to recover a few personal effects from its clothing. Following the normal custom of the time, the badly decomposed body was quickly returned to the sea. The personal items were brought to the Dutch port of Vlissingen, where they were subsequently identified by Diesel's son, Eugen, as having belonged to his father.

The exact circumstances surrounding Diesel's death will never be known. There was no official investigation or trial, not even a hearing by the ship's company. Since the body was never recovered again, there was no autopsy and no official coroner's report. Diesel did not leave a suicide note. He did not leave a will. And he apparently never mentioned any intention to kill himself. He did, however, leave an enigmatic cross penciled in his notebook after the date for September 29.

Shortly after Rudolf Diesel disappeared, his wife, Martha, opened a bag that her husband had given to her just before his ill-fated voyage, with directions that it should not be opened until the following week. She discovered 20,000 German marks in cash and a number of financial statements indicating that their bank accounts were virtually empty. At the time, the general public had assumed that the Diesel family was enormously wealthy, but in fact they were almost broke.

HIS EARLY YEARS

Rudolf Christian Karl Diesel was born in Paris on March 18, 1858. His parents, Theodor and Elise Diesel, were both originally from the Bavarian city of Augsburg. Rudolf had an older sister, Louise, born in 1856, and a younger one, Emma, born in 1859. Theodor, a leatherworker who had emigrated from Germany around 1850, ran a small leather shop in Paris. Rudolf was a shy but intelligent child who spent a lot of time drawing. Early on, Rudolf exhibited an interest in things mechanical. He dismantled the family's cherished cuckoo clock but was embarrassed when he was unable to reassemble it.[1] As he grew older, Rudolf excelled in his studies, learned to speak three languages—German, French, and English—and spent much of his spare time in the Conservatoire des Arts et Metiers, the oldest technical museum in Paris, which housed an interesting collection of mechanical inventions. Rudolf completed his elementary schooling and was awarded a bronze medal for academic excellence. His future looked extremely promising.

Then fate stepped in and severely disrupted the family's normal routine.

On July 19, 1870, France declared war on Prussia. Ironically, from the start the war did not go well for the overconfident French, and as one humiliating military defeat followed another, waves of refugees began to flow into Paris. The Diesels (and anyone else of German descent) were increasingly viewed with suspicion, and they were soon expelled from France. On September 6, the family boarded a steamer from Dieppe to England. A few days later the refugees arrived in London, where, after much difficulty, Theodor Diesel finally managed to find a low-paying job. Rudolf, now 12 years old, was able to resume his studies in a London school. In his spare time, he visited the British Museum and the South Kensington Museum, where he often lingered in the science and engineering exhibits. His previous interest in machinery was strongly reinforced. The family, however, struggled to make ends meet, and when Christoph Barnickel, the husband of a cousin of Theodor's named Betty Barnickel, who lived in Augsburg, Germany, offered to take Rudolf in, his parents quickly agreed.

After his arrival in Augsburg, Rudolf was enrolled in a three-year program at the local technical school, where he happily immersed himself in chemistry, art, and—especially—a machine shop with a forge. He was in mechanical heaven. At the age of 14, Rudolf wrote to his parents to tell them that he had decided to become an engineer. After the Franco-Prussian War ended in early 1871, the rest of the Diesel family returned to Paris. Rudolf remained in Augsburg, where he subsequently graduated at the head of his class, and then journeyed to Paris to be with his family. But his stay there was cut short by the sudden death of his older sister, Louise, possibly from heart failure. Grief-stricken, the Diesels eventually agreed to let Rudolf return to Augsburg when Professor Barnickel and his wife, who had become quite fond of their young relative, again offered to take him in.

In October 1873, Rudolf enrolled in a mechanical engineering program in Augsburg, and as usual he ended up at the top of his class. It was in the physics laboratory at this school that Rudolf saw a mysterious device that almost certainly had a profound impact on his life. The device, a pneumatic lighter, was a small cylinder constructed much like a bicycle tire pump, but with a barrel made of glass, in which air was greatly compressed, causing a substantial rise in temperature—and a hot spark that could be viewed through the glass. The seeds of an idea had been planted. But it would take many years before those seeds germinated and grew into a practical invention.

THE "HEAT ENGINE"

Upon graduation, Rudolf was awarded a scholarship to attend the Technische Hochschule München (literally the Technical High School of Munich, but generally known as the Munich Institute of Technology), where he studied thermodynamics under Professor Carl von Linde (the inventor of the first reliable and efficient compressed-ammonia refrigerator). By the time Diesel arrived in Munich, his interests had expanded to include the arts, linguistics, and social theories of the day. It was in Munich that Diesel first began to seriously consider the possibility of developing what he called a "heat engine," one that would not need a spark plug to create combustion (the general idea perhaps inspired by the pneumatic lighter he had seen in Augsburg, and unquestionably by some of Professor von Linde's lectures). At the time, the steam engine was the main source of industrial power, but the engine's poor heat-utilization efficiency—which ran around 10 percent or less—bothered Diesel, who abhorred even the slightest waste. Steam engines were also extremely expensive and consequently tended to benefit large and wealthy companies that could afford them over small businesses that could not. This deeply troubled Diesel's newfound social conscience. From this time on, Diesel combined his interests in engineering and thermodynamics with his desire to help small business owners, which became one of the motivations for his subsequent work in developing his own engine.

After graduating from the Munich Institute of Technology in 1880, Diesel became an apprentice at the large and successful Gebrüder Sulzer Maschinenfabrik (Sulzer Brothers Machine Works) in Winterthur, Switzerland, where he gained a good deal of practical mechanical experience. Diesel immediately impressed his supervisors, and it wasn't long before he was sent to Paris, where Carl von Linde (who was also a businessman) helped arrange a job for him in a new refrigerator manufacturing plant that was under construction. Diesel quickly became a refrigeration expert and was soon promoted to plant manager. In 1881, Diesel obtained his first patent (for the production of ice in glass containers), and a short time later he met his future wife, Martha Flasche, an attractive, blue-eyed German blonde who was a governess for a wealthy Parisian family. With his patent in hand, Diesel searched for a manufacturer for his ice-making machine. He soon came to an agreement with the Maschinenfabrik Augsburg (Augsburg Machine Works, known today as Maschinenfabrik Augsburg-Nuerenberg, or MAN AG), which began producing parts for his invention in 1883, the same year he married Martha in Munich. The next year, their first child, Rudolf Jr., was born, followed by a daughter, Heddy, and a few years later another son, Eugen.

In 1885, Diesel began to experiment with an ammonia-fueled engine in his own laboratory. Although the experiments were ultimately a failure, these early trials set the stage for his later engine development work. In 1890, Carl von Linde helped Diesel obtain a new franchise to distribute and sell Linde's refrigerators in northern Germany. As a consequence, the young Diesel family moved to Berlin. Not long after arriving in Berlin, Diesel came up with a concept for a new engine design, which was finally patented on February 28, 1892. The following year, he published his now-famous paper "Theory and Construction of a Rational Heat Engine to Replace the Steam Engine and Contemporary Combustion Engine." The paper described his invention as a "compression ignition engine" that could burn virtually any fuel, ignited not by a spark but by the extremely high temperature caused by highly compressing the air before the injection of fuel into the cylinder. It was a radical idea. Now the challenge was to turn his idea into an engine that actually worked.

A PROTOTYPE

Diesel searched for someone to help him build a prototype. He again turned to the Augsburg Machine Works, which agreed to assist in exchange for future engine sales rights for most of Germany. He also received the backing of the Friedrich Krupp Werke at Essen (the German steel producer better known for locomotive and arms manufacture in later years) in exchange for all German sales rights not already given to the Augsburg company. In addition, Diesel received financial backing from the Sulzer Brothers company in Switzerland, in exchange for an option on Swiss patent rights. The first prototype, constructed in 1893, consisted of a single 10-foot black iron cylinder with a large flywheel at its base. The engine, which was initially fueled by kerosene and later by gasoline, ran briefly, but it had to be shut down when the pressure gauge exploded.[2] Diesel and one of his assistants narrowly escaped injury. Although the engine had run, it had not generated enough power to sustain its own operation. Diesel went back to the drawing board and extensively redesigned the engine. He also continued to experiment with various fuels, and a second prototype actually ran briefly under its own power on February 17, 1894. But numerous technical problems still needed to be resolved. During this period, Diesel, who had always been prone to headaches, began to suffer severe migraine symptoms.

"When I began constructing my engine in the early nineties, the existing method was a total failure," Diesel later admitted. "The enormous pressures generated in my machine, the friction between moving parts, the magnitude of which had not been seen before, forced me to minutely examine the stress on each single organ and to delve extensively into material science."[3] This examination process delayed Diesel's project for more than a year. In addition to making major redesigns of numerous components, he also continued his fuel experiments with gasoline, kerosene, lighting gas, and heavy oils.[4] Low-grade kerosene was the final fuel of choice. But the time he took to fine-tune his engine design was well spent. On the last day of 1896, a third prototype engine, containing many refinements over Diesel's earlier models, was started up and thoroughly tested. The 600-pound, 57-inch-high engine ran smoothly and was considered a success.

As word of the invention spread, Diesel began to sell licenses to entrepreneurs who wanted to manufacture his engine outside of Germany.

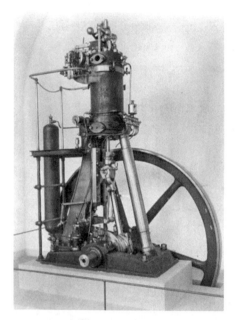

Figure 2. Rudolf Diesel's 1897 Engine Photo
Deutches Museum

Most of these manufacturers added technical refinements of their own to Diesel's basic design, and they were supposed to share their improvements with the other licensed manufacturers. But, in practice, this often did not occur due to competitive pressures.

One of the most prominent of these businessmen was Adolphus Busch, the U.S. beer baron of St. Louis, Missouri. After hearing about Diesel's new engine, Busch sent a trusted friend and technical advisor, Colonel Edward D. Meier, to Augsburg to study the engine and report his findings. Meanwhile, Busch settled into a luxury hotel in Baden-Baden, where he awaited the outcome of the investigation. Meier's eventual report was extremely favorable, and Rudolf Diesel was invited to Baden-Baden. Diesel quickly sensed that the high-spending Busch wanted the manufacturing license for his engine in the United States badly and wouldn't be stingy about paying for it. Diesel boldly suggested a fee of one million German marks (roughly equivalent to US$5 million today). Busch didn't bat an eyelash and calmly wrote out a check on the spot. Busch subsequently had the first commercial engine built in the United States on the Diesel patent installed in his St. Louis brewery.

THE MUNICH EXHIBITION

In 1898, four of Diesel's engines—each one built by a different German manufacturer—were put on display at the Munich Power and Machinery Exhibition. The special pavilion housing the exhibit was one of the highlights of the fair. Although the engines, which were constantly attended by technicians, ran without serious difficulties most of the time, there were bad days. One of the worst occurred when Prince Ludwig (the future King Ludwig III) of Bavaria visited the exhibit. The prince had heard that Diesel's engines were considerably less noisy than other engines of the day and wanted to see (and hear) them for himself. Unfortunately, none of the engines were working that day. The prince entered the pavilion and inspected the silent, inoperative engines. After a lengthy pause, he commented that the new engines were *indeed* quiet.

As more and more companies lined up to buy licenses, Rudolf Diesel quickly became a wealthy man. To help manage the exponential growth of the business, Diesel established the Allgemeine Gesellschaft für Dieselmotoren (General Society for Diesel Engines) on September 17, 1898. The new company paid him 3.5 million marks in cash and stocks for the rights to his engine and assumed control of all future development work. The following month, still suffering from severe headaches and exhaustion from his years of constant work, Diesel was urged by his doctors to enter a private sanitarium in Munich for a rest cure. But he was still too close to his business, and after the unsuccessful stay at the sanitarium, his physicians recommended a further course of treatment in a remote castle in the Alps. Although the second rest cure was a pleasant diversion in spectacularly beautiful surroundings, it was not entirely successful either.

It was during this turbulent period in Diesel's life that he had a sumptuous mansion built for his family. Located at Maria-Theresia-Strasse, 32 in a fashionable suburb of Munich, this extravagant indulgence featured decorated vaulted ceilings, the most modern plumbing and electrical wiring, a marble fireplace in every room, and the finest French and Italian furnishings. It cost a fortune. Under normal circumstances the cost would not have been a problem for a multimillionaire, but Diesel, who had been so successful with his own business, seemed to have an uncanny talent for losing money in other people's ventures. He managed to lose 300,000 marks in a questionable Balkan oil scheme that went bad. He also was frequently engaged in expensive legal proceedings to protect his patents or defend his business interests. One legal judgment (involving some of his real estate speculations) cost Diesel 600,000 marks.

THE PARIS EXPOSITION

In 1900, a smaller version of Diesel's engine manufactured by the French Otto Company was shown at the Paris Exposition. To demonstrate the engine's ability to operate on various fuels, it ran on peanut oil. Approximately 50,000 people attended the exposition, and some of them were undoubtedly confused when they followed the scent of what they thought was a food concession and ended up standing in front of Diesel's engine chugging away on the peanut oil. Some people cite this exhibit as evidence of Diesel's early support of renewable biofuels. But the idea for the use of peanut oil appears to have come from the French government, which had its peanut-growing African colonies in mind. Diesel did conduct similar vegetable-oil tests in later years, however.[5] In any case, the exhibit was extremely popular, and the engine won the exhibition's coveted Grand Prize. If nothing else, the prize was good publicity. By the end of 1901, some 31 companies were licensed to build and sell diesel engines and sales franchises covered 11 countries. A survey conducted the following year counted 359 diesel engines in use worldwide.

Unfortunately, Diesel's vision of an engine that would empower small businesses to compete with large industrial companies didn't come to pass in his lifetime. His early models required a separate air compression unit to inject the fuel, and the engines themselves were too large and expensive for all but the heaviest of industrial uses. If anything, Diesel's engine simply increased the competitive advantage of the big companies. However, oceangoing ships had plenty of space in their engine rooms, and Diesel's engine design was adopted for marine use as early as 1903. Despite these early successes, Diesel's persistent bad luck with investments continued to multiply. By 1905, his estimated losses came to more than 3 million marks.

Ironically, Diesel came to hate his extravagant mansion in Munich, which he began to describe as an "outrageous mausoleum." But he did make it somewhat more practical by converting the entire upper floor into an engine research and construction office. Martha Diesel was not thrilled. Nevertheless, it was in his home office that Diesel planned a smaller version of his engine that would be suitable for automotive use. The "petite model," as he called it, earned a Grand Prize at the Brussels World's Fair in 1910. But this engine still had many shortcomings and was not immediately adopted for its intended use.

VEGETABLE-OIL VISION

In his later years Diesel was unquestionably a strong advocate for the use of renewable fuels such as seed oils, and this almost certainly was an outgrowth of his original desire to create an engine that could be operated almost anywhere on almost any fuel. In 1911 he said, "The diesel engine can be fed with vegetable oils and would help considerably in the development of agriculture of the countries which use it." The following year, despite the growing dominance of petroleum-based fuels, Diesel continued to make the case for vegetable-oil fuels. In an April 13, 1912 speech in St. Louis, Missouri, Diesel described the recent changes to the fuel atomizers in his engines that allowed the use of castor oil, palm oil, lard, or other natural fuels. He went on to say, "The use of vegetable oils for engine fuels may seem insignificant today, but such oils may become, in the course of time, as important as petroleum and the coal-tar products of the present time. . . . Motive power can still be produced from the heat of the sun, always available, even when the natural stores of solid and liquid fuels are completely exhausted." Prophetic words indeed.

In November 1912, Diesel published a book, *Die Entstehung des Dieselmotors*, describing the development of his engine. The income from the book helped offset a little his mounting debts from continuing bad investments and various lawsuits, which by this time totaled nearly 10 million marks. Nevertheless, in early 1913 Rudolf and Martha vacationed in Italy. After their return, the summer of 1913 was not distinguished by any dramatic events, except that Rudolf seemed to be sinking into a deepening depression over his financial difficulties. On September 26, Diesel took a slow train to Belgium. He spent several days at the elegant Hotel de la Poste in Ghent, where he wrote a confused letter to his wife.

Unfortunately, he misaddressed it, and she did not receive it until about five days after he disappeared. Then, on the evening of September 29, he boarded the steamship *Dresden* at Antwerp for his ill-fated voyage to England.

What really happened in the middle of the English Channel on that dark night in September 1913? No one will ever know for sure. Perhaps Diesel, distraught over his impending financial calamity, decided to take his own life. Diesel had been suffering from heart trouble for a number of years, and it's possible that he was struck by a sudden heart attack and fell overboard. Or, perhaps, the dark speculations about embittered enemies or international intrigue were correct. . . . Whatever happened, with Diesel's passing the world lost one of the most brilliant engineering minds of the day—and the father of what has become the biodiesel industry as well.

Vegetable-Oil Revival

In 1912, a year before Rudolf Diesel's death, there were more than 70,000 diesel engines operating around the world. But because of their size and weight and the fact that they ran best at a relatively constant speed, about 95 percent of these engines were restricted mainly to stationary uses in factories and electrical-generation plants.[1] Nevertheless, the use of diesels to power ships was beginning to gain some momentum. By 1912, approximately 365 ships and at least 60 heavy merchant vessels were equipped with diesel engines.[2] In the last five years of his life, Diesel had focused much of his research and development work on trying to build a successful diesel-powered locomotive—with little success. Nevertheless, he was convinced that his engine would revolutionize the railroad industry around the world. Diesel was right, but he didn't live to see it happen. The first diesel-powered locomotive ran on the Prussian and Saxon State Railways in 1914,[3] and early diesel locomotives in the United States finally made their debut around 1925. However, diesels did not seriously displace steam engines on most railroads until after World War II.

The automotive industry was particularly slow to adopt Diesel's engine. But eventually smaller, lighter engines with self-contained direct-injection pumps were developed, and in 1924 Germany's MAN AG was the first company to power a truck with one of these new engines. In the United States, the Mack Truck Company began experimenting with diesel engines around 1927, and by 1935 Mack was producing and selling their own diesel-powered trucks.[4] In the early 1930s, diesel trucks were soon followed by diesel-powered public buses in Berlin and many other cities. In the fall of 1936, Daimler-Benz unveiled their first large-scale-production diesel passenger car at the International Automobile Exhibition in Berlin. Although diesel-powered taxis subsequently became popular, particularly in Europe, due to their low fuel consumption and rugged dependability, the general driving public found diesel-powered pleasure cars of this era to be sluggish and unexciting. But for heavy commercial purposes, such as factories, power plants, oceangoing ships, railway locomotives, and construction equipment, diesel engines quickly became the motive power of choice—and remain so to this day.

Although Diesel had originally intended that his engine would be able to run on a variety of fuels, including whale oil, hemp oil, and coal dust, early on he opted for less expensive petroleum-based kerosene, which was plentiful and relatively cheap in the late nineteenth century. In the early twentieth century, as diesel engines came into wider use, the petroleum industry developed so-called diesel fuel, a lower-grade by-product of the gasoline refining process. Diesel engines were modified to burn this new fuel. But the more Diesel's engine was altered to burn this dirtier, petroleum-based fuel (petrodiesel), the further it diverged from its inventor's vision of biofuel flexibility, which he had especially promoted in the last few years of his life. And as the number of diesel engines that were manufactured to run on petrodiesel grew exponentially, the less likely it seemed that Diesel's vegetable-oil vision would ever be realized.

EARLY EXPERIMENTS

Despite the development of an almost total reliance on petrodiesel, the idea of using vegetable oil as an alternative source of diesel fuel was not completely forgotten. There were a number of experiments on the use of vegetable oils beginning in the 1920s and continuing through the early 1940s, particularly among European nations with tropical colonies—especially African colonies. In 1937, a Belgian patent was granted to G. Chavanne of the University of Brussels for the use of ethyl esters of palm oil (which would now be described as a type of biodiesel). The following year, a commercial passenger bus operated between Brussels and Louvain on palm-oil ethyl ester. The test was reportedly a success.[5] Shortly before World War II, there were some experiments in South Africa in which heavy farming equipment was operated with fuels derived from sunflower seed oil. But these promising experiments were abandoned in favor of making synthetic liquid fuel from coal, due to South Africa's abundant coal reserves.[6]

During World War II, vegetable oils were used as emergency fuels by various nations when normal supplies of petroleum-based fuels were disrupted. Brazil utilized cottonseed oil in place of imported diesel fuel. Argentina, China, India, and Japan all used some form of vegetable oil during the war years. The Japanese battleship *Yamato* is reported to have used refined, food-grade soybean oil as bunker fuel.[7] But after the war, with the return of steady supplies of cheap petroleum oil, virtually all research on vegetable-oil fuels ceased.

THE OIL EMBARGO

Fast-forward to 1973. On October 6 of that year, Egyptian military forces attacked Israel across the Suez Canal, while in the north Syrian troops simultaneously attacked from the Golan Heights. Initially taken by surprise, the Israeli army pulled back, but then with assistance from the United States and other Western nations, the Israelis counterattacked and eventually reversed all the early Arab gains and ended up occupying additional territory. Although a cease-fire was concluded in November, the Arab nations were humiliated by their losses—and angry. In retaliation for the United States's and other Western countries' support for Israel, the Organization of Petroleum Exporting Countries (OPEC) initiated an oil embargo against the West in general, and the United States in particular.

The embargo sent shock waves through the global economy and dramatically inflated energy prices worldwide. Overnight, the price of a barrel of oil rose from $3 to more than $5. By the end of 1974, the price was more than $12 a barrel. In the United States, President Richard Nixon instituted voluntary gasoline rationing in December 1973 and urged homeowners to turn down the thermostats on their heating systems. In response to the crisis, the U.S. Congress approved the construction of the trans-Alaskan oil pipeline (completed in 1977 at a cost of $8 billion), which was designed to supply 2 million barrels of oil a day. Much of the Western world was hit by a severe economic recession, but Western Europe was especially hard-hit because it had very limited domestic supplies of oil at the time.

In 1979, the revolution in Iran that resulted in the ousting of the U.S.-backed shah precipitated yet another global energy crisis. Oil prices again doubled, sending the industrial world into an economic tailspin.

The price increases and fuel shortages of the 1970s and early 1980s spurred interest in the development of alternate fuels around the world, but especially in Western Europe and the United States, where the economic damage from the turmoil in the petroleum markets was most severe. One of these alternate fuels was ethanol. During both World War I and World War II, the United States and European nations used alcohol fuels as supplements to petroleum-based fuels. In the 1920s and 1930s, a considerable amount of ethanol, fermented from corn and generally referred to as "gasohol," had been produced in the United States for use as a vehicle fuel. But this effort ended with the low petroleum prices of the 1940s, and as ethanol production facilities closed, gasohol disappeared from the market. But in 1979, in response to the ongoing international oil crisis, ethanol-gasoline blends were reintroduced to the U.S. market by several oil companies, which promoted the ethanol blends as "gasoline extenders" and octane enhancers. And although ethanol was being promoted as a gasoline additive, a separate but related line of research into alternate diesel fuels made from vegetable oils was being conducted in a number of countries.

Some of the earliest documented experiments with vegetable oils as diesel fuels in the United States took place in the early 1950s at Ohio State University, where a "dual fuel" project was conducted with cottonseed oil and corn oil blended with petroleum diesel.[8] But with the resumption of seemingly endless supplies of petroleum fuels after World War II, these studies soon gathered dust. After the shocks to the oil industry of the 1970s, however, interest in alternate diesel fuels was revived, especially in Europe and the United States. Two long-term research programs stand out in particular: one in Austria, the other in Idaho. A third program in South Africa showed a lot of initial progress, but was not sustained.

THE AUSTRIAN CONNECTION

About a year after the energy crisis of 1973, representatives from the Austrian Federal Institute of Agricultural Engineering (Bundesanstalt für Landtechnik, or BLT) began preliminary discussions on alternate biofuels for diesel engines (particularly for farm tractors) with the diesel engine developer AVL-List GmbH in Graz, Austria. Those discussions were the impetus for a series of experiments that mixed various vegetable oils with petrodiesel. In 1975, Manfred Wörgetter, who had just graduated from the Mechanical Engineering Department of the Technical University of Graz, took a job at BLT in Wieselburg, Austria, and a short time later began the fuel experiments. "In my work I concentrated on bench and field tests of farm diesel engines, while my boss, Josef Pernkopf, focused on production issues," Wörgetter (who is now deputy director and head of Research Agricultural Engineering at BLT) recalls. "The main aim of our work was to ensure the supply of fuel for farm tractor engines in the event of another oil crisis."[9]

In 1976, and again in 1978, a small, old tractor was bench- and field-tested with various mixtures of linseed oil and petrodiesel. The initial results from the mixing tests showed that the viscosity (thickness) of the vegetable oil needed to be reduced and that the use of the oil would considerably increase engine maintenance costs due to excessive deposits on a number of internal engine components. In virtually all cases, if the vegetable oils were used over a long period of time, the engine could be seriously damaged. In early 1979, BLT published an article in a regional journal recommending that engine tests with linseed oil be discontinued. "We concluded that either the fuel needed to be adapted to the needs of the engine, or the engine needed to be adapted to the fuel," Wörgetter says. "We decided to adapt the fuel."

At this point the BLT researchers reviewed plant-oil production world-wide and concluded that a vegetable-oil strategy needed to be oriented toward local climate and agricultural conditions. "We concentrated our efforts on bench tests with different mixtures of rapeseed and sunflower oil with fossil diesel," Wörgetter says. "After four hundred operating hours with a farm tractor using 50 percent rapeseed oil in diesel we had to stop because of an engine failure." During this time, a copy of a South African newsletter containing information about work that had been done with sunflower methyl ester (a form of biodiesel) was brought to Wörgetter's attention. When he met some of the members of the South African research team at an energy conference in Berlin, Germany, in October 1981, Wörgetter's discussions with them convinced him that producing methyl ester from vegetable oil was probably the best approach for adapting the fuel for a diesel engine. The report of the use of sunflower oil methyl esters at this conference is generally viewed as marking the redis-covery of what came to be known a few years later as biodiesel.[10] In 1981, BLT published an article in a highly regarded Austrian journal on the idea of chemical modification of vegetable-oil fuels. At about the same time, some early feasibility studies on the use of fatty acid methyl esters as fuel in diesel engines and heating boilers were conducted in France, but most observers credit the sustained Austrian research work as being the main foundation for the subsequent European biodiesel industry.

Independently, and at the same time, a number of chemists at the University of Graz were having discussions in their laboratory about the possibility of using vegetable oil as fuel. They wanted to start a research project on that topic and contacted the Austrian Ministry of Agriculture to see if it was interested. The ministry informed them that BLT in Wieselburg was already working on the same topic, so a decision was made to contact BLT. Professor Martin Mittelbach, an organic chemist from the university, accompanied by his boss, Professor Hans Junek, traveled to

Wieselburg. "When we arrived, they told us they were having a bad day because the engine in one of their test tractors had stopped running," Mittelbach recalls. "They said, 'Maybe you have come at the right time; you are chemists, tell us what's going on.'"[11] Mittelbach says that he offered some immediate observations about the probable causes of the engine difficulties. He also agreed to investigate the problem further and promised to let the BLT researchers know if he discovered anything interesting. This meeting marked the beginning of a long-lasting friendship between the researchers. Mittelbach returned to Graz and gave a good deal of thought to the best ways of modifying the vegetable oil chemically to make a fuel that would not cause operational problems in diesel engines.

Mittelbach, who had received his PhD in 1979 at the University of Graz, says that shortly before he visited BLT in Wieselburg, someone in his department had mentioned the idea of using vegetable oil as a diesel fuel. "If you are an organic chemist you know that vegetable oil and diesel fuel have a totally different chemical structure, so it was hard to believe that vegetable oil could be used as a fuel," he recalls. "For nonchemists all oils may sound like they are the same, but they really aren't, so this looked like an interesting idea." Mittelbach checked the available literature to see what was available and didn't find much. (Mittelbach heard about Rudolf Diesel's work with vegetable oils a few years later, but admits that this did not influence his early experiments because he was unaware of it at the time.)

THE PROCESS

Mittelbach persisted in his research and then conducted a series of chemical laboratory experiments and tests using rapeseed oil (also known as canola oil), which turned out to be successful. In chemical terms, vegetable oil normally is composed of three fatty-acid molecules linked to a glycerol molecule; combined in this way, the fatty acids and glycerol are referred to as a triglyceride. Mittelbach relied on a standard chemical process known as *transesterification*, in which the vegetable oil, an alcohol, and a catalyst were mixed, resulting in the removal of glycerin from the vegetable oil, to make the oil thinner. The products of the reaction are alkyl esters (biodiesel) and glycerin. The main point of the process was to produce alkyl esters that would flow through and combust properly in modern diesel engines without leaving damaging internal-engine deposits. "The fatty acids and fatty acid esters were not new," Mittelbach notes. "They were well known in chemical literature for making detergents, and then nonionic detergents, which are used for dishwashing and so on. What was new was their use as a diesel fuel."

Another goal of this early research was to change the production process, because the existing oleochemical route for the production of fatty acid methyl esters was too complicated and expensive. (Oleochemicals, which are used in a wide range of products, are derived from biological oils or fats.) The researchers wanted to find a low-temperature and low-pressure process that would perform the reaction without all the expensive equipment used by the traditional oleochemical industry. The research work finally resulted in a patent application that described the first method for cheap biodiesel production.

After their initial success with the experiments, Mittelbach and Wörgetter agreed that they needed a larger supply of biodiesel in order to conduct additional tests on the fuel. "At the university we were experienced chemists, but we were not set up to produce large quantities in the laboratory," Mittelbach says. "We would make a small batch and then have to clean everything up and do it again and again. It just was not practical." After receiving a small grant for the necessary equipment, the researchers were able to produce biodiesel in 30- to 40-liter (8- to 10- gallon) batches. When they had accumulated several hundred liters, they were ready for more extensive laboratory and field tests. The researchers looked for a financial partner for the field testing. The Austrian tractor manufacturer Steyr stepped forward. Steyr agreed to run the field tests while BLT would conduct the engine bench tests. The subsequent tests were successful, with no engine damage detected, and Wörgetter and Mittelbach published their findings in 1982.

Although the tests using rapeseed oil had been a technical success, it was quickly realized that the oil was too expensive as a feedstock to be used for biodiesel, since petrodiesel prices were much lower. At the time, very little rapeseed was being grown in Europe and most rapeseed had to be imported from abroad. Consequently, in 1983 Mittelbach began to look for less expensive feedstocks, and he soon discovered that there was a fairly large supply of used cooking oil available. He conducted some preliminary experiments using the waste cooking oil and found it to be a viable source for making biodiesel.

Mittelbach says that the most exhilarating moment for him, personally, was when he decided to run his own car on biodiesel. "During the early tests we made the biodiesel and sent it out to be tested; I didn't actually see the tests," Mittelbach recalls. "The tests were working, and so I thought, 'Well, what about passenger cars?' At the time I had a diesel car, and the most exciting moment for me was when I put around 20 liters of biodiesel in the fuel tank. I started the engine and checked the exhaust gases, which smelled like burned fat. It was very exciting. At that time, nobody could give a guarantee that your car would run on biodiesel. So, the next day, when it started and ran normally, I could see that it was going to work. We could have found a test car somewhere, I suppose, but I just decided to use my own car. That was the best proof that it worked. It was easier to convince other people that it worked because I could say that I was using it in my own car without any problems."

REAL-WORLD TESTS

Now that the viability of biodiesel for use in a diesel engine had been demonstrated, the researchers decided that it was time to expand the tests to larger numbers of tractors in real-world settings. The idea was to persuade local farmers to use biodiesel in their tractors, creating a closed energy loop in the agricultural community. The farmers could grow and use their own fuel and be independent from the fluctuations in the international oil market. But in order to do that, much larger quantities of biodiesel needed to be produced. The solution was a pilot plant.

But the researchers didn't have the money for a venture of this size and were forced to look for an outside source of funds. The Wieselburg and Graz groups began searching for funding. Finally Mittelbach was successful, finding support in an unlikely place: the petroleum industry. Ironically, the main financial backing came from OMV AG, Austria's largest oil and petrochemical industry, according to Mittelbach. At the time, OMV was building a new oil pipeline from the Adriatic Sea to Vienna, and the money they paid for the acquisition of the land traversed by the pipeline was put into a fund to be used for the benefit of the farmers whose land was taken for the right-of-way. "The money was to be used for some sort of energy projects, and the farmers decided that the biodiesel pilot plant was one of the projects where the money should be spent," Mittelbach says. The pilot plant (which no longer exists) was constructed in 1985 at the Silberberg Agricultural College in Styria, Austria, and it was capable of producing approximately 500 tons (142,500 gallons) of biodiesel from rapeseed oil annually.

Once the pilot plant was up and running, the researchers discovered they had another problem. The farmers who were supposed to run the field tests had suddenly become skeptical. "We wanted to conduct fleet tests, and trying to find ten or twenty farmers who were willing to make the tests was not easy," Mittelbach says. "They were very anxious that something might happen to their tractors." But eventually, with a lot of gentle persuasion, the farmers agreed to the tests, which, happily, turned out to be successful.

In 1986, BLT was contacted by an official from Gaskoks, a large Austrian energy company, who wanted to help finance a pilot project, titled "Biodiesel," designed to create a basis for supplying the agricultural sector with fuel produced from rapeseed oil. The project, which was subsequently carried out at BLT in Wieselburg, also was financed and assisted by the Austrian Federal Ministry for Agriculture. During the project (which cost in excess of US$1 million), fleet tests were conducted on 35 farm tractors from 10 different producers between 1987 and 1989. A series of bench tests, extensive emissions measurements, and investigations of the engine oil, along with the results of other Austrian research programs, finally led to a proposal for a biodiesel quality standard.

Thanks to the cooperative efforts of all the biodiesel researchers, the Austrian Standardization Institute (Osterreichische Normungsinstitut, or ON) established a working group that succeeded in creating the first biodiesel standard in the world (ON C 1190). "All subsequent standards have been based on the foundation set by this groundbreaking work," says Wörgetter.[12]

TURNING WASTE INTO FUEL

In 1987, building on the previous successes with rapeseed oil and Mittelbach's earlier experiments with used cooking oil, a series of engine and emissions tests using methyl ester from waste cooking oils was conducted in cooperation with AVL-List GmbH in Graz. "We used the same chemical reactions that we had used with the rapeseed oil," says Mittelbach. "We also measured the emissions, because if you are using an alternative fuel this is an important issue. The soot and particulate emissions with the vegetable-oil biodiesel were far lower—about 50 percent lower—than with the petroleum diesel. And the used frying oil was even better than rapeseed oil biodiesel. That was really surprising; we had not anticipated that. This was a very important step for the further development of biodiesel." In 1988, the results of the tests were published in *Science News* and the *Journal of the American Oil Chemists' Society*. The international scientific community began to sit up and take notice. In the same year, the first Austrian patent for the transesterification process was taken out by Mittelbach and several collaborators. Although the research teams in Graz and Wieselburg worked well together in the early years, eventually a certain amount of competition developed between them, especially from around 1987 to 1990, according to Wörgetter. But since then, a good, cooperative relationship has been reestablished, according to members of both teams.

In 1989, an Austrian-government-supported research project, "High-Quality Fuel from Waste Cooking Oils," was launched. A short time later, large-scale esterification experiments followed using 100 percent waste cooking oil, conducted by Mittelbach and others, in cooperation with the Technical University of Graz. But not all the emphasis was on used cooking oil. In 1990, the first farmers' cooperative in Asperhofen (near Vienna), with approximately 290 farmer members, began commercial production of biodiesel made from rapeseed as well as sunflower oil. In the same year, large-scale tractor engine tests were finally completed by BLT at Wieselburg, which convinced major tractor manufacturers such as John Deere, Ford, Massey-Ferguson, Mercedes, Same, and others to issue engine warranties for biodiesel use.[13] The warranties were a major step forward in the development of a viable European biodiesel market.

Although a group of Austrian farmers were participants in some of the earliest large-scale field trials of biodiesel, the farming sector was generally slow to adopt biodiesel for regular use. "The idea sounded very attractive and was well accepted by politicians, scientists, the general public, and the tractor industry, but less well accepted by the farmers, who were just looking at the cost," says Werner Körbitz, the chairman of the Austrian Biofuels Institute (Osterreichisches Biotreibstoff Institut, or OBI) in Vienna. "If it was one penny more expensive, they were not interested. If it was one penny less expensive, then they were willing to try it. There was very little market research done about customer behavior and attitudes at this early stage. It was thought that biodiesel would be produced by farmers for farmers, but in actual practice, this was not true."[14] In later years, however, even the reluctant agricultural community began to embrace biodiesel.

SOUTH AFRICA

The story of biodiesel research in South Africa is a tale of missed opportunity—on more than one occasion. As mentioned previously, there were some early experiments in South Africa prior to World War II using vegetable-derived fuels, which were abandoned in favor of synthetic liquid fuels produced from coal. Then, around 1980, the idea of using vegetable oil as a diesel fuel was revived at the Council for Scientific and Industrial

Research (CSIR) in Pretoria, South Africa. Lourens du Plessis, now a semiretired special research scientist in the food science group at CSIR, recalls what happened.

"One of our engineers did some early vegetable-oil tests on a diesel engine on our campus," he says. "The agricultural engineers quickly realized that this was a good idea that could benefit the agricultural community."[15]

The CSIR research project began with investigations into the use of straight sunflower oil as a fuel. But the experiments soon ran into problems with improper fuel vaporization and the leaking of fuel into the lubricating oil inside the engine. "So, we thought about how we could overcome these problems, and we quickly switched over to producing ethyl esters of sunflower oil," du Plessis continues. "I was in favor of methyl esters, but the engineers wanted to stick to ethyl esters because they regarded it as an agricultural product. The whole issue at the time was for the agricultural sector to completely produce the diesel fuel." Du Plessis, who had degrees in chemistry and botany as well as a PhD in plant biochemistry, was asked to be part of the research team because of his expertise in oils and fats.

The initial test results were positive, and du Plessis and his coworkers produced about 500 liters (132 gallons) of the biodiesel fuel in a processor housed in a small steel shed. "It wasn't a technically designed facility," he recalls. "We built it ourselves in the backyard; it was actually quite a lot of fun to do it all ourselves." With a relatively large supply of fuel on hand, the agricultural engineers ran the engine tests, mostly on diesel tractor engines, in their laboratories. The results of the tests were encouraging, with very few problems of any sort. An additional series of tests was run on the stability of the biodiesel fuels. The results of all the tests were published in various journals and presented at a number of international conferences. Then, around 1984 or '85, the South African Department of Agriculture pulled the plug on the program, according to du Plessis. "They decided it was not economic or worthwhile, and they stopped the whole process," he says. "And that was the end of it."

Du Plessis looks back on his early biodiesel research with mixed emotions. "What was really rewarding was the fact that as a food chemist and

scientist, I worked with the mechanical engineering people; it was a mul-
tidisciplined team," he says. "But the main thing that was so exciting for
us was that the fuel was running in the engines without any problems. We
actually tried to patent our ideas in the early 1980s, but we found that
there was some early Belgian work already registered, so we couldn't get
a patent. That was a pity, because we spent a lot of time and made a lot
of effort to get the fuel well tested." But the saddest part of all was that
the research program, which had made such progress and showed such
promise, was canceled altogether.

EARLY EXPERIMENTS IN IDAHO

The oil shocks of the 1970s sparked considerable interest in alternate
fuels on the other side of the Atlantic Ocean as well. The initial research
work with vegetable-oil diesel fuels in the United States began at about
the same time as the Austrian and South African studies were getting
under way. There were a number of early studies in the late 1970s, but one
of the main pioneers in this work was Dr. Charles Peterson, an agricultural engineer at the University of Idaho. In 1979, the dean of the
College of Agriculture at UI told Peterson he had heard that vegetable
oil could be used as a diesel fuel. Peterson was intrigued, and he agreed to
give it a try. He went to a grocery store and bought some sunflower oil,
which he mixed with regular petroleum-based diesel fuel. "We had a Ford
tractor that we used in class experiments, so we had an easy way to switch
fuels," Peterson, who is now emeritus professor of biological and agricultural engineering at UI, recalls. "We had a dynamometer that we put it
on, but I think the biggest thing we learned was that the vegetable oil
blended very well with the diesel fuel."[16]

The following summer, encouraged by the initial tests, Peterson
decided to try running the tractor on straight safflower oil in demonstra-
tions at a number of local county fairs. "We used safflower oil because in
testing the viscosity of the oils that we grew in Idaho, safflower oil had
the lowest viscosity, and we thought it might perform a little better,"

Peterson says. That experiment worked too—at least initially. "When we got done, however, the tractor engine was completely shot," he admits. "At the end of the test we had severe polymerization of the piston rings and the engine wouldn't start."

Peterson's experiment with straight vegetable oil didn't work any better than other similar experiments of that period because modern diesel engines simply weren't designed to run on vegetable oil. Consequently, in order to use vegetable oil successfully as a fuel, either the oil or the engine had to be modified. Peterson and his colleagues, like their counterparts in Austria and South Africa, quickly came to the conclusion that modifying the oil was the preferred approach. Trying to develop the best ways to do that, and then testing the performance of the resulting fuels, became the main focus of UI's subsequent, highly regarded biodiesel research program.

Around 1982, the university bought a new Sato tractor for its farm operation. "We started out using a 50 percent rapeseed oil/fuel blend for the tractor, but shortly afterward we switched it to run on 100 percent esterified rapeseed oil," Peterson recalls. The switch to esterified oil (biodiesel) marked the real beginning of biodiesel experiments at the university. The UI Chemical Engineering Department helped Peterson come up with his early biodiesel recipe. But even the best-designed plan can go astray. "I remember when I tried to make the very first batch of esterified oil, I miscalculated the catalyst and ended up with a big glob of soap," Peterson recalls, laughing. "People still do that sometimes, I guess." Undaunted, Peterson persisted, and soon he and his coworkers were regularly turning out high-quality biodiesel from rapeseed oil. The process was facilitated with the help of a small mechanical screw press for extracting the seed oil and a 200-gallon batch reactor to make the biodiesel. Since then, the university has produced thousands of gallons of biodiesel for its many research projects.

Another early "disaster" from the mid-1980s that eventually led to a research breakthrough was an engine test with raw vegetable oil conducted by one of Peterson's graduate students. "He did a power test with raw vegetable oil in an engine connected to an electric dynamometer," Peterson recalls. "In a very short time it polymerized the piston rings and the engine seized up. This gave us the idea of running what we called the 'injector coking test' as a screening method for evaluating alternative fuels. We would run the engine through a torque test, and at the end of that period we pulled the injectors and evaluated the coking [carbon deposits] on them. The coking on the injectors could then be related to the coking that was going on inside the rest of the engine. This was a screening test that was a lot easier on the engines. We felt that the fuels that performed better in the test would do better on longer-term tests as well. There have been a lot of people who have adapted the test in different ways, and now they use machine vision to do the evaluation. When we did the tests we used photographs to determine the extent of the coking. It's interesting that this test came out of what was a failure and pretty severe damage to that engine."[17]

THE RESEARCH EXPANDS

Before long, the rapeseed biodiesel project became a joint effort with the Idaho Department of Water Resources, the U.S. Department of Energy, and the U.S. Department of Agriculture. Over the years, Peterson, who is widely acknowledged as a leader in the field, has headed numerous biodiesel research projects for various local, state, federal, and private agencies. During that time, UI has conducted biodiesel research on rapeseed oil, soybean oil, hydrogenated soybean oil, tallow, and a number of other feedstock sources.

The University of Idaho has used different forms of biodiesel to run many diesel engines in various types of farm machinery, stationary installations, Dodge and Ford pickups, and a long-haul Kenworth truck. The 200,000-mile, over-the-road test on the Kenworth tractor-trailer, operated by Simplot Transportation of Caldwell, Idaho, was completed in 1999. The truck ran on a blend of 50 percent biodiesel and 50 percent petrodiesel. The biodiesel was made from waste vegetable oil from the Simplot Inc. french-fry plant located in Caldwell. After the test, the Caterpillar engine was removed from the truck and sent to the manufacturer for evaluation.

The entire test was considered to be a success. "I think that test, which involved Caterpillar, really helped get their attention focused on biodiesel, and it probably was as important as anything in making people aware that waste oil could be used for biodiesel," Peterson says.[18]

Another research initiative that Peterson is especially proud of is the "Truck in the Park Project" conducted in Yellowstone National Park. The project had two main goals: to provide data on emissions and performance and to define a niche market for biodiesel in an environmentally sensitive area. The project also developed partnerships among the U.S. Department of Energy, the states of Montana and Wyoming, the National Park Service, regional businesses, and regulators. In 1994, Peterson and Daryl Reece, a graduate student working on his master's degree in agricultural engineering, drove a new Dodge pickup fueled with 100 percent rapeseed ethyl ester (REE) to Mammoth Hot Springs. Due to the park's extremely cold winter-weather conditions, the truck was equipped with a standard winterization package for diesel engines, but no modifications were made to the truck's engine or fuel system.[19] Initially there were some concerns that the odor of the biodiesel exhaust (which smelled like french fries) might attract bears. A special bear study was conducted, and it was found that the bears couldn't have cared less. The truck then was driven by park employees, who accumulated more than 130,000 miles running on the biodiesel fuel. In September 1998, the truck's engine was completely torn down and thoroughly inspected, revealing very little wear and no carbon buildup. The engine was reassembled and put back in the truck for the second phase of the test, designed to accumulate 200,000 miles of use. Overall, the project has been a great success with park employees and visitors alike, and the National Park Service has since introduced biodiesel to more than 50 other parks through the Green Energy Parks Program.

"That certainly was a dramatic setting for the tests," Peterson notes. "I think our Yellowstone Park project, where we used biodiesel in such an environmentally sensitive area, is probably the project that made biodiesel as well known in this country as it is." Looking back on his quarter century of work with biodiesel, Peterson is somewhat bemused by the present scope of the program. "I always tell people that this started out more as a hobby than anything," he says. "In the early years, I never ever thought that this would develop into our principal research program that I was involved with. But after a while, we were able to get more funding for this project than for others, and it kept on getting larger and larger." Although it didn't occur until about 90 years after his death, Rudolf Diesel's vegetable-oil vision was finally being realized.

3

Biodiesel 101

As most of the early biodiesel researchers quickly discovered, using straight plant oil as a fuel substitute in diesel engines was not especially good for the engines. Numerous test engines around the world were undoubtedly ruined in many of those early experiments. The problem was that, for almost a century, diesel engines had been gradually developed, adapted, and fine-tuned to burn petroleum-based diesel fuel. Faced with the prospect of having to modify the millions of diesel engines in use around the world in a wide range of different types of vehicles, researchers in the late 1970s and early 1980s opted to modify the vegetable-oil fuel instead. This was actually a fairly easy choice, according to Werner Körbitz, the chairman of the Austrian Biofuels Institute in Vienna. "That's because, even then, it was clear that biodiesel would not be able to replace more than about 10 percent of the petroleum diesel market share," he says.[1] And trying to retool the diesel engine for such a small potential part of the market didn't generate much enthusiasm among engine manufacturers.

MAKING BIODIESEL

As mentioned previously, the oil transformation process the researchers selected was *transesterification*, or the transformation of one form of ester into another. (Esters are naturally occurring compounds such as oils and fats, or any of a large group of organic compounds formed when an acid and alcohol are mixed.) In order to understand the process (without getting too technical), we need to take a closer look at vegetable oil. One of the main problems with vegetable oil compared to diesel fuel is that it's thicker, or more viscous. This is due to the fact that vegetable oil contains glycerin—a thick, sticky substance—in its chemical structure. Every vegetable-oil molecule is composed of three fatty acid chains attached to a molecule of glycerin. Picture a microscopic three-legged creature with a round glycerin head and three long, dangling legs. This is why vegetable oil is described technically as a *triglyceride*, or *three* fatty acid chains and glycerin.

Although the exact percentage varies somewhat depending on what kind of plant the oil comes from, approximately 20 percent or less of a vegetable-oil molecule is composed of glycerin. Transesterification involves breaking every oil (triglyceride) molecule into three fatty acid chains and a separate (or free) glycerin molecule. During the process, alcohol is added, and each of the fatty acid chains attaches to one of the new alcohol molecules, creating three mono-alkyl esters. This process makes the esters thinner and more suitable for use as diesel fuel. Once separated from the glycerin, the alkyl ester chains then become what we call biodiesel.

The alcohol used in the process can be either ethanol (made from grains) or methanol (made from wood, coal, or natural gas). Methanol is usually preferred because it's cheaper and tends to produce a more predictable reaction. On the downside, methanol dissolves rubber, can be fatal if swallowed, and must be handled with extreme caution. Ethanol, on the other hand, is generally more expensive and may not always produce a consistent, stable reaction. On the upside, ethanol is less toxic and is made from a renewable resource. If biodiesel is produced with methanol, it is referred to as *methyl esters* and if it is made with ethanol, it is referred to as *ethyl esters*. A more generic term, *alkyl esters*, refers to any alcohol-produced vegetable-oil esters.

But as a good high-school chemistry teacher would point out, there is still one more ingredient needed to make the process work—a catalyst. The catalyst is the substance that initiates the reaction between the vegetable oil and the alcohol by "cracking" the triglycerides (vegetable oil) and releasing the alkyl esters (biodiesel). There are two main catalysts that can be used, sodium hydroxide (NaOH) and potassium hydroxide (KOH). Sodium hydroxide, which is less expensive, is commonly referred to as lye or caustic soda (the same chemical used to unclog kitchen or bathroom drains). If sodium hydroxide is not available, potassium hydroxide can be used instead, but a larger quantity is required. Sometimes a third catalyst, sulfuric acid, is used by commercial biodiesel producers as a pretreatment for waste cooking oils to prevent excessive soap formation. All of these chemicals are dangerous, however, and must be handled carefully.

These traditional catalysts are known as "homogeneous catalysts" because they dissolve in the reaction solution. But there is another catalyst—new to the market—that is quite different. This new catalyst is referred to as a "heterogeneous catalyst" because it does not dissolve in solution, and can easily be removed, and then reused, after the reaction takes place. This so-called "solid catalyst" strategy is being developed by several companies, and if it proves to be commercially viable, it could have a big impact on the industry. This new catalyst appears to be appropriate for use in both large and smaller biodiesel refineries.[2] You can expect to hear more about this development soon.

THE PROCESS

The transesterification process is initiated by adding carefully measured amounts of alcohol mixed with the catalyst to the vegetable oil. How much catalyst is used depends on the pH (acidity) of the oil. When used cooking oil is chosen as the feedstock for biodiesel, one additional factor comes into play—free fatty acids. Because they are considered acids, oils and fats are sometimes referred to as fatty acids. When vegetable oil is fried, free fatty acids (which are not bound or attached to other molecules) are released and end up floating around among the triglycerides. These free fatty acids can use up too much catalyst and result in the formation of excess amounts of soap (an unhelpful trait), so they need to be eliminated. The way to accomplish this is to add more catalyst to the mix; the exact amount is determined either by the pH of the used cooking oil or by the trial-and-error method.

One of the main advantages of biodiesel is that the transesterification process used to produce it can be conducted at almost any scale—from a kitchen blender that makes a few liters on up to a large industrial facility capable of producing millions of gallons per year. Although an industrial-size biodiesel facility uses a lot of high-tech equipment to wring every last productive ounce out of all the ingredients (and recycles many of them for reuse), the basic transesterification process is more or less the same as that used in a small-scale facility located in a garage or backyard shed. The main difference is that very large-scale operations often are designed to produce biodiesel on a continuous basis—the *continuous-flow process*—while the small processor normally produces smaller, individual batches at a time—the *batch process*. In the batch process, the reaction and subsequent settling procedure takes place in a single tank or container over a period of time. In the continuous-flow process, however, there is a constant movement of feedstock and other ingredients through the system, resulting in finished biodiesel at the end of the process.

Here's how the basic process works (using methanol and sodium hydroxide as an example). Carefully measured quantities of methanol and sodium hydroxide (lye) are mixed to create sodium methoxide, which is then mixed with the vegetable oil and stirred or agitated (and sometimes heated) for a specified length of time. If used vegetable oil is the feedstock, the process requires a bit more testing, lye, and filtration, but it is otherwise essentially the same. During the mixing, the oil molecules are split or "cracked" and the methyl esters (biodiesel) rise to the top of the settling/mixing tank while the glycerin and catalyst settle to the bottom. (The separation process can be speeded up with the use of a centrifuge.) After about eight hours, the glycerin and catalyst are drawn off the bottom, leaving biodiesel in the tank. In most cases the biodiesel needs to be washed with water to remove any remaining traces of alcohol, catalyst, and glycerin. In this procedure, water is mixed with the biodiesel, allowed to settle out over several days, and then removed. The wash process can be repeated if needed, but it is time-consuming. Not everyone agrees on whether this water-wash step is necessary. Some smaller producers who are making biodiesel for themselves skip the process, while commercial producers usually must perform it to meet industry standards. In the case of some larger, more sophisticated manufacturing facilities, the transesterification process itself is so carefully controlled and refined that the water wash may not be needed. There are, of course, quite a few technical variations on this entire process for large-scale industrial operations, but the general transesterification procedure is similar.

Because making biodiesel is relatively simple and can be very low-tech (an old 55-gallon drum often is used as the settling/mixing tank), it has attracted an enthusiastic community of backyard enthusiasts or "homebrewers" around the world. For those who want to make their own biodiesel, *From the Fryer to the Fuel Tank: The Complete Guide to Using Vegetable Oil as an Alternative Fuel* by Joshua Tickell is one of the older and more popular books on the subject. A newer reference that goes into more accurate detail is available from Maria "Mark" Alovert, and is titled the *Biodiesel Homebrew Guide 2004*. And Bill Kemp's 2006 book *Biodiesel Basics and Beyond* is another excellent source for the homebrewer (see the bibliography).

BIODIESEL FEEDSTOCKS

Another remarkable feature of the transesterification process is that it can use a wide range of feedstocks—virgin vegetable oils, used fryer oil, animal fats, even pond algae—to produce the same basic biodiesel end product (with minor differences in fuel characteristics). These feedstocks can be used individually or blended to produce biodiesel with specific traits. The ability to adapt the production process to locally available feedstocks and end-user needs is one of biodiesel's most attractive advantages.

There are hundreds of oil-producing plants that can be used as feedstocks for biodiesel, from corn to soybeans and sunflower to oil palm. Even avocado and industrial hemp will work. Here is a brief overview of some of the most common crops, listed in descending order of oil production in liters per hectare (one hectare equals 2.47 acres) and in U.S. gallons per acre.[3] The production figures given are conservative averages and can vary widely depending on specific plant variety, cultivation practices, and weather and soil conditions. Other products made from these plants and their oils are included in the listings to highlight existing competitive markets. Figure 3 shows the main feedstock crops used worldwide for biodiesel production, based on a rough estimate by the author.

Oil Palm

The African oil palm is at the top of the list of oil-producing plants, with a remarkable yield of up to 5 metric tons of oil per hectare every year (about 5,950 liters per hectare, or 635 gallons per acre). The African palm produces two types of oil, palm oil and palm kernel oil. Palm oil is extracted from the fleshy part of the fruit, which contains 45 to 55 percent oil, and it is used mainly in the manufacture of soaps, candles, margarine, and cooking oils. Palm kernel oil, as its name implies, comes from the kernel of the fruit, which contains about 50 percent oil. Nearly colorless, palm kernel oil is solid at room temperature and is used in making ice cream, mayonnaise, baked goods, and soaps and detergents. The pressed "cake" (pulp) remaining after the oil has been extracted is used as an animal feed. The African palm can be found along the coast of West Africa from Liberia to Angola and eastward to the Indian Ocean islands of Zanzibar and Madagascar. The African palm is also sometimes grown as an ornamental tree in other subtropical locations, such as Florida and Southern California.[4] Palm oil is the main biodiesel feedstock in Malaysia, but palm-oil biodiesel has a high *cloud point* (the temperature at which the first wax crystals appear in biodiesel), making it less desirable for use in colder climates in its pure form. Despite its top ranking in terms of oil yield, palm oil at present represents only about 10 percent of total global biodiesel raw material sources.[5]

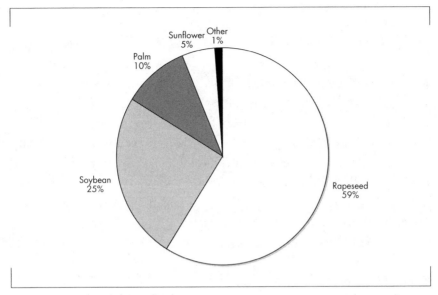

Figure 3. Raw Materials for Biodiesel Author's estimate for 2007

Coconut

As a source of oil, the coconut tree produces only around half as much oil as the oil palm, which is still an impressive 2,689 liters per hectare (287 gallons per acre) annually. Coconut is the third-most-produced oil in the world, after peanut and soybean. To make coconut oil, the meat of the coconut is peeled from the husk, dried, and pressed (or the meat can be shredded fresh and cold-pressed). The residual cake is used as an animal feed. Among the many products made from various parts of the coconut tree are twine or rope, mattress padding, mats, rugs, brushes, filler for plastics, charcoal filters, vinegar, soaps, lubricants, hydraulic fluid, paints, synthetic rubber, and margarine; it also can be used in making ice cream. The coconut palm may have originated in northwestern South America, but it is now found in all tropical regions of the world, especially along coastlines, and increasingly in plantations (which may have some negative environmental implications in certain locations).

Jatropha

Jatropha is a versatile bush or tree that has many uses. It has garnered increased interest recently because it adapts well to semiarid, marginal locations and can be grown as a hedge for erosion control, property boundaries, and animal fencing. Jatropha is used to make lamp oil, soap, candles, poisons, and a wide range of folk remedies. Jatropha can produce about 1,590 liters of oil per hectare (202 gallons per acre) annually. Widely grown as a medicinal plant, jatropha establishes itself easily and is found in Brazil, Fiji, Honduras, India, Jamaica, Nicaragua, Panama, Puerto Rico, Mexico, El Salvador, and much of Africa. Jatropha is presently a very minor source of biodiesel, but it has a lot of potential and is the object of intense study around the world.

Rapeseed/Canola

Sometimes cultivated in small quantities as a potherb, this yellow-flowering plant is more commonly grown in Europe as a forage feed for livestock and as a source of rapeseed oil (known in North America as canola oil). Rapeseed/canola produces 1,190 to 1,500 liters per hectare (127 to 160 gallons per acre), giving it the highest yield of any conventional oilseed field crop. The residual seed is used as a high-protein animal feed. Rapeseed is cultivated in most European countries, China, Canada, India, Australia, and Russia, but it will grow in most temperate regions. Rapeseed, which represents about 59 percent of global biodiesel raw material sources, is the principal feedstock for biodiesel produced in Europe, and most biodiesel research has been based on rapeseed methyl ester or rapeseed ethyl ester.

Camelina

Camelina, whose common names are gold-of-pleasure or false flax in English, is a member of the mustard family. Native to Northern Europe and parts of Central Asia, camelina has been cultivated in Europe for more than 3,000 years. Camelina is grown mainly as an oilseed crop for vegetable oil and animal feed. Camelina oil is also naturally high in omega-3 fatty acids, making it a valuable food-grade oil for cooking and other purposes. Oil yield from camelina is similar to rapeseed, but with lower costs due to lower fertilizer and pesticide requirements. Due to these traits, camelina has been the object of growing interest in various countries, but especially in the western United States as an alternate feedstock crop for biodiesel. Some studies indicate that the use of camelina as a biodiesel feedstock could reduce the cost of production about $1 per gallon.

Peanut

Peanuts produce an edible oil that can be used for cooking and deep-frying; in margarines, salad dressings, and shortenings for pastry and breads; and for the manufacture of pharmaceuticals, soaps, and lubricants. The seeds (peanuts) are also eaten raw, roasted and salted, chopped in confectioneries, or (especially in the United States) ground into peanut butter. In most other countries, peanuts are mainly processed for oil. The peanut produces about 1,059 liters per hectare (113 gallons per acre). Native to South America, peanuts are now widely cultivated in warm climates and sandy soils throughout the world. Although one of Rudolf Diesel's engines used peanut oil as fuel at the 1900 Paris Exposition, today it is a relatively minor source of feedstock for biodiesel.

Sunflower

The sunflower is cultivated mainly for its seeds, which yield the world's second-most-important type of edible oil, and it produces around 952 liters per hectare (102 gallons per acre). Sunflower oil is used for cooking, in margarine and salad dressings, and in lubricants, soaps, lamp oil, and a variety of paints and varnishes. Sunflower kernels are eaten raw or roasted and salted; they also can be made into flour. The cake (with seed hulls removed) is used as a high-protein animal feed. Native to western North America, the sunflower is the only major oil crop to have evolved in what is now the United States. Introduced early to Europe and Russia, sunflowers now grow in many countries, in both temperate and tropical regions. Sunflowers represent about 5 percent of global biodiesel raw material sources.

Safflower

An annual thistlelike herb, safflower is grown mainly for its edible-oil-producing seeds, which yield about 779 liters per hectare (83 gallons per acre) of safflower oil. High in essential unsaturated fatty acids, safflower oil is light-colored and easily clarified. It is used in salad and cooking oils, margarine, and candles, and as a drying oil in paints, linoleum, and varnishes. Safflower is believed to have originated in southern Asia and was cultivated in China, India, Persia, and Egypt from prehistoric times. Later, safflower was grown in Europe and subsequently was introduced to Mexico, South America, and the United States. Safflower, which does well in the same areas that favor the growth of wheat and barley, can be planted, cultivated, and harvested with standard farm machinery used for small grains. Safflower oil is not a major biodiesel feedstock at the present time.

Mustard

Several varieties of mustard are grown as vegetable greens and as cover crops, but they are especially valued for their seeds, which can produce about 572 liters per hectare (61 gallons per acre) of mustard oil. The seeds also can be processed into a wide range of products, including various types of commercial mustards, lubricants, and hair oil. The seed residue is fed to animals, used in fertilizers, or used to make a valuable organic pesticide. Widely grown in many countries, mustard is considered a weed on cultivated lands in some locations. Most mustard crops can be planted, cultivated, and harvested using ordinary farm machinery. Although mustard seed oil is not widely used as a biodiesel feedstock, quite a lot of research is presently focused on its use, particularly in the United States.

Soybean

The soybean produces one of the world's most important sources of oil and protein. Soybeans yield around 446 liters per hectare (48 gallons per acre) of edible oil, which is used as a salad oil and for the manufacture of margarine and shortening. The oil also is used industrially in the manufacture of paints, linoleum, printing inks, soap, insecticides, disinfectants, and a wide range of other products. The most commonly raised crop in the United States (around 3.1 billion bushels were produced in 2006), the soybean has been grown in East Asia for thousands of years and is widely cultivated in subtropical and tropical regions such as Brazil and Argentina. Although most biodiesel in the United States is made from soybean oil (virgin or used), soybeans are not the best crop for oil production. Worldwide, soybeans probably represent about 25 percent of total global biodiesel raw material sources.

Hemp

As *Cannabis sativa*, hemp has been cultivated for more than 4,500 years for many different purposes, including fiber, oil, medicine, and narcotics. Native to Central Asia, hemp has also long been cultivated in Asia, Europe, and China. Hemp can produce around 363 liters of oil per hectare (39 gallons per acre).

Drug-producing varieties (those containing high levels of the psychoactive compound THC) thrive in the tropics, while oil- and fiber-producing plants do better in temperate and subtropical areas. When grown for nondrug purposes, this plant usually is referred to as "industrial hemp." Licenses for hemp cultivation have been issued in the European Union and Canada. The U.S. government still considers industrial hemp and marijuana to be identical from a regulatory perspective and little progress toward greater use of industrial hemp in the U.S. is expected until hemp is no longer classified as an illegal drug. As of 2006, China produced about 40 percent of global hemp fiber.

Corn (Maize)

Corn (also known as maize) is believed to have originated in Mexico in prehistoric times. Few plants are grown as extensively or put to more uses. Corn is a staple cereal food in Central and South America and throughout much of Africa. The oil produced, which amounts to only about 172 liters per hectare (18 gallons per acre), represents about 7 to 8 percent of the grain, placing corn at the bottom of the oil-production spectrum. Nevertheless, processed corn oil is used extensively as a frying oil in the United States. Ethanol, a renewable alcohol fuel, is made from cornstarch in the United States.

Used Cooking Oil

The United States is almost as addicted to fast food as it is to petroleum. Every year, more than 3 billion gallons (11.4 billion liters) of used cooking oil is drained from deep-fat fryers in this country. Most restaurants and other frying-oil users have to pay a rendering or waste company to haul the greasy stuff away. Some used cooking oils are processed into animal feed, makeup, or fertilizer, but for many years much of it has been dumped in landfills or sewers in some locations, causing a lot of headaches for sewage treatment plants. But used vegetable oil is a great low-cost resource for making biodiesel, and it is increasingly used for that purpose in numerous countries. However, because it is of lower quality due to repeated heating at high temperatures and possible contamination with animal fats and food particles, used cooking oil requires additional treatment, both before and after the transesterification process. This is an important issue if the biodiesel made from it is to meet stringent government standards for use as a vehicle fuel. Despite this fact, turning what would otherwise be a waste product into usable fuel makes a lot of sense from a variety of environmental perspectives. And in locations where large quantities of used cooking oil are available, such as in major cities, there is a lot of potential for making substantial amounts of biodiesel. For example, according to the *New York Times*, it would take just one-fifth of the waste cooking oil produced in New York City to run its entire public transit bus system. However, due to recent price increases in virgin oils as well as the growing popularity of biodiesel, demand for used cooking oil has grown dramatically, and free sources of this feedstock are disappearing rapidly in many parts of the United States.

Animal Fat

Animal fat, a by-product of the animal rendering process, is the least expensive feedstock currently available for biodiesel production. Beef tallow, poultry fat, fish oil, and other types of animal fats will all work. In recent years, the United States has generated around 11 billion pounds of animal fats, according to U.S. Department of Agriculture averages, which theoretically is enough to produce roughly 1.5 billion gallons of biodiesel. Substantial quantities of animal fats are available in many other countries for potential biodiesel production as well. Admittedly, not all of this feedstock is available for biodiesel production, but it does present an opportunity to lower input costs substantially.

However, biodiesel made from animal fats tends to have poor cold-weather properties compared with biodiesel made from virgin oils such as soybean or rapeseed. Nevertheless, in warmer climates around the world, there is considerable potential for using animal fats as a feedstock. What's more, when biodiesel is made from animal fats, it generally has lower tailpipe nitrous oxide (NO_x) emissions than other biodiesels. The growing global concern about how to dispose of cattle infected with bovine spongiform encephalopathy (BSE, or mad cow disease) or otherwise suspect cattle has provided another opportunity for the biodiesel industry, especially in the European Union, where feeding ground-up animal parts to other animals has been banned since January 2001. As a result, a market for the production of biodiesel from rendered animal fats has developed in a number of European nations (see Chapter 6).

Algae

There is one additional biodiesel feedstock that potentially dwarfs all the others in terms of oil production: algae. Although pond algae may seem like a bizarre source of diesel fuel, it's not as far-fetched as it may sound. That's because much of the original organic matter that formed the basis of the world's present petroleum resources was algae—vast amounts of algae—in shallow, prehistoric bodies of water. These deposits subsequently were buried by sediments and later transformed by pressure and heat over millions of years into petroleum. Today the process can be shortened to just a few days simply by growing the algae in ponds and extracting the oil directly from the harvested algae.

From 1976 to 1998, the National Renewable Energy Laboratory (NREL) conducted a $25 million program funded by the U.S. Department of Energy called the Aquatic Species Program: Biodiesel from Algae. The main focus of the program was the production of biodiesel from algae grown in ponds utilizing waste CO_2 from coal-fired power plants. Although the program's use of one of the most environmentally unfriendly fossil fuels—coal—in its experiments seems incongruous, it was at least utilizing the carbon dioxide emissions in a productive way. (This is very much in line with the current idea of "waste exchange," where the waste from one industry is used as a feedstock for another.)

One of the main accomplishments of the program was to locate, catalog, and study more than 3,000 strains of algae from all across the continental United States and Hawaii. After preliminary testing in California and Hawaii, a large-scale pond facility was built in the desert near Roswell, New Mexico. The quantity of oil that the algae in Roswell could produce was remarkable. A series of 1,000-square-meter ponds were filled with floating algae. Up to 50 grams of algae were produced per square meter daily.[6] Thus, each pond had a potential daily yield of 50,000 grams (50 kilograms, or 110 pounds). Oil concentrations between 30 and 40 percent are not unusual for some types of algae, so using a conservative 30 percent as a multiplier yields approximately 15 kilograms (33 pounds) of oil per pond per day. This means that one pond could theoretically produce 5,475 kilograms (12,045 pounds) of oil per year. Assuming that 7.5 pounds (roughly) equals a gallon, one algae pond could possibly yield 3,418 liters (roughly 903 gallons) of oil per year (or about 6,499 gallons per acre per year). However, in actual practice, the Roswell researchers were unable to maintain consistent production levels due to cold overnight desert temperatures. Other researchers have estimated that one acre of algae could produce anywhere from 500 to 30,000 gallons of algal oil every year, so there is clearly a very wide range of opinion on this issue.

The NREL researchers concluded that "algal biodiesel could easily supply several 'quads' of biodiesel—substantially more than existing oilseed crops could provide." (A quad represents one quadrillion Btu of energy.) Two hundred thousand hectares (494,000 acres), representing less than 0.1 percent of suitable land area in the United States, could produce one quad of fuel.[7] Two quads would require an area the size of the state of Rhode Island, the smallest state in the union. Regardless of whose production statistics you use, figures like this clearly demonstrate the potential for huge quantities of oil production from algae. Although the NREL tests were reluctantly terminated in 1998 due to budget cuts (prompted by political pressure from large petroleum companies, according to some observers), the researchers hoped that their work would be used as a foundation for the future production of biodiesel from algae. In the past few years, their hopes have been realized as researchers all over the world have resumed studies and trials of a wide range of algae strains in a variety of different settings.

PRODUCTION POTENTIAL

In 2006, total global output of biodiesel was about 5.7 million metric tons (1.7 billion gallons). About 85 percent of that production is generated by Europe, with the United States and the rest of the world making up the balance. By comparison, the United States alone consumes approximately 60 *billion* gallons of middle-distillate fuels annually, according to the U.S. Energy Information Agency. (Middle-distillate fuels include diesel fuel, heating oil, kerosene, jet fuels, and gas turbine engine fuels.) Of that total, about 35 billion gallons (58 percent) is for on-highway vehicle use, while the rest is divided among a wide range of residential, commercial, and industrial purposes. How much of this consumption can be met by biodiesel? This is a question that has sparked quite a lot of debate. As noted earlier, the United States generates about 3 billion gallons (11.4 billion liters) of used frying oil every year. Assuming that half of this oil could be converted to biodiesel, that would represent perhaps 2.5 percent of the current petrodiesel market. If half of all 11 billion pounds of animal fats produced in the United States could be processed into biodiesel, that might yield roughly 750 million gallons of biodiesel, or 1.25 percent of the petrodiesel market. If all the fallow cropland in the United States (about 60 million acres) were planted to rapeseed (admittedly a large assumption) and yielded 100 gallons of oil per acre (not such a large assumption), that could produce roughly another 6 billion gallons, or about 10 percent of the current petrodiesel market. The total estimated production from all of these sources comes to 8.25 billion gallons, or about 14 percent of the U.S. petrodiesel market. All of these sources theoretically could be used without displacing any current food crop production, an important consideration. This projection admittedly does not take into account possible biodiesel production from algae, which, at the present time, is not yet occurring on a large commercial scale and would require massive investment in production facilities. Optimistically, several quads of production could be added for algae, which would boost total U.S. biodiesel production by about 20 billion gallons to roughly 28 billion, approximately half the present petrodiesel market. Although production capacities in other countries vary quite a bit, in Europe it is estimated that biodiesel has the potential to replace somewhere between 10 and 15 percent of the current petrodiesel market given present technology.[8]

How do these admittedly sketchy (and optimistic) estimates compare with "official" projections for the United States? The National Biodiesel Board of Jefferson City, Missouri, says that, as of the end of 2007, there was over 1.85 billion gallons of biodiesel production capacity per year. (The *actual* production for 2007 was only about 450 million gallons due to a variety of reasons we'll discuss later in the book.) Much of this production capacity is modular, which means that it could be doubled in 12 to 18 months. The NBB recently set a production goal for the U.S. biodiesel industry of 2 billion gallons by 2015.

Looking beyond production facilities, some observers in the farming sector say that the upper limit for biodiesel production from soybeans in the United States would be about 600 million gallons. The National Renewable Energy Laboratory, on the other hand, estimated that around 1 billion gallons of biodiesel could be produced annually in the United States, and that production could be boosted to 2 billion gallons by 2010. (This prediction was made a number of years ago, but is more or less in line with the NBB's more recent 2 billion gallon goal mentioned above.)

Another 4 to 10 billion gallons might be produced using mustard seed, according to the Department of Energy Biomass Program. These official estimates come to a total of 12 billion gallons, or about 20 percent of the present U.S. petrodiesel market. This is a very optimistic projection. Whether the correct figure is 14 percent or an even more optimistic 20 percent, it's clear that future U.S. biodiesel production capacity (in terms of both facilities and feedstocks) falls far short of the current petrodiesel market, unless algae were used as a biodiesel feedstock in a massive, industrial-scale initiative that would take many years and a lot of money to develop. This would be difficult, but not impossible given the recent increase in algae-to-biodiesel research and development activities around the nation. This more optimistic scenario might be possible, especially if a *real* national energy plan (rather than the usual handouts to the fossil fuel and nuclear industries), one that strongly emphasized renewables, were finally implemented in the United States. Also difficult, but, again, not impossible.

HOW DOES BIODIESEL COMPARE?

By now it should be obvious that biodiesel has a lot going for it. It can be made almost anywhere by almost anybody from a wide range of ingredients. But how does it really stack up when compared to petrodiesel? In most respects, biodiesel is better, while in others, it's not quite as good. First, biodiesel can be used in any modern diesel engine without any modifications to the engine. (This is not the case with some other fuels, such as compressed natural gas, which require major capital expenditures for engine retrofits or new engines as well as new infrastructure.) Biodiesel has excellent lubricating properties and will lubricate many moving parts in the engine, increasing engine life. It also has a higher cetane number than petrodiesel (cetane is a measure of the ignition quality of diesel fuel), indicating better ignition properties. On the downside, the energy content of biodiesel is about 10 to 12 percent lower than that of petrodiesel (about 121,000 Btu compared to 135,000 Btu for Number 2 diesel fuel). Biodiesel has an oxygen content of around 10 percent, more than that of petrodiesel fuel. The oxygen results in more favorable emission levels, but also results in reduced energy content. Although this causes a slight reduction in engine performance, the loss is partly offset by a 7 percent average increase in combustion efficiency. Generally speaking, the use of biodiesel results in about a 5 percent decrease in torque, power, and fuel efficiency in diesel engines. However, some recent performance results in school bus and other fleets have indicated an *increase* in fuel efficiency, so the jury is still out on this performance factor. The bottom line, though, is that most people can't detect any noticeable difference in the performance of their vehicles, and some even claim that overall performance is actually better.

Biodiesel is free of lead, contains virtually no sulfur or aromatics (toxic compounds such as benzene, toluene, and xylene), and results in substantial reductions in the release of unburned hydrocarbons, carbon monoxide, and particulate matter (soot), which has been linked to respiratory disease, cancer, and other adverse health effects. The production and use of biodiesel results in a 78 percent reduction in carbon dioxide emissions, according to a joint U.S. Department of Energy (DOE) and U.S. Department of Agriculture (USDA) study published in 1998.[9] This is mainly because the plants used to grow biodiesel feedstock absorb most of the CO_2 emissions from biodiesel combustion. On the downside, emissions of nitrogen oxides (NO_x, a contributing factor in the formation of smog and ozone) are usually slightly greater with biodiesel according to a 2002 EPA study. This can be partly offset by tuning the engine specifically for biodiesel. Recent additive tests show promising results for reducing nitrogen oxide emissions as well. However, a recent NREL study published in 2006 appears to contradict the earlier EPA study and shows no consistent effect of biodiesel use on NO_x emissions. It is still unclear which of these studies is more correct. Last but not least, biodiesel replaces the typically noxious exhaust smell of petrodiesel with an odor faintly like that of french fries or doughnuts, especially if it has been made from recycled cooking oil.

Life-Cycle Studies

Biodiesel has other positive features as well. The overall life-cycle production of wastewater from the production of biodiesel is 79 percent lower than the overall production of wastewater from petrodiesel. What's more, the overall life-cycle production of hazardous solid wastes from biodiesel is 96 percent lower than overall production of hazardous solid wastes from diesel, according to the DOE/USDA study.

Another important factor when comparing biodiesel with other fuels is what is known as the *energy-efficiency ratio*. This is a numerical figure that represents the energy stored in the fuel compared to the total energy required to produce, manufacture, transport, and distribute the fuel. The total energy-efficiency ratios for biodiesel and petrodiesel are very similar. However, the total *fossil* energy-efficiency ratio for biodiesel shows that biodiesel is about four times as efficient as petrodiesel in utilizing fossil energy, according to the DOE/USDA study cited above. Biodiesel has a positive fossil energy-efficiency ratio of 3.2 to 1. (A more recent study puts the ratio at 3.5 to 1, reflecting a number of improvements in the overall process.) Petrodiesel, on the other hand, has a negative fossil energy-efficiency ratio of 0.83 to 1, according to the study.[10] Just for the sake of comparison, ethanol has an energy-efficiency ratio of 1.34 to 1, according to a 2001 USDA study.[11] Other studies say that biodiesel has an energy-efficiency ratio of around 2.5 to 1, while still others say it can be as high as 7 to 1. Energy-efficiency ratios and similar statistics are tricky and can vary widely depending on what factors are included in the studies, so it's not unusual to see a wide range of figures. In any case, biodiesel generally provides more energy with far less negative environmental impacts on a life-cycle basis than petrodiesel.

Biodegradability and Toxicity

When it comes to biodegradability and toxicity, biodiesel wins the contest over petrodiesel hands down. Biodiesel fuel is not especially harmful to humans or the environment. One hundred percent biodiesel (also called B100 or "neat" biodiesel) is as biodegradable as sugar and up to 10 times less toxic than table salt. One study showed that biodiesel rubbed on the skin was less irritating than a 4 percent water and soap solution. Although biodiesel smells and feels somewhat, well, greasy, otherwise it's relatively benign. On the other hand, just the fumes from petrodiesel are toxic and dangerous.

Although ingesting carefully measured amounts of biodiesel (not recommended) caused no major problems for a group of laboratory rats in a 1996 Ohio laboratory study, drinking even a small amount of petrodiesel could prove extremely dangerous, even fatal for humans (or rats). Studies also have shown that if spilled accidentally, biodiesel will biodegrade up to four times faster than petrodiesel fuel. In about three to four weeks' time, biodiesel will have almost entirely biodegraded.[12] This is an especially important advantage in environmentally sensitive areas such as national parks and forests or waterways. Although accidental spills of biodiesel will still cause some environmental problems, they are minimal when compared to an equivalent spill of petroleum diesel or crude oil.

Transportation and Storage

Biodiesel is much safer to transport and store than petrodiesel. That's because, aside from its low toxicity, biodiesel also has a high *flash point*, or ignition temperature. This means that biodiesel needs to be above 260°F (126°C) before it will ignite. The flash point of petrodiesel, by comparison, is around 125°F (52°C), which is one reason why it is considered a hazardous material. Biodiesel, on the other hand, is so safe that it can be shipped by common carrier, or even UPS or FedEx. Tests also have shown that the flash point of biodiesel blends increases as the percentage of biodiesel increases. Consequently, biodiesel and blends of biodiesel with petrodiesel are safer to ship and store than straight petrodiesel.

In general, the same storage and handling procedures used for petrodiesel can be used for biodiesel, meaning that no major changes to infrastructure are needed. Biodiesel should be stored in a clean, dry, dark environment in tanks made from steel, aluminum, fluorinated polyethylene, fluorinated polypropylene, or Teflon. Copper, brass, tin, lead, and zinc should be avoided, according to the National Biodiesel Board.

Due to its organic nature, biodiesel is somewhat more susceptible to the growth of bacteria and mold in the presence of moisture than petrodiesel. Consequently, it is sometimes necessary to add small quantities of biocides to stored biodiesel, especially in warmer climates, to reduce the growth of these contaminating organisms. It is also important to eliminate as much moisture from biodiesel storage tanks as possible to minimize the growth of bacteria and mold. Ideally, the storage time for biodiesel (as with petrodiesel) should be limited to six months.

BIODIESEL BLENDS

In addition to its many environmental advantages, another extremely useful characteristic of biodiesel is that it can be blended with petrodiesel in any percentage. A B20 blend (20 percent biodiesel and 80 percent petrodiesel) has demonstrated significant environmental benefits with a minimum increase in cost to consumers. In recent years, B20 has become a popular fuel with many fleets in the United States, encouraged in part by the 1998 Congressional approval of B20 as a compliance strategy for fleets under the Energy Policy Act of 1992. (For more on the EPAct, see Chapter 11.) It is biodiesel's ability to be blended with petrodiesel at any percentage from B1 to B99 that makes it such a flexible fuel that can meet a wide variety of different needs, as we will see in the next chapter.

Biodiesel's Many Uses

Although Rudolf Diesel's early designs were for large, industrial-scale engines in stationary settings, he believed that his invention would be used eventually to power a wide range of vehicles of virtually every description. His vision eventually came to pass (after his death), but by then his engine design had been modified to burn petroleum-based diesel fuel. There were, however, a few exceptions. In the 1950s, Germany's MAN AG developed its "salad oil engine" design (named for its ability to run on a wide variety of different fuels), but the design was mainly confined to limited numbers of larger stationary installations.[1] The vast majority of diesel engines in use around the world, however, burned (and still burn) petrodiesel fuel, which is similar to kerosene, jet fuel, and home heating oil. Consequently, early biodiesel researchers were focused on creating a renewable fuel for diesel engines that were optimized to run on petrodiesel. As we have seen, there are many feedstocks that can be used to make biodiesel. But the different ways of producing biodiesel are exceeded by the many ways of using it. In this chapter we'll explore those applications. First, though, we'll take a quick look at the modern diesel engine and a few issues related to using biodiesel as a fuel.

HOW THE DIESEL ENGINE WORKS

Although there are many variations in size and usage, most diesel engines share the same basic features they inherited from Rudolf Diesel's original designs. Like their gasoline-powered relatives, diesels have a number of cylinders (usually four, six, or eight) containing pistons that are connected to a crankshaft. When the pistons move up and down in the cylinders, they cause the crankshaft to turn. The crankshaft's rotational force passes through a transmission to a driveshaft that ultimately turns the wheels of the vehicle (or, in the case of ships, a propeller). One of the main differences between a spark-ignition (gasoline) engine and a compression-ignition (diesel) engine is in the fuel system and how the fuel is combusted. In a gasoline-powered engine, the spark plug causes the fuel to ignite. In a diesel engine, the extreme high pressure and temperature (around 1,000°F/540°C) of the air in the top of the cylinder causes the fuel to ignite spontaneously, without the use of an electric spark. One key advantage of diesel engines, therefore, is that they don't need spark plugs, an ignition coil, a distributor, or a carburetor. Most diesel engines do, however, need what is called a "glow plug," which preheats the combustion chamber and helps the fuel ignite when the engine is being started. Because a diesel engine compresses air at a much greater force (with a compression ratio of as much as 24 to 1), its engine block and many other internal components are stronger than in a gasoline-powered engine. This results in a heavier, more expensive, but longer-lasting engine. It is not unusual for diesel vehicles to run for 200,000 to 300,000 miles or more before needing an engine overhaul.

Although there have been many advances in diesel engine design, some of the most significant progress over the years has been related to improvements in fuel-injection technology. Germany's Robert Bosch is generally credited with developing the world's first commercial diesel-injection pump. Prototypes were tested as early as 1923, and the pump was mass-produced beginning in 1927.[2] The pump was the key development that made it possible to produce smaller, lighter diesel engines for use in trucks and automobiles. The fuel system on a diesel engine consists of a fuel tank, an injector pump, and the injectors, which, as their name implies, inject fuel into the engine under high pressure. Older diesel engines use what is called indirect injection, in which the fuel is injected into a prechamber, where it is partially combusted before it enters the cylinder. Most newer diesels use what is called direct injection (DI), in which the fuel is injected directly into the cylinder. The older engines are tough and extremely reliable but noisier, and they tend to emit more soot (partially burned fuel). The newer, direct-injection engines are quieter and cleaner and have better acceleration. Modern diesel engines come in a wide range of sizes and designs for many different uses, but their rugged dependability makes them the motive power of choice for heavy commercial and industrial uses and other applications.

COLD-WEATHER ISSUES

Some diesel engines are notoriously hard to start in cold weather. The reason for this is because petrodiesel can begin to "cloud" at about 32°F (0°C), the temperature at which paraffin wax crystals begin to form. Cloudy diesel fuel can clog fuel filters and keep the engine from starting or cause it to stall. When temperatures fall further, diesel fuel reaches its *pour point* (the temperature below which it will not pour). At this point diesel fuel generally stops flowing through fuel lines and diesel engines stop running. Normally the cloud point and the pour point are about 15 to 20 degrees Fahrenheit apart. When the temperature drops even further, to about 10°F (–12°C), diesel fuel reaches its *gel point*, when it becomes the consistency of petroleum jelly. Biodiesel, unfortunately, suffers from all of these problems, but at higher temperatures, making the situation even worse than it is with petrodiesel. The actual cloud point for biodiesel varies depending on what kind of material was used for the feedstock. Biodiesel made from used cooking oil or animal fats will cloud at higher temperatures than biodiesel made from virgin rapeseed/canola oil. But there are even different cloud points among various types of virgin oils. As noted in Chapter 3, palm-oil biodiesel has a high cloud point and generally is not suitable for use in extremely cold climates.

During the winter months in cold regions, petrodiesel fuel is normally altered with special winter formulations to help it perform better. There are also winterizing agents, antigel formulas, and other additives that can lower the cloud point of petrodiesel. Fortunately, these same agents and formulas can be added to biodiesel blends to improve their winter performance as well, although these additives tend to work primarily on the petrodiesel portion of the blend. Another widely used strategy is to use a lower biodiesel blend, such as B5 or B20, during the winter and a higher biodiesel concentration, such as B50 or B100, during the warmer months. For hardcore biodiesel purists who live in cold climates, special cold-weather heating kits are available for most diesel cars and trucks.

SLUDGE AND SLIME

One characteristic of biodiesel that can cause some problems in older diesel engines is its high solvent potential. On engines manufactured before 1994, the rubber seals, hoses, and gaskets will be degraded with the use of high concentrations of biodiesel, especially B100. If these rubber parts are replaced with biodiesel-resistant materials such as Viton, B100 can be used without any problems. But biodiesel's solvent properties can also cause problems with old fuel tanks and fuel lines, which are typically coated with sludge. The biodiesel dissolves the sludge, which then can end up in the fuel filter, causing the engine to malfunction. However, once the sludge has been cleaned out of the fuel system, the engine should run without further trouble on biodiesel.

In warm climates, diesel engines—and especially their fuel tanks—are susceptible to bacteria growth, which can clog the fuel system and cause engine failure. This problem can occur with either petrodiesel or biodiesel fuels. The typical greenish to black bacterial slime grows in the absence of light but in the presence of moisture in the fuel tank as it feeds on the hydrocarbons in the fuel. The bacteria can be eliminated with the use of biocides, which are widely available at automotive parts stores, fuel dealers, and other retail outlets. Keeping the fuel tank as full as possible will minimize the amount of condensation, oxygen, and bacteriological activity.

ENGINE WARRANTIES

Although biodiesel will run in virtually any diesel engine, not all engine manufacturers will honor their engine warranties if biodiesel has been used as a fuel. Today this is less of a problem than it used to be, because more experience in the use of biodiesel has been gained. In the past, the main problem with using biodiesel was its solvent properties, which softened rubber engine parts. Most diesel engines manufactured after 1994 use components that are biodiesel-resistant. In the United States, diesel engines are manufactured for petroleum diesel fuels that meet the requirements set by the American Society for Testing and Materials (ASTM). For petrodiesel fuel, the ASTM standard is ASTM D 975. In 2001, ASTM approved a new standard for biodiesel, D 6751, covering pure biodiesel (B100) for blending with petrodiesel up to 20 percent by volume. Most major U.S. diesel engine manufacturers and some carmakers now say that the use of blends up to B5 that meet the ASTM standard will not void their warranties. Many now approve the use of B20, and a few, such as Case IH, Fairbanks Morse, and New Holland have given the nod to B100 in some or all of their equipment. For blends above B5, the individual engine manufacturer should be contacted just to be sure.

Diesel engine manufacturers guarantee their engines against defects in "materials and workmanship." They do not warrant the fuels. Consequently, if there is a fuel-related problem with an engine (whether it is petrodiesel or biodiesel), the fuel manufacturer is responsible. This is why it is important to purchase biodiesel fuel from a commercial supplier who will certify that its product meets the ASTM standard (unless you are making fuel for your own vehicle and don't care about warranty issues). The National Biodiesel Board has formed the National Biodiesel Accreditation Commission (NBAC) to support biodiesel fuel quality standards in the United States. The NBAC issues a "Certified Biodiesel Marketer" seal of approval to biodiesel marketers that have met its standards. People who make their own biodiesel have only themselves to blame if they damage their engines with off-spec fuel. This seems to be a risk most of them are willing to take, but it is something that large fleet operators simply cannot afford to do.

In Europe, biodiesel standards were set by individual countries for many years. Austria's first standard was issued in 1991 and was the basis for numerous engine warranties. A German standard, DIN 51606, was formalized in 1997 and has been widely used as a guide for other European national standards. After many years of effort, a new Europea-wide biodiesel standard, DIN EN 14214, was published in October 2003 that essentially superseded previous European national standards in March 2004. Many European engine and car manufacturers' warranties cover the use of biodiesel in various blends.

WHERE THE DIESEL ENGINE IS USED

Now it's time to take a look at where diesel engines are used today. The transportation sector is responsible for more than 70 percent of the petroleum consumed in the United States and one-third of U.S. carbon dioxide emissions. The same general statistics apply to Europe as well. Over-the-road vehicles account for the vast majority of diesel fuel use (35 billion gallons, or about 59 percent of middle-distillate fuels) in the United States. When most Americans think of transportation, they generally think first of automobiles. But the transport sector extends well beyond cars to include trucks, buses, trains, boats, and planes. And the vast majority of these other types of transport are powered by diesel engines or, in the case of commercial aircraft, on diesel-like aviation fuels. It's no exaggeration to say that the vast majority of the world's heavy transport sector is diesel-powered. This offers obvious potential for a wide range of uses for biodiesel. Here is a brief overview of those applications.

Automobiles
Diesel-powered cars were first introduced to the European market in 1936, and their use in most European countries has grown steadily ever since. Diesel cars became popular as alternate-fueled vehicles in the

United States during the oil crises of the 1970s. The main reasons for this were that diesel vehicles got better mileage and diesel fuel was relatively inexpensive in those days. (The fact that diesel was still a petroleum-based fuel seems to have escaped the attention of a lot of people initially.) But most American drivers were not impressed by the slower acceleration and noisier operation of diesel automobiles, some of which were poorly designed. By the mid-1980s, demand for diesel cars had fallen off, and production declined as well. With the return of cheap gasoline after the oil crisis ended, U.S. drivers went back to driving their gas-guzzlers. Meanwhile, much of the rest of the world continued to drive diesel-powered cars and light trucks, due mainly to higher fuel prices. In the late 1990s, diesel cars and pickup trucks returned to the U.S. market with new DI engines. Volkswagen and Mercedes-Benz both offered diesel-powered cars, while Dodge and Ford offered diesel-powered pickups. Despite the reintroduction of diesels to the United States, they represent only about 3.5 percent of total personal vehicles (mainly pickup trucks) driven in this country, while in Europe diesel cars represent about 50 percent. Diesel cars in Europe are increasingly popular and their numbers continue to climb. Using biodiesel as a fuel for automobiles in countries that have larger percentages of diesel-powered cars obviously has more potential for reducing petrodiesel fuel consumption in the automotive sector there than in the United States.

Fleets

In the United States, biodiesel is being used in more than 700 fleets across the country, and that number is growing rapidly. All four branches of the U.S. military, NASA, the Postal Service, L.L. Bean, and dozens of school districts and municipal fleets have been jumping on the biodiesel bandwagon in recent years. The Postal Service alone used half a million gallons of biodiesel fuel in 2002. In 2000, Lambert International Airport in St. Louis switched to a B20 blend in 300 of its vehicles, including deicer tankers, snowplows, high-speed runway brooms, lawn equipment, passenger shuttles, and aircraft rescue fire trucks. "The reliability is great," says Frank Williams, fleet maintenance foreman. "We've had sustained

25-below wind chill factor for multiple days and never had a problem with the B20. Most of the vehicle operators didn't even know we had switched to biodiesel."[3] Fort Wayne, Indiana, has switched to a biodiesel blend for about 300 of its municipal vehicles. In California, the city of Berkeley runs its trucks on B20. The city of Boulder, Colorado, has been testing biodiesel in its tractors, dump trucks, and fire trucks, and as of the end of 2007, it was powering about 100 vehicles or other equipment with B20. In April 2007, San Francisco announced that it was switching to B20 in its entire diesel fleet of more than 1,500 vehicles, and achieved its goal by the end of the year. This dramatic shift makes San Francisco the largest city in the United States to use B20 fleet-wide.

Unlike compressed natural gas, biodiesel does not require expensive engine retrofits, making the switch to biodiesel attractive to large fleet owners. And since many municipalities across the United States have been suffering from severe budget limitations in recent years, they can keep their older vehicles running longer while lowering emissions at the same time by using biodiesel.

Mass Transit
Some of the most highly visible users of diesel engines are mass transit fleets. Diesel engines power about 80 percent of the more than 81,000 active transit buses in the United States, according to the American Public Transportation Association. Since most large urban bus systems typically have hundreds or even thousands of buses on the road, they can be significant contributors to urban air pollution. An increasing number of transit fleets are reporting positive experiences with biodiesel. Among the many city bus fleets in the United States using biodiesel are those in Cedar Rapids, Iowa; Bloomington, Indiana; St. Louis, Missouri; Oklahoma City, Oklahoma; Olympia and Seattle, Washington; Knoxville, Tennessee; Raleigh, North Carolina; and Springfield, Illinois.

One of the most successful early biodiesel initiatives in a city transit fleet has taken place in Graz, Austria. Beginning in 1994, the Grazer Verkehrsbetriebe (GVB) began field tests with two of its city buses running on biodiesel made from recycled frying oil. After many years of continued positive test results, the GVB converted its entire fleet to biodiesel in 2005.[4] Many other bus fleets around the world are experimenting with (or regularly using) biodiesel. In Italy, public transit buses in Florence, Gorgonzola, Padua, and Perugia now run on biodiesel. French buses in Paris, Bordeaux, Dijon, Dunkirk, Grenoble, and Strasbourg are using a biodiesel blend. Biodiesel systems in Canada include those of Brampton, Ontario; Halifax, Nova Scotia; and Saskatoon, Saskatchewan. The use of biodiesel in urban transit systems around the world is growing exponentially.

School buses are one of the largest mass-transit programs in the United States. Approximately 460,000 school buses transport more than 24 million children to and from schools and school-related activities every school day.[5] The vast majority of these buses are powered by diesel engines. The use of biodiesel in school bus fleets is growing not only due to environmental awareness but also because of health concerns about students. Several thousand school buses in more than 100 fleets are currently operating successfully on biodiesel blends in the United States, including those in school districts in Olympia and Chicago, Illinois; Clark County, Nevada; Denver, Colorado; and Medford, New Jersey. "It's been absolutely fantastic," said Joe Biluck, Jr., director of operations and technology for the Medford district. "We've had no downtime as a result of this fuel. We've never had a fuel system gel up on us and we've run down to temperatures of 11 degrees below zero and haven't experienced any problems."[6] Biodiesel-powered school buses can be found throughout many parts of Europe as well.

Trucks and Heavy Equipment
Commercial trucks and heavy industrial equipment are almost entirely powered by diesel engines. The use of biodiesel in heavy construction and mining equipment has been growing in the United States in recent years. Carmeuse Lime Mines, which operates two of the nation's largest lime mines in Maysville and Butler, Kentucky, runs about 150 pieces of underground equipment on a biodiesel blend and is now one of the largest single users of biodiesel in the state. In 2002, Alcoa Davenport Works in Davenport, Iowa, switched all of its diesel-powered equipment to a B20

blend. This aluminum manufacturing plant used approximately 250,000 gallons of biodiesel in 2003, roughly the equivalent of all the soybeans from a 1,000-acre farm, according to the National Biodiesel Board. However, except for a number of tests conducted by various research organizations, the use of biodiesel by the commercial, over-the-road trucking industry initially made very slow progress in the United States due to the highly competitive nature of the industry and the higher cost of biodiesel fuel. This situation has begun to change recently with growing customer demand for "green" shippers and increased acceptance of biodiesel by more and more truckers around the nation. Although the cost of biodiesel is still an issue, the trucking industry will almost certainly use more biodiesel in the years ahead because of its many other benefits, especially reduced maintenance costs and increased engine life. In Europe and elsewhere around the world, many trucking companies and heavy equipment operators are using biodiesel as well.

Farm Equipment

Most of the early biodiesel experiments were conducted on diesel farm tractor engines. The main idea was to guarantee the ability to grow food in the face of a future oil crisis. That rationale is just as valid today. But diesel engines power much more than tractors on most large modern farms. Biodiesel can be used as a fuel or fuel additive in virtually any diesel-powered farm equipment, such as trucks, harvesters, balers, irrigation pumps, and other machinery. This provides farmers with the opportunity to help create demand for biodiesel by using it themselves in their day-to-day operations while they are growing feedstock to make more biodiesel. Biodiesel blends offer improved lubricating properties over straight petrodiesel, as well as lower maintenance costs, less downtime, and increased equipment life.

One of the principal feedstocks for biodiesel in the United States is soybeans, a major crop produced by almost 400,000 farmers in 29 states. Total acreage planted to soybeans in 2006 was 75.5 million acres. A U.S. Department of Agriculture study completed in 2001 found that an average annual increase of the equivalent of 200 million gallons of soy-based biodiesel demand would boost total crop cash receipts by $5.2 billion cumulatively by 2010.[7] The farming sector itself in the United States consumes about 3.2 billion gallons of diesel fuel every year, representing about 5 percent of the middle-distillate fuel market, according to the Energy Information Administration. Biodiesel use in farming equipment around the world is growing steadily.

Boats

Rudolf Diesel's engine was adapted for marine use as early as 1903. Since then, diesel engines have spread to virtually every corner of the world's marine environments. Unfortunately, diesel engines can cause considerable environmental damage, especially in the case of a petrodiesel fuel spill. But it is precisely this environmental fragility that makes marine use of biodiesel so attractive. Tests have concluded that biodiesel is not harmful to fish, and that when spilled in water, biodiesel will be 95 percent degraded after 28 days as compared with only 40 percent for petrodiesel in the same time period.[8] Diesel engines are used in a wide range of marine applications, including merchant ships, cruise ships, ferryboats, and powerboats, and even in sailboats as auxiliary engines. Electrical generators, bilge pumps, and other onboard equipment also can be diesel-powered. One of the reasons petrodiesel fuel is popular in boats is its low risk of spontaneous combustion when compared with gasoline. Biodiesel, with its higher flash point, is even safer than petrodiesel.

Vessel operators report a noticeable improvement in the odor of engine exhaust when biodiesel is used instead of petrodiesel, making it less objectionable for engine crew and passengers alike. Anyone who has sat on deck downwind of the exhaust stack on a diesel-powered cruise ship knows that this is a less-than-relaxing experience. Another benefit of marine use is that biodiesel tends to keep boats cleaner. Diesel exhaust leaves a film of soot on the stern, and biodiesel minimizes or eliminates that problem. The marine industry presently accounts for about 10 percent of the petrodiesel consumed in the United States, so the potential for the increased use of biodiesel, especially in sensitive or protected waterway areas, is fairly substantial. Currently in the United States much

of the emphasis on biodiesel is focused on recreational boats, which consume about 95 million gallons of diesel fuel every year, according to the
National Biodiesel Board. However, a number of companies that operate
ferries have been testing various blends of biodiesel. Marine charter boat
operators and other maritime businesses around the world are beginning
to use biodiesel as well.

Trains

The testing of biodiesel as a fuel for diesel locomotives has been conducted on a limited basis in various countries. In December 2001, the Tri-
County Commuter Rail Authority in southern Florida began running one
of its locomotives on B100. The locomotive operated for three months in
regular passenger service without any problems, and Tri-County continues to consider the future use of biodiesel. India Railways conducted
its first trial run for B5 in a locomotive on the Delhi-Amritsar Shatabdi
Express, one of its high-speed passenger trains, in December 2002. The
railway plans to use locally produced biodiesel on more of its trains in the
future. In December 2003, the transportation company América Latina
Logística (ALL), with 15,000 kilometers of railroad in Argentina and
southern Brazil, decided to replace about a quarter of the petrodiesel fuel
it consumes with B20. Preliminary tests were reportedly conducted on
two trains in early 2004, with subsequent tests of B100, and there are
plans to expand biodiesel use to the entire system with a total of 580
trains.[9] In October 2004, the Minnesota Prairie Line Railroad began
fueling its locomotives with a B2 biodiesel blend. The 94-mile short-line
railroad runs through five southern Minnesota counties, from Hanley
Falls to Norwood. And in June 2007, Virgin Trains in the UK announced
a B20 trial with one of its Voyagers in regularly scheduled service. During
the trial the biodiesel train will run across much of Britain, from
Birmingham to Scotland, in South Wales, North East England, the North
West, Lake District, West Country, the South West, and the South Coast.
There is a good deal of potential for additional expansion of biodiesel in
the railroad industry worldwide, especially in congested urban areas
where air quality is a particular concern.

Electrical Generators

Most larger electrical generators, whether for primary or standby generation, are powered by diesel engines. Even individual homeowners can use
smaller diesel-powered generators in conjunction with off-grid energy systems or as emergency backup. Isolated communities in wilderness areas or
on islands frequently rely on diesel generators as their sole source of electricity. Importing petrodiesel fuel for generator use in isolated locations
can be very expensive (and potentially hazardous for sensitive environments). Large institutions such as hospitals, universities, military installations, and some businesses also use diesel-powered generators for
emergency backup when grid power fails. All of these installations offer
potential for expanded biodiesel use, especially in cities, where backup
generators can be significant sources of air pollution if they burn
petrodiesel fuel.

Aircraft

In the United States, a number of studies have been conducted on the
possible use of biodiesel as an aviation fuel for both military and civilian
use. A 1995 study at Purdue University funded by the Indiana Soybean
Growers Association found that biodiesel blended with jet fuel showed
potential for use in aircraft with jet turbine engines, and that further
testing was warranted.[10] The Renewable Aviation Fuel Development
Center at Baylor University also has initiated research, development, and
testing programs of blends of biodiesel and Jet A fuel for turbine engine
aircraft. In October 2007, the world's first biodiesel-fueled jet using B100
was test-flown from the Reno-Stead airport in Nevada. The jet, a retired
former Czechoslovakian L-29 military jet, was originally designed to run
on a variety of fuels, making it ideal for the experiment. The jet, piloted
by Carol Sugars and Douglas Rodante, flew up to 17,000 feet and no significant difference in performance was observed. Additional flights are
planned.[11] And in February 2008, Virgin Atlantic Airways successfully
flew a 747 from London to Amsterdam partly on a B20 blend with
kerosene-based aviation fuel.

BIODIESEL AS LUBRICANT AND SOLVENT

Potential markets for biodiesel extend beyond the transportation and electrical-generation sectors. Even when used in low concentrations such as B2 or B5, biodiesel can offer a significant (up to 65 percent) lubricity advantage in any diesel engine. The use of low blends of biodiesel in combination with the new ultra-low-sulfur diesel (ULSD) fuel in the United States adds the needed lubricity that has been removed as part of the refining process for the new ULSD fuel. Biodiesel can even be used straight as a machinery lubricant. It is also possible to use biodiesel instead of kerosene in some camping lanterns and stoves. Biodiesel's solvent properties may be used to clean dirty or greasy engine or other machine parts; left in a bucket of B100, dirty parts are usually clean by the next morning.

But biodiesel can be used to clean more than dirty machinery. A series of laboratory experiments were conducted at the School of Ocean Science at the University of Wales to test the potential of biodiesel as a cleaning agent for shorelines contaminated by crude-oil spills. Pure vegetable-oil biodiesels (rapeseed or soybean) were shown to have a considerable capacity to dissolve crude oil. In a separate study in Texas, a commercial biosolvent, CytoSol, based on vegetable-oil methyl esters similar to biodiesel, was shown to be effective in coagulating the crude oil and allowing it to float to the surface of the water, where it can be collected. CytoSol was licensed by the California Department of Fish and Game as a shoreline cleaning agent in 1997.[12]

HEATING WITH BIODIESEL

In the United States, until fairly recently, biodiesel has been promoted mainly as a fuel for diesel-powered vehicles. But many people don't realize that biodiesel also can be used as a heating fuel additive or replacement in a standard oil-fired furnace or boiler. That's because Number 2 heating oil (another middle-distillate fuel) is virtually the same as standard petrodiesel vehicle fuel, and biodiesel can be mixed at any percentage with Number 2 oil. When used for space heating, biodiesel is sometimes referred to as "biofuel" or "bioheat" in the United States. The conversion process for an oil-fired furnace or boiler is just as simple as the conversion for a diesel engine; just add the biodiesel to the fuel tank. No new heating appliance or expensive retrofitting is required as long as the blend is B20 or less. For higher concentrations of biodiesel, a small rubber seal in the fuel pump (and possibly gaskets, O-rings, or hoses on some models) needs to be replaced with a synthetic seal made out of Viton or Teflon. What's more, since it's located in a protected space, using biodiesel as a home heating fuel has virtually none of the cold-weather operating problems that are associated with using biodiesel in vehicles during the winter. Using biodiesel as a heating fuel is such a simple idea, you have to wonder why nobody thought of it sooner. Actually, someone did, because biodiesel has been used as a heating fuel in Italy, France, and a number of other European countries for many years. Now the United States is beginning to catch up. "Bioheat," as the heating industry now officially refers to it, is available from a rapidly increasing number of fuel distributors, especially in the Northeast.

Special Considerations
As mentioned previously, biodiesel has high solvent properties and tends to dissolve the sludge that often coats the insides of old fuel tanks and fuel lines. When used in a heating system, this can potentially cause a clogged fuel filter or burner head, so biodiesel should be added carefully at first to old heating systems. Until all the sludge in the fuel tank has been dissolved, keeping an extra fuel filter on hand also might be a good idea for the first heating season.

Biodiesel should be stored in an indoor (or underground) storage tank because biodiesel, like Number 2 heating oil, will gel if stored outside in extremely cold weather. The pour point (the temperature below which the fuel will not pour) must be kept in mind if biodiesel is used. The pour point for Number 2 fuel oil is around $-11°F$ ($-24°C$). Although the actual pour-point temperature for biodiesel varies, depending on its concentration and original feedstock, it is consistently higher than that of Number 2 fuel oil. Consequently, biodiesel fuel should be stored at temperatures above its pour point.

Some people actually have used B100 to heat their homes with no problem. However, other people burning high percentages of biodiesel have experienced seal failures in their fuel pumps or ignition failures in the burners. The leaky rubber seal (or pump) usually can be repaired or replaced by a heating service technician in a short time, but the potential for this problem should be kept in mind if a high concentration of biodiesel is used. Some oil burner manufacturers are testing new seal materials to eliminate this problem in future burner models. In addition to occasional ignition failures with B100, some burners malfunction upon startup because some makes of CAD cells used to ensure that the burner ignites properly can't "see" the biodiesel flame (which is in a slightly different part of the visible light spectrum from Number 2 oil) and shuts the burner off as a safety measure. An oil burner conversion kit may be needed to solve this problem.

Great Potential

The potential for reduced reliance on imported oil with the increased use of biodiesel as a heating fuel additive is substantial. In fact, if everyone in the northeastern United States used just a B5 blend, it could save 50 million gallons of regular heating oil a year, according to officials at the USDA Agricultural Experiment Station in Beltsville, Maryland. The experiment station has been heating its many buildings successfully with a biodiesel blend since 1999. At first the station staff used a B5 blend, but in 2001, encouraged by the test results, they switched to B20 and haven't experienced any problems. "Using biodiesel offers an opportunity to reduce emissions, especially particulate matter and hydrocarbons, and that's a great advantage," says John Van de Vaarst, deputy area director, who is responsible for facilities management and operations. "I used to refer to biodiesel as an alternative fuel, but now I call it an 'American fuel, made by American farmers.' I think it's an obvious strategy to help clean up the environment and reduce our dependency on foreign oil."[13]

LIBRARY
WAUKESHA COUNTY TECHNICAL COLLEGE
800 MAIN STREET
PEWAUKEE, WI 53072

WITHDRAWN

Another series of tests on the use of biodiesel for space heating was conducted at the Brookhaven National Laboratory on Long Island. Sponsored by the National Renewable Energy Laboratory and the U.S. Department of Energy, the 2001 test report found that biodiesel blends at or below B30 can replace fuel oil with no noticeable changes in performance. Burning of the blends also reduced emissions of carbon monoxide and nitrogen oxide.

"There has been a lot of interest, particularly in the Northeast, in using biodiesel as a home heating oil," says Jenna Higgins, director of communications for the National Biodiesel Board in Jefferson City, Missouri. "Bioheat is really starting to take off and we're seeing more and more retailers offering it."[14] Roughly three out of four homes in this country that use oil heat are located in the Northeast, so the potential for expanding the use of biodiesel in the region is substantial. But can biodiesel meet the increased demand for the heating market? Residential consumption of Number 2 heating oil in 2002 was around 6 billion gallons nationwide, according to the Energy Information Agency. That figure has not substantially changed since then. Assuming that every homeowner in the United States currently heating with oil switched to B20 (admittedly a large assumption), that would require about 1.2 billion gallons of biodiesel. United States production of biodiesel could probably cover this amount in the near future, assuming that this was the only use for biodiesel, which, of course, is not the case. Still, bioheat is an excellent strategy to clean up fairly dirty emissions quickly and easily without large retrofit costs.

Real-World Tests
But is bioheat really safe and effective? Bob Cerio, the former energy manager for the Warwick, Rhode Island, school district says it is. The district ran biodiesel-fuel tests, originally sponsored by the National Renewable Energy Lab, in three of its schools beginning in 2001. During the first heating season, the district ran three different percentages of biodiesel (B10, B15, and B20), as well as a Number 2 fuel-oil control in a fourth school. "It just worked very, very well for us," Cerio reports. "We

had three different types of burners, three different types of boilers, and three different sizes, so we had an opportunity to test a wide spectrum of capacities. With the smaller boilers, we were able to get similar test data to what people would be experiencing in their homes," he adds.[15]

After a successful first season, Cerio switched to a B20 blend in the test schools for the 2002–03 heating season without any problems, and he stopped experimenting with any other concentrations. A number of tests for boiler efficiency and emissions were conducted. "We did not see any change in efficiency, but we saw a reduction in sulfur dioxide, nitrous oxides, carbon monoxide, and carbon dioxide," he reports. "We also discovered that our boilers were running much cleaner, so that saved us quite a lot of work cleaning them." Cerio is enthusiastic about the use of biodiesel as a heating fuel. "It's a very easy match for home heating, particularly if you have an indoor storage tank," he says. "Other than that, there really isn't anything that has to be done in order to use it."

Another biodiesel field trial involving about 100 homes, sponsored by the U.S. Department of Energy, the New York State Energy Research and Development Authority (NYSERDA), and the National Oilheat Research Alliance, was conducted by Abbott & Mills Inc., a fuel-oil dealer in Newburgh, New York. The tests with B20, which ran for about three years, went well, according to Ralph Mills, the company's general manager. "We have no news to report, which is good news," he says. "We've had no service problems associated with the fuel at all. The conclusion that we've come to is that B20 is a viable replacement for traditional fuel oil."[16] Continuing the experiment at the next level, Abbott & Mills has been heating its office with B100.

Clearly, bioheat works. But there are two main obstacles to heating a home in the United States with biodiesel: price and availability. Biodiesel generally costs more than Number 2 heating oil. How much more depends on who the supplier is and the quantity purchased. A federal biodiesel blender's credit, which became available in January 2005, allows blended bioheat fuel to be sold at roughly the same price as conventional heating oil, although not all dealers have passed those savings on to their customers. Nationally, the price of pure biodiesel ranges from about $3.50

to $5.00 per gallon, depending on the time of year and supply/demand. Some fuel dealers have offered up to a B5 blend to their customers at no increase in price, but in any case a B5 blend should be only a few cents per gallon more than regular Number 2 heating oil. The price for B20 can be around 20 cents per gallon more than Number 2 oil.

Finding a local source of biodiesel for home heating can be a problem in some areas. Although there are now more than 170 major producers (and numerous small producers) of biodiesel scattered around the country, as well as hundreds of local distributors, the vast majority of the distributors are clustered in the Midwest, where biodiesel feedstocks are grown. Finding a biodiesel source in some parts of the country can still be a chal-lenge, but in the past few years it definitely has gotten easier. And locating a fuel-oil dealer that offered biodiesel home deliveries was a real challenge until fairly recently, even in New England, where 2.2 billion gallons of heating oil are consumed every winter. But that's changing dramatically as bioheat becomes available in an increasing number of communities.

Catching On

The state of Maine is well known for its rugged winters, and bioheat would seem to be an obvious choice for the state's many environmentally conscious citizens. One industry pioneer capitalizing on this is Frontier Energy Inc. of South China, Maine. In 2002, sensing a new market opportunity, this offshoot of Frontier Oil Company was the first in the nation to offer biodiesel to homeowners in its regular delivery area between Augusta and Waterville. The company currently is offering—and actively promoting—a B5 "Basic Bioheat" blend as well as a B20 "Premium Bioheat" blend. For those who want it, B100 is also available, although the company doesn't recommend using it as a heating fuel at that concentration. "It's going very well so far," says Joel Glatz, Frontier Energy's vice president. "We're probably selling about the same amount for vehicular use as we are for heating use at this point, but I think the heating application is what is really going to catch on in this state. We use about 300 million gallons in Maine for heating oil and about 150 mil-lion gallons for transportation annually, so, obviously, there is a much larger market for heating in this state."[17] Frontier's bioheat marketing strategy has worked; homeowner response has been extremely positive. "It's been fantastic," Glatz reports. "Those who have used it love it. The comment I usually get is, 'I can't tell the difference,' which is exactly what you want to hear." In early 2004, Frontier Energy reached an agreement with the C. N. Brown Company of South Paris, Maine, to provide the state of Maine with 40,000 gallons of B20 to heat state office buildings. The state capitol building, the governor's mansion, and state offices that rely on Number 2 heating oil now use bioheat. Approximately six fuel dealers presently offer bioheat across the state.

In Vermont, also famous for frigid winters and many environmentally aware citizens, the bioheat sector is growing rapidly. A number of businesses, but especially established fuel dealers, in cooperation with the Vermont Biofuels Association, the Vermont Sustainable Jobs Fund, and other key players, have been working hard to help facilitate the delivery of biodiesel home heating blends across the state.[18] Just a few years ago, there were virtually no fuel dealers who carried biodiesel blends. As of January 2008, there were approximately 20, and the number continues to grow along with demand.

According to the U.S. Census, more than half of households in New Hampshire use Number 2 oil to heat their homes. Early on, the Granite State Clean Cities Coalition played a key role in encouraging a number of bioheat pilot projects across the state. In 2004, sensing a business opportunity, Rymes Propane & Oils in Concord, New Hampshire, introduced biodiesel and bioheat to its customers. In the following years, a number of other fuel dealers followed Rymes' lead, and bioheat is now available to many homeowners in the state.

In Massachusetts, bioheat is coming on strong. In the fall of 2003, Alliance Energy Services of Holyoke, Massachusetts, began to offer biodiesel as a heating fuel in a B20 blend. "Biofuel is readily available, and it makes sense for a lot of people," says Stephan Chase, the company's president. Alliance, which has been actively promoting its biofuel, has a growing number of customers. "It will be interesting to see what happens," Chase says. "The biofuel is a good product, and the Pioneer Valley has a lot of residents who are concerned about the environment, so it's a good combination; we should do very well with it here."[19] In 2004, Alliance began adding B3 to *all* its home heating oil at no additional cost to its customers. And in July 2007, a new state mandate was announced that requires a minimum blend of 3 percent bioheat for all state agencies that use Number 2 heating oil in their operations. The minimum level will rise to B10 by 2012.[20] This initiative is expected to increase both use and availability of bioheat across the state.

Farther south on the East Coast, where winter weather is somewhat milder, Tevis Oil of Westminster, Maryland, delivered its first B20 blend of heating oil to a residential customer in Upperco, Maryland, in November 2003. "This was the first delivery of soy biodiesel for use in home heating that we know of in this area," said Jack Tevis, president of S. H. Tevis and Son, which operates Tevis Oil. "Soy and other biodiesel fuels are used in the Midwest and New England to heat homes and run farm equipment, but it's still a fairly new concept here in the Mid-Atlantic region."[21] Since then, customer demand for bioheat has grown in Maryland, and as of October 2006, there were approximately six fuel dealers offering various blends of biodiesel to homeowners across the state, and the number continues to grow. More and more fuel dealers in the surrounding states of Virginia, Pennsylvania, and New Jersey are now offering bioheat as well.

The future of biodiesel as a home heating fuel looks good. It can provide farmers with a steady cash crop, help boost the economy, reduce dependence on foreign oil, and benefit the environment all at the same time. What's more, the use of bioheat is far beyond the experimentation phase, and many people are now asking for bioheat from their local fuel dealers. Those dealers are responding to the increased demand.

PART TWO

Biodiesel
around the
World

5

Europe, the Global Leader

With a combined output of 4.9 million metric tons (1.5 billion gallons), Europe is the global leader in biodiesel production, accounting for about 80 percent of the total world market as of 2006. (Output for 2007 was about 5.4 million tons.) Germany, France, and Italy combined produced around five times more biodiesel than the United States in 2006, although the U.S. is gaining rapidly and has moved into the number two position after Germany. As of 2006, 55 percent of European Union (EU) transport needs were met with diesel fuels, offering a huge opportunity for additional growth in the biodiesel sector.

Biodiesel has been manufactured on an industrial scale in Europe since 1992, mainly in response to the supportive actions of various European Union (EU) institutions. Today, there are more than 185 plants in the EU (up dramatically from approximately 50 in 2004) with a production capacity of about 11.2 million metric tons (3.3 billion gallons) annually as of 2007. There are an additional 58 plants under construction as well, offering the potential for continued increases in production capacity. These facilities are located primarily in Germany, France, Italy, the UK, Austria, Poland, the Czech Republic, Spain, Portugal, Slovakia, and Denmark, according to the European Biodiesel Board (EBB). Germany accounts for about half of the total EU capacity, with more than 4.4 million metric tons. Italy is now second with 1,366,000 tons; France is third with 780,000 tons; the UK fourth with 657,000 tons; and Spain is fifth with 508,000 tons capacity. These rankings could change substantially if all the plants under construction come online in the next few years. With all this activity, however, Europe has managed to build a large overcapacity for biodiesel production, especially in Germany, where the industry has run into serious problems.

Actual biodiesel production is roughly half of these figures. Nevertheless, production increased a dramatic 54 percent between 2005 and 2006, continuing a strong growth pattern of about 50 percent annually in recent years. Germany produced about 2,662,000 tons, France about 743,000, Italy about 447,000, and the UK 192,000 tons, according to EBB statistics for 2006.

After the initial successes of the first biodiesel pilot plant at the Silberberg Agricultural College in Styria, Austria, in 1985, some small-scale commercial facilities began biodiesel production in other locations, but the quality of their output was uneven. Other early plants capable of more consistent quality followed, and in 1991 the first industrial-scale biodiesel plant with a production capacity of 10,000 metric tons per year was started in Aschach, Austria, along with some small-scale plants in France and the former Czechoslovakia. This was followed in 1992 by a much larger plant in Livorno, Italy, with an annual capacity of 80,000 metric tons.

THE COMMON AGRICULTURAL POLICY

In 1992, a number of changes made to the European Union's Common Agricultural Policy (CAP) had a dramatic impact on the biodiesel industry. In response to increasing surpluses of grains and other crops, the new policy established a set-aside program that prohibited farmers from growing food or feed crops on 10 percent of their arable land in the EU, while simultaneously allowing them to grow crops such as rapeseed, sunflowers, or soybeans on the set-aside lands for "industrial purposes." The production of biodiesel feedstock was one possible option under this provision, and beginning the following year, biodiesel production began to grow dramatically. In 1992, Germany produced only 5,000 metric tons of

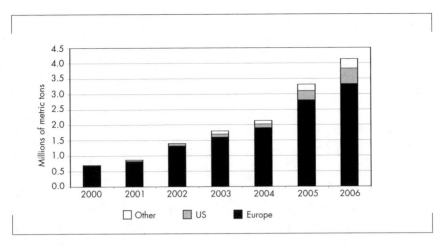

Figure 4. Shares of World Biodiesel Production (2000–2006) European Biodiesel Board

biodiesel, while France produced virtually none. By the next year, German production had doubled to 10,000 metric tons, while French production had soared to 20,000 metric tons.[1] As demand grew, additional plants were constructed all over Europe in locations such as Leer, Germany (80,000 tons per year) in 1994, and Rouen, France (120,000 metric tons per year) in 1996, as well as in Sweden and elsewhere.

In 2004, the new, expanded EU-25,[2] including the so-called new accession countries, offered the potential to increase the set-aside lands from previous levels of about 7 million hectares (17.3 million acres) to about 12 million hectares (29.6 million acres), opening up the possibility for a large increase in biofuels feedstock production. Actual acreage planted to biofuel crops in 2005, however, was 2.6 million hectares; 2.4 million hectares to produce biodiesel (95 percent rapeseed, 5 percent sunflower), and 0.2 million hectares for ethanol (51 percent sugar beets, 49 percent wheat).[3]

However, by 2007 the surplus grain situation had changed dramatically, and in September the EU agriculture ministers decided to suspend temporarily the set-aside program for 2008 in an attempt to bring down soaring wheat prices. This was in recognition of the fact that the EU farm policy was helping to drive prices for cereals up to historic highs and increasing the cost of bread, pasta, and meat products for consumers, sparking angry protests. Poor harvests and increased demand in recent years had reduced grain surpluses to almost nothing.[4] The future fate of the set-aside program and the long-term effect this will have on biofuels crop production, is unclear. But in the short term, this will definitely reduce the amount of domestic agricultural land available for biofuels crop production. The EU Commission expects that there will be an increase in oilseed prices of between 8 to 10 percent for rapeseed and about 15 percent for sunflower seed through 2020.

The European nations have not, however, relied entirely on virgin vegetable-oil crops for all of their biodiesel feedstock. Used frying oils (UFO)[5] have played an increasingly important role in the picture. Although there were a number of limited experiments in prior years, the start of industrial biodiesel production from 100 percent UFO began in Mureck, Austria, in 1994. Subsequently, the UFO sector of the biodiesel market made especially impressive gains between 1998 and 1999. This was a time when set-aside lands for industrial crop production had been cut back to 5 percent, oilseed costs were high, and petrodiesel prices were at record lows, causing substantial losses for many biodiesel producers and prompting them to search for lower-cost feedstocks. They found what they were looking for with UFO. Despite the fact that it is limited in supply, UFO does offer a relatively low-cost feedstock option for many producers, although recent high demand has prompted price increases even for this commodity.

But beginning in 1999 and continuing into mid-2002, vegetable-oil prices dropped again while petrodiesel prices were, on average, relatively high, causing a boom in the growth of biodiesel manufacturing plants, especially in Germany, where the number of plants increased by a remarkable 200 percent. Numerous construction projects were approved, especially in 2000, dramatically increasing the country's production capacity. Biodiesel production in Germany catapulted from 130,000 metric tons in 1999 to around 500,000 metric tons in 2002, with much of that increase taking place in the former East Germany. But by 2004, with vegetable-oil prices on the international market again moving higher, German biodiesel production capacity outstripped demand, causing considerable concern in the industry. "The current high vegetable-oil prices in Europe are definitely causing problems for biodiesel producers," says Raffaello Garofalo, secretary-general of the European Biodiesel Board. "This is happening at the same time that we have legislation in the member states promoting biofuel. With the right foot we are going forward, but with the left foot we are lagging behind, and the gap between the two is getting wider and wider. It's an awkward situation."[6] The strong upward trend in vegetable-oil prices from 2005 through 2008, reflecting growing demand for food *and* biofuel uses around the world, have continued to challenge the biodiesel industry in the EU and beyond.

PRODUCTION ISSUES

Since feedstock generally represents about 85 percent[7] of the direct production cost of biodiesel, it's clear that anything that affects the cost of vegetable oils will have a significant impact on the price of biodiesel made from these oils. The price of vegetable oil is driven by international markets beyond the control of biodiesel producers, and the inevitable ups and downs in those markets have caused the biodiesel industry significant challenges during periods of high vegetable-oil prices, like those that occurred in 1997–98 and again in 2004–08. Nevertheless, in August 2000, for the first time in history, the crude petroleum oil price climbed above the price for vegetable oil on the world market. Whether this phenomenon will be repeated in the future as the world supply of petroleum oil continues to decrease and demand continues to increase is hard to predict. But regardless of the fluctuations in the international commodities markets, the historical disparity between the cost of production for biodiesel and that for petrodiesel has been substantial, putting biodiesel at a significant price disadvantage. For example, the cost of producing biodiesel in Europe was around 500 euros per 1,000 liters in 2001, while the cost of petroleum-based diesel was 200 to 250 euros per 1,000 liters.[8] More recently, the cost of production for biodiesel in Germany has risen to 640 euros per 1,000 liters, indicating that this relative disadvantage remains a problem. This general price disparity is expected to continue for the foreseeable future, and it is the main rationale behind various tax strategies that are aimed at leveling the playing field.

There has been one development, however, that has offered biodiesel producers a little more flexibility in recent years. Initially, most of the biodiesel research in Europe was focused on rapeseed methyl ester (RME) and, later, on used frying oils (UFO). But after a lot of additional research and screening of potential feedstocks, it was determined that good biodiesel could be made from a wide range of feedstocks and even from multifeedstock blends (MFB). This discovery was one of the key factors in the subsequent development of modern biodiesel production facilities that can take advantage of the lowest-cost feedstocks or blends of feedstocks. Since the price and availability of various feedstocks can change rapidly (the ups and downs in the international vegetable-oil market in recent years are a prime example), designing manufacturing facilities that can alter their ingredients mix quickly has been extremely important. This design goal has been achieved in some of the most up-to-date facilities, which can select the desired "recipe" from a range of options stored in their process-control systems. "The selection of an efficient process technology is very important," says Werner Körbitz, chairman of the Austrian Biofuels Institute. "Fortunately, there are many on the market right now, so one can choose."[9]

The other key factor in profitable biodiesel plant operation is yield. In most early plants, the yield from the transesterification process was between 85 and 95 percent. Although this may sound fairly impressive, it's important to understand that a drop in yield of only 10 percent can reduce profits by about 25 percent. Modern biodiesel plants convert all the free fatty acids and triglycerides in the vegetable oil and achieve a 100 percent yield.[10]

In a related issue, since the transesterification process produces roughly 10 percent glycerin as a by-product, the value of this substance also plays a role in the profitability of any biodiesel plant. The current world demand for glycerin is around 950,000 tons per year in a niche market that is highly sensitive to oversupply. Recently, this market has not been able to absorb the huge increases in production resulting from the exponential growth in EU biodiesel production, which has triggered a price collapse for this commodity. Trying to figure out what to do with the thousands of tons of excess glycerin from the continuing expansion of the biodiesel industry around the world will be keeping quite a few people awake at night for the foreseeable future. Nevertheless, the huge quantities of low-priced glycerin currently on the market also offer huge opportunities for people who can figure out what to do with the stuff.

THE EUROPEAN BIODIESEL BOARD

In 1997, as a sign of continuing development within the industry, the European Biodiesel Board (EBB) was formed. A nonprofit organization with the aim of promoting the use of biodiesel in the European Union, the EBB also serves as an organizational structure and collective voice for its member-producers. Located in Brussels, Belgium, the EBB has achieved a high degree of visibility and earned the confidence of EU institutions as well as nongovernmental organizations through its many activities. The organization has come a long way since its founding. "In the early 1990s, biodiesel was really an outsider in the fuel industry and was not well known," says Raffaello Garofalo, secretary-general of the EBB. "But today, with about 4.9 million tons of annual production, the biodiesel industry is well established in the EU, and we now have the involvement of big multinational corporations. That's a major achievement."[11] Despite its many achievements, in 2007 and 2008 the industry has been struggling to deal with a wide range of problems, which the EBB has been grappling with.

REGULATION AND LEGISLATION

The regulatory and legal frameworks covering biodiesel in Europe have developed gradually since the 1980s but have followed different paths in the various nations (even after the founding of the European Union).

Austria developed the first RME biodiesel standard (ON C 1190) in 1991, followed by others in France and Italy in 1993, the Czech Republic in 1994, and Germany, with the most elaborate standard of all (DIN E 51606), in 1997. These standards were crucial in obtaining warranties for biodiesel use from numerous diesel vehicle manufacturers in these countries, resulting in increased demand for biodiesel as a vehicle fuel.

However, an even larger stimulus for biodiesel use was the Kyoto Protocol. Signed in 1997, the protocol committed the European Union to reducing greenhouse gas emissions by 8 percent from 1990 levels by a series of target dates from 2008 to 2012. But it wasn't until May 14, 2003, that the European Commission adopted the Directive for the Promotion of Biofuels (2003/30), which called for the increase of biofuels' market share to 2 percent by 2005 and 5.75 percent by 2010. The directive was motivated by the need to reduce greenhouse gas emissions in the transport sector in response to the Kyoto Protocol, and to enhance energy security by reducing European dependence on imported oil. Transportation is responsible for approximately one-third of carbon dioxide emissions in the EU, and road vehicles are almost entirely dependent on oil for their fuel. Consequently, one significant provision of the directive (among many others) encouraged greater use of biofuels in public transport and taxi fleets.

Overall, the directive gave the increased production of biodiesel feedstocks a significant boost, especially in the accession countries in the expanded EU-27. In most of the EU, these goals were viewed as achievable, but not without considerable effort and commitment. The French Agency for Environment and Energy Management (ADEME) estimated that reaching the 2010 objective would require an increase in industrial rapeseed plantings from 3 million hectares (7.4 million acres) in the EU to 8 million hectares (19.8 million acres). Since the directives involved specify only sales rather than production, if local supply within any nation cannot meet supply, it is possible to import the difference.[12]

However, in March 2007, the European Council adopted a new, even more aggressive Energy Action Plan, also referred to as the Renewable Energy Road Map. The new plan mandates that 10 percent of all fuels used for transport in the EU will be biofuels by 2020. This represents a significant increase over the original targets set in 2003. The original targets were voluntary, and it was up to the individual member states to establish their own standards; however, many of the states missed their early targets. The new plan was intended to make up for those shortfalls and to place even greater emphasis on meeting the new targets.

The new plan is mandatory for all member states as well as the EU as a whole. The plan also hopes to achieve a 500-million-ton reduction in carbon dioxide by 2020, as well as promote the development of so-called second-generation biofuels, including cellulosic ethanol, biodiesel from algae, animal waste, and other sources not currently being used for biofuels production. These second-generation biofuels may replace current production methods within the next decade or so and are expected to more than double current supplies. However, even with these additional supplies of biofuels, it is unlikely that they will be able to replace more than 12 to 18 percent of liquid EU transport fuels by 2015, according to some observers.[13] It also is hoped that these second-generation biofuels will reduce the growing competition with food production. Nevertheless, the EU recognizes that there is not enough domestic agricultural land available to meet both food and fuel requirements by the 2020 target, and it is now looking overseas for biodiesel feedstocks to make up the difference.[14]

In addition to the Directive for the Promotion of Biofuels, another main force driving increased biodiesel use in the EU is the Directive on Fuel Quality. The EU Directive on Fuel Quality (and a number of voluntary agreements under other programs) has resulted in significant advances in diesel engine technology that have improved energy efficiency and reduced emissions. These improvements require high standards for the fuel used by these engines, and the EU CEN (European Committee on Standardization) fuel standard EN 14214—developed in cooperation with automotive, oil, and biodiesel industries in the member nations—ensures biodiesel's continued consistent high quality. First proposed in the late 1990s, the standard finally came into effect in late 2003. The standard is being used as a basis for the replacement of the various individual national standards mentioned earlier. "The new standard is a major achievement," says Raffaello Garofalo of the EBB. "It's a reference point for our entire industry."[15]

Since biodiesel has consistently been more expensive to produce than petrodiesel, national fuel-taxation policy has also played a key role in the development of the industry in Europe. But up until fairly recently, these decisions have been made on a nation-by-nation basis. Germany, France, Italy, Sweden, Austria, and the Czech Republic all have had some form of biodiesel tax exemption. These same countries, as well as Denmark, Finland, the Netherlands, and Norway, also have had various types of "carbon taxes" on the books that have been designed to reduce the use of fossil fuels gradually and encourage the use of renewables. But the patchwork of individual laws has created a bureaucratic nightmare as well as substantial differences in taxation levels among the various nations. Trying to resolve this tangled mess with a common European standard has been a long-standing goal.

This seemingly impossible task was finally realized with the EU Directive on Energy Taxation (2003/96). Since the late 1990s, the EU had tried to develop an EU-wide energy tax strategy for its member states. After years of wrangling and numerous attempts—especially on the part of the United Kingdom—to block the initiative, the European Council of Ministers finally unanimously adopted a new directive on energy taxation on October 27, 2003. The directive established, among other things, rules for the detaxation of biodiesel and biofuels. Article 16 of the directive enables member states to apply an exemption or a reduced rate of excise duty to all biofuels (in pure form or in blends) sold in the EU for a period of six years, beginning in 2004. In addition, the directive makes it much easier for individual countries to amend their fuel tax laws. The UK, Germany, and Austria took advantage of this provision in 2004; the Netherlands, Lithuania, and the Czech Republic in 2005; and Sweden and Spain in 2006.[16] Nevertheless, it's important to understand that the decision to offer tax breaks for biofuels still rests with the individual EU member nations. Recently, there seems to be a trend toward establishing mandatory biofuel content in liquid fuels, rather than tax incentives.

The directive also was designed to reduce the distortions in competition that existed between energy products. Previously, only petroleum-based oils had been subject to EU tax legislation, and not coal, natural gas, or electricity; these latter forms of energy are now included. The directive also was intended to increase the incentive to use energy more efficiently and to allow member states to offer companies tax incentives in exchange for specific initiatives to reduce emissions.

"In practice, what the directive says is those member states of the European Union who intend to detax biodiesel can do so without asking for prior authorization from Brussels," explains Raffaello Garofalo. "This is very important because, until the directive was published and adopted, the member states had to pass through a very long procedure in Brussels in order to have biodiesel and biofuels detaxation." This procedure required the approval of all the other states, lengthy debates, and inevitable delays, according to Garofalo.[17] The new directive eliminates all of that—at the EU level, at least. In January 2004, in response to concerns raised by some of the new accession countries, the European Commission proposed a series of transitional measures to help these nations gradually come into harmony with the new EU-wide tax directives.[18]

GROWING PROBLEMS

Despite its many remarkable achievements, the European biodiesel industry has faced some serious challenges, especially in 2007 and 2008. Due to a number of reasons, there has been growing negative media and political attention focused on biofuels in the past few years. Much of the attention has been centered on the food-versus-fuel issue, higher food prices, food security, effects on biodiversity, and wildlife habitat destruction caused by the clearing of tropical rain forests for oil-palm plantations in some countries such as Indonesia, Malaysia, and Thailand. These plantations now cover millions of hectares in these countries, and additional conversions are planned or under way. This practice results in serious degradation of soils, to say nothing of the loss of the forests, which traditionally have supplied local communities with a wide variety of services and products. It's important to note, however, that there are a complex series of reasons for this deforestation and that these general trends were already ongoing before the more recent rise in demand for biodiesel feedstocks. (For more on these issues, see Chapter 13.)

Nevertheless, the EU's growing need to import substantial quantities of biodiesel feedstocks from overseas raises some difficult issues for the industry. This growing trend in "outsourcing" of biofuel feedstock production to third world nations—some of which can barely feed their own populations—is beginning to raise serious concerns in the environmental community, which would otherwise generally support the shift from petroleum to biofuels. A September 2007 report by the Organisation for Economic Cooperation and Development (OECD) was highly critical of the potential social and environmental effects of large-scale biofuels production, particularly in tropical regions.

Another problem has been the growing doubt among European governments about the true environmental benefits of some biofuels, resulting in reassessments of some long-standing, across-the-board subsidies and tax breaks. There has been a realization that different biofuels have different impacts on the environment based on their feedstocks and how those feedstocks are produced. U.S.-produced corn-based ethanol, in particular, has come in for serious criticism. As a result, the United Kingdom, France, Germany, the Netherlands, and Switzerland have removed or are revising various incentives for farmers, biofuel companies, and distributors.[19] Although public opinion in Europe is now divided on biofuels, support for biodiesel generally remains fairly strong in some EU institutions, the World Energy Council, and even the World Wildlife Fund (WWF).

Overall, the WWF is supportive of the EU initiative for the greater use of biofuels; however, that support is conditional on a number of provisos, including legally binding environmental certification for both imported and domestically produced biofuels covering such issues as carbon balance and greenhouse gas life-cycle analysis, land use, and neutral impacts on water, soil, and biodiversity. Among other issues, the fund also calls for assistance for developing countries, the promotion of second-generation biofuels, and the use of careful rural development plans.[20]

A related initiative to develop global standards for sustainable biofuels production and processing is being spearheaded by The Energy Center at the Swiss Federal Technical Institute in Lausanne (EPFL). This multi-stakeholder effort from around the world will host a series of meetings, teleconferences, and online discussions through June 2008 and hopes to develop draft standards that will "ensure that biofuels deliver on their promise of sustainability."[21] Another Swiss-headquartered organization that is focused on a wide range of environmental concerns related to biodiesel feedstock production is the Roundtable on Sustainable Palm Oil. Founded in 2004, the group is a collaboration between various companies and NGOs that support the sustainable production of palm oil. In November 2005, the RSPO, which also has offices in Kuala Lumpur, Malaysia, agreed to a set of principles and criteria governing palm-oil-plantation development.[22]

Another issue that is causing problems for the industry is that a number of EU member nations have been slow to implement promises to increase their biofuel use. "We have been promised a market, but it is not there yet," says Raffaello Garofalo of the EBB. In addition, recent decisions by several EU nations to tax biodiesel at a time when the EU is promoting greater biodiesel use would seem to be counterintuitive. Garofalo says that these nations are putting their own financial considerations above larger environmental concerns. "If there is no legislative support on taxation or on binding targets, then there is no real market for biodiesel," he adds.[23]

There is yet another issue that has created a good deal of controversy in the EU that has nothing to do with the environment—the so-called "splash and dash" problem. In 2004, the U.S. Congress passed a provision for a tax credit for biodiesel that is blended with petrodiesel (see Chapter 10). The tax credit amounted to a one-cent credit per percent of biodiesel blended with petrodiesel. Under the plan, a gallon of B2 would earn a two-cent credit while B20 would receive a 20-cent credit, and so on.

Although the credit was intended mainly to stimulate biodiesel production and consumption in the United States, there was what some (especially in Europe) have described as a "loophole" that included foreign exports large enough to sail a supertanker through, and that is exactly what has been happening.

Beginning in early 2007, EU nations began to notice tanker loads of B99, some of which had paused just long enough in U.S. ports to add one percent (or less) petrodiesel to the cargo to qualify it for the blending tax credit. When the shipment arrived in a European port, it qualified for yet another tax credit, this time from the EU. Although a lot of biodiesel actually produced in the United States is also finding its way to European markets, the "splash and dash" practice has caused a good deal of consternation by many in the industry, since this was not the original intent of the legislation. By the end of 2007, the import flood was estimated to be around 700,000 tons (210 million gallons). Nations such as Germany, Spain, Italy, Austria, and the UK have been especially hard-hit. Countries such as France that have import quotas in place have been somewhat insulated from the recent import rush.

The European Biodiesel Board, representing EU biodiesel producers, protested the rising tide of cheap imports from the United States and called for duties to be placed on them. In October 2007, the EBB turned up the heat by threatening legal action with the possibility of a formal World Trade Organization complaint, unless the U.S. Congress closed the loophole. Congress didn't. Two months later the EBB voted unanimously to initiate that legal action.[24] The official complaint was filed in April 2008, followed by a 45-day examination period to determine its validity. In June, the European Commission launched an investigation and could impose duties on imports from the United States.[25]

GERMANY, THE EUROPEAN LEADER

While Europe is the global leader in biodiesel production, Germany, until very recently, was the undisputed highest-producing nation in the EU

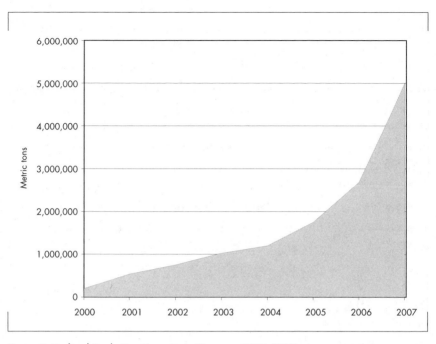

Figure 5. Biodiesel Production Capacity in Germany, 2000–2007 Various sources

(and the world), with an annual production of 2,662,000 tons (798.6 million gallons) in 2006. By the end of 2007, total German production *capacity* reached a record 5 million tons (1.5 billion gallons). Although it's very hard to make long-term predictions, this may prove to be the high-water mark for the industry.

At first, Germany's biodiesel was intended for use primarily in German agricultural machinery, especially farm tractors, in a closed-loop cycle. In 1982, some early tests were conducted on farm-tractor diesel engines at the Institut für Biosystemtechnik (Institute for Agricultural Engineering) in Braunschweig. But the uses for biodiesel soon multiplied well beyond the agricultural community, and today biodiesel is used in Germany to power a wide range of vehicles as well as forestry machinery, various kinds of boats, and other types of equipment.

The first supplies of rapeseed methyl ester for early German trials were manufactured by the Henkel Company in Düsseldorf in a nondedicated chemical plant. A pilot plant then was established at the Connemann oil mill in the port city of Leer, located on the North Sea coast (Oelmuhle Leer Connemann GmbH & Co. has been in the seed oil business since 1750), and in June 1991 it produced its first 10 tons of biodiesel. Three years later the pilot plant was replaced by an 80,000-ton-capacity (later increased to 100,000 tons), industrial-scale facility at the same location that used a so-called continuous deglycerolization (CD) process. This facility is now owned by the U.S. multinational Archer Daniels Midland (ADM). The following year a small-scale, 2,000-ton plant was opened by a farmers' cooperative in Grossfriesen in the former East German state of Saxony, and a short time later a 5,000-ton plant owned by a farmers' cooperative in Henningsleben in Thuringia began production.[26] In 1999, a 50,000-ton-capacity plant was started up at Ochsenfurt in Bavaria, and another facility began operating in Wittenberg in the former East German state of Mecklenburg-Vorpommern (Mecklenburg–West Pomerania), with a production capacity of 60,000 tons per year.

The Building Boom
The post-1999 boom in German biodiesel plant construction mentioned earlier really hit its stride in 2001 with the construction of Oelmuhle Hamburg, a 100,000-ton-capacity plant owned by ADM based on the same CD production process used in ADM's Leer facility. Also in 2001, a 100,000-ton plant operated by Mitteldeutsche Umesterungswerke GmbH opened in Bitterfeld, as well as three smaller plants by various companies in Rudolstadt, Malchin, and Oranienburg.

In 2002, the construction boom continued with the opening of three 100,000-ton plants in Marl, Neuss, and Schwarzheide, as well as four smaller facilities in Sohland, Bokel, Südlohn, and Harth-Pöllnitz. The building binge persisted in 2003 when a 50,000-ton facility was opened in Magdeburg and five smaller plants owned by various companies opened in Falkenhagen, Kyritz, Schleswig, Wiedemar, and Uckerland. In 2004, the rate of the expansion finally began to slow as German production capacity far outpaced demand. As of mid-2004, Germany had a total of 25 biodiesel-manufacturing plants.[27]

Since then, a smaller number of plants, but generally much larger in size, have been built in various locations. In 2006, ADM Mainz GmbH opened a new 275,000-ton-capacity plant in Mainz. And in 2007, the company also completed a huge expansion project at its Hamburg plant, which now has a 1-million-ton (300-million-gallon) capacity, making the plant the largest in Europe—and ADM the largest biodiesel producer in Germany.[28] Several other large projects also were completed recently in various locations, but the "gold rush" mentality that had predominated in the industry for years has definitely come to an end, and has recently been replaced by a serious overcapacity crisis.

Pure Success

Biodiesel was traditionally marketed in Germany mainly as a vehicle fuel (for both cars and trucks) at a "pure" B100 concentration; however, this policy was changed in late 2003 to include various blends of biodiesel as well. Until recently, this strategy was successful from a price standpoint because biodiesel was exempt from fuel taxes, making it less expensive than petrodiesel for drivers of diesel cars and other vehicles. (The 2003 approval of biodiesel blending altered the tax reduction to correspond with the amount of biodiesel in the blend.) Unlike France and Italy, however, Germany did not establish any upper limit on the amount of biodiesel produced that was eligible for tax exemption, which partly explains the huge expansion of production capacity until fairly recently.

Biodiesel production was given a boost in 1996 when the sale of leaded gasoline was banned in Germany. As a result, about 1,000 fuel pumps at filling stations suddenly became available, and within a few months about 600 of them had been converted to biodiesel use. In 1999, the German government introduced an additional ecological tax (eco-tax) on fossil fuels that increased in a series of steps over the following four years, making biodiesel even more attractive as an alternate fuel. The idea was to further reduce greenhouse gases while transferring the costs to polluters.

One of the earliest major successes for biodiesel took place when the German Taxi Association adopted the use of biodiesel nationwide after a successful 1991–92 test in the city of Freiburg. The taxi association's practice of ordering large numbers of new taxis at one time helped spur

Mercedes-Benz, Volkswagen, Audi, and Volvo to offer warranties for the use of biodiesel in their vehicles. The early collaboration of biodiesel producers with German automakers Mercedes-Benz, Volkswagen, BMW, and MAN has been a significant factor in the success of biodiesel in Germany. A number of public transport fleets in German cities also have switched to biodiesel. One of the best examples has been in Heinsberg (located near the German border with the Netherlands), where the city works department, Kreiswerke Heinsberg GmbH, began its first tests with biodiesel in 1996. During the next few years, the tests continued while the city also considered converting its buses to run on compressed natural gas. After all the pros and cons were weighed carefully, biodiesel won the contest. By 1998, the entire fleet of 130 buses (Mercedes-Benz and MAN) had been switched over to biodiesel, making Kreiswerke Heinsberg the first local public-transport enterprise in Germany to operate its entire fleet on biodiesel.

The city of Neuwied, located beside the Rhine River, was the first municipality in the Rhineland-Pfalz region to convert its entire bus fleet to biodiesel. But unlike similar biodiesel projects in other cities, the Neuwied conversion involved the transformation of a mostly older fleet, in this case 26 1982–83 vintage buses. Although the age of the vehicles required that various seals and hoses on the diesel engines be replaced in order to make them serviceable for use with biodiesel, it still proved to be a cost-effective program. During the course of the project, which ran from January 1997 to October 2000, a number of the older buses were replaced with new models. One of the main criteria of selection for the new buses was that they be suitable for use with biodiesel. By the end of the study, it was clear that the buses worked well on biodiesel, and the project was judged to be a success.[29] Other German cities are experimenting with biodiesel in their municipal fleets as well.

But biodiesel isn't being used just on dry land. Since the early 1990s some tourist boats have been using B100 as their fuel on the Bodensee (Lake Constance, located along the German-Swiss border), Europe's second-largest drinking-water supply. The yacht clubs and individual boating enthusiasts using the lake also have embraced biodiesel and have included its use as part of their efforts to keep the lake clean and free from toxic fuel spills.[30]

Support Organizations

Biodiesel also has benefited from promotional efforts by organizations such as the Union for the Promotion of Oil and Protein Plants (UFOP) that have emphasized biodiesel's locally grown feedstocks, environmental advantages, and high quality standards. As a result, biodiesel is widely associated with the brilliant yellow rapeseed fields that produce the feedstock. The UFOP's "Germany's most beautiful oil fields" publicity has helped propel biodiesel's popularity, and the fuel now accounts for nearly 5 percent of the German diesel market. There are currently more than 1,900 fueling stations offering biodiesel across the country. About 40 percent of the nation's biodiesel is sold at these fueling stations, while the remaining 60 percent is marketed through fleet operations for public transportation and commercial freight haulers.[31]

Germany's first preliminary fuel standard for biodiesel made from plant oil methyl ester (PME), DIN (Deutsches Institut für Normung) V 5 1.606, was published in 1994 and included a wider range of possible oil feedstocks than the earlier 1991 Austrian standard. The first German standard was followed by a second, improved standard (DIN E 51.606) for fatty acid methyl ester (FAME) in 1997. The importance of maintaining high quality standards for biodiesel production was clearly demonstrated around 1999. Due to the rapid expansion of both production and demand, biodiesel of inferior quality began to enter the market, causing engine problems and complaints from both consumers and auto manufacturers. In response, with the assistance of the UFOP, biodiesel producers created an organization called the Association for Quality Management of Biodiesel (Arbeitsgemeinschaft Qualitatsmanagement Biodiesel, or AGQM) to ensure that only high-quality fuel would be sold. The group developed a quality seal that is displayed on fuel pumps containing approved biodiesel. There are many other organizations that support the German biodiesel industry as well, but they are simply too numerous to mention.

Crop Issues

As in most of Europe, rapeseed is the main feedstock for biodiesel in Germany, accounting for more than 70 percent of total biodiesel production. (Imported soybean and palm oil make up another 20 percent.) Around 1974–75, the first varieties of so-called 0-rapeseed were introduced, followed in 1986–87 by improved 00-rapeseed (canola) varieties, which turned out, more or less by accident, to have very favorable characteristics for biodiesel. Initially the oleochemical industry consumed most of the annual German rapeseed crop, but after 1999 the biodiesel industry became the most important processor. German rapeseed breeders continue to work on additional improvements to enhance certain genetic traits by breeding for new oil properties and increased resistance to disease; they also hope to increase the oil yield to 3.5 to 4 tons of oil per hectare (1.4 to 1.6 tons per acre). The best reported yield for rapeseed of 2.9 tons of oil per hectare was reported in 1999 by a farmer in Schleswig-Holstein. Improvements in precision plowing, soil preparation, fertilizer application, pest control, and harvesting will make the most of the genetic advances.

The former East German states account for about 60 percent of the annual German rapeseed harvest. In terms of the production of nonfood crops, rapeseed is the largest crop, while sunflower is a distant second. In 2006, rapeseed was grown on about 1.5 million hectares (3.7 million acres) in Germany, representing about 11 percent of total agricultural land. This compares with about 7 million hectares (17.3 million acres) used for the production of food crops.[32] In the past few years, demand for rapeseed oil in Germany has exceeded the supply, and imports of rapeseed oil from Canada and even China have been used to fill the gap. Since much of this imported oil does not meet strict EU food-grade oil standards, the imported oil has been mainly used in the production of biodiesel, leaving the non-genetically modified EU oil for the domestic food market.[33]

An Industry in Crisis

In a dramatic turn of events, the German biodiesel industry recently has gone from being the world leader to finding itself in the midst of a serious crisis bordering on total collapse. This unprecedented situation is the result of a convergence of a number of different factors—high rapeseed oil prices, overexpansion of production capacity, the ending of the tax breaks for biodiesel by the German government and resulting falling demand—that has created a "perfect storm" that seriously threatens the entire industry.

This storm has been developing gradually in recent years. The over-building of production capacity began to be apparent around 2004 and became worse with the opening of a number of very large new facilities since then. This was taking place while vegetable-oil prices continued to rise as the demand for food and fuel expanded simultaneously with poor harvests in some countries due to climate change. (Rapeseed prices in the EU increased about 20 percent between March 2006 and March 2007.) But the most dramatic trigger event was the imposition of a 9-euro cent-per-liter tax (nearly 36 cents per gallon) by the new German government headed by Angela Merkel. This tax, which went into effect on August 1, 2006, is designed to increase by six euro cents in a series of additional steps beginning in 2008 until biodiesel reaches parity with petrodiesel at a 45-cent tax in 2012. "It's death by installments," said Karin Ratzlaff of the Association of the German Biofuel Industry (Verband der Deutschen Biokraftstoffindustrie—VDB).[34]

Although the biodiesel industry has repeatedly called for relief on the tax issue, the government appears to be unwilling to change its position. The Merkel government did, however, mandate that a B5 blend of biodiesel and petrodiesel be sold at all filling stations throughout the country, helping to offset at least some of the pain of the new tax. (German biodiesel representatives warned this would only encourage a flood of cheaper imports from foreign producers.) According to the gov-ernment, one rationale for the new tax (in addition to raising tax rev-enues) was to help accelerate the development of second-generation biofuels. "We would like to see the second-generation biofuels developed as soon as possible," said Tobias Dunow, a spokesperson for the German environment ministry. Nevertheless, "first-generation" biodiesel is still seen by most observers as an important part of Germany's energy mix until these second-generation fuels enter the market, sometime after 2015.[35]

The new tax has had the most devastating effect on B100 sales, the very heart of the German industry until fairly recently. With the price advantage for biodiesel removed, many large German fleets that had been using biodiesel suddenly switched back to petrodiesel, causing a dramatic drop of 25 percent in biodiesel sales. In response, EOP Biodiesel, which recently opened a 100,000-ton plant, has almost completely withdrawn from the B100 market and now has shifted its focus to blended biodiesel. Other companies are making similar moves. Adding to the German industry's woes, the flood of less expensive imports from the United States (the "splash and dash" problem mentioned previously), coupled with increased imports from non-EU nations such as Indonesia, Malaysia, and Brazil, where feedstock costs are much lower, has caused most German biodiesel plants to cut back on their production or shut down altogether.[36] As a result of these combined challenges, at the beginning of 2008, after a second round of tax increases went into effect, it was estimated that the industry was producing at only 10 percent of its capacity. It also was estimated that imports now account for about 90 percent of biodiesel used for blending in Germany, confirming the industry's earlier predictions—and fears.[37]

Most participants in the German biodiesel industy feel that they have been betrayed by the government. "We are not just in a crisis, we are in a state of collapse," says Peter Schrum, head of the German Association for Biogenic and Regenerative Fuels (Bundesverband Biogene Kraftstoffe e.V.). "Many small companies have reduced production to zero; production is largely being continued by larger companies with large financial resources."[38] In a sign of the times, a brand-new, 100,000-ton plant at the North Sea port of Emden owned by Petrotec delayed its opening when construction was completed at the end of 2007.[39] Many investors in German biodiesel plants are losing money, and there are a growing number who are talking about getting out of the business altogether. In fact, several of the shut-down German biodiesel plants have already been sold to buyers in the United States and are scheduled to be dismantled and shipped overseas.[40]

The shortsighted changes in the Merkel government's tax policy regarding biodiesel have been an unmitigated disaster. In just over a year, the government has managed to transform an industry that was once the global leader into almost a total ruin. Whether the industry will be able to recover from its current dilemma is unclear. Unless there is a reversal in government tax policy, or perhaps an antidumping import tariff imposed by the EU against the "splash and dash" flood, the German biodiesel industry's prospects do not look good, at least in the short term.

Straight Vegetable Oil

In recent years there has been some use of virgin rapeseed oil as a diesel-fuel substitute in Germany. The oil either is used in vehicles with modified engines or is mixed with petrodiesel in the fuel tank. It is estimated that about 5,000 vehicles with modified fuel systems operate on straight vegetable oil in Germany, and that number has been relatively stable for some time.[41] Engine manufacturers have serious doubts about this practice, however, and virtually all of the big players in the biodiesel industry take a dim view of this strategy. "The position of the EBB is very clear on this," Raffaello Garofalo says. "One barrel of bad biodiesel is enough to spoil the reputation of the entire biodiesel industry. The auto industry doesn't want to build special engines, which means that straight vegetable oil cannot be an option for the main market, except for those individuals who want to convert their engines. If you use straight vegetable oil in a normal diesel engine it will work for about six months if you are lucky, but then the engine will be ruined. We have had big problems with this in Germany. Some people were putting straight vegetable oil in their cars, and then after the engine seized up, they went to the manufacturer claiming that biodiesel had caused the problem. This was having a tremendous negative impact on the biodiesel industry."[42] (See "Straight Veg" in Chapter 11 for more on this issue.)

FRANCE

After Germany, France is the second-largest producer of biodiesel in the EU, with an annual production of approximately 743,000 metric tons

(223 million gallons) as of 2006. (Between 1993 and 1999, France was the leading European producer of biodiesel, but it was overtaken by Germany in 2000.) Two main biofuels have been promoted in France to meet biofuels targets set by the European Commission for transportation fuels: ethanol and biodiesel.[43] Taken together, these two biofuels represented about 1 percent of total fuel consumption in France in 2005. (Of that total, biodiesel represented about 1 percent of diesel use.) More recent statistics indicate that biofuels represented about 1.6 percent of total fuels in 2006, with a projected 3.5 percent in 2007.

France produces roughly five times more biodiesel than it does ethanol. This is mainly in response to the fact that about 75 percent of cars and trucks in France are now diesel-powered. France's strategy for implementing greater biodiesel use has been to blend it with petrodiesel fuel and heating oil, usually at a B5 concentration, although a new B10 blend standard was launched in early 2007. Consequently, unlike their counterparts in Germany and Austria, where biodiesel has been promoted and sold mainly as a differentiated product, many biodiesel users in France may not even be aware of the presence of biodiesel in their fuels. Biodiesel is also sometimes blended at rates between B5 and B30 for use in French vehicle fleets. Thousands of vehicles in various communities are involved in municipal fleet programs across the country. Public buses in Paris, Bordeaux, Dijon, Dunkerque, Grenoble, Strasbourg, and elsewhere are now using a biodiesel blend.

Early Activity
In 1981, feasibility studies for the use of fatty acid methyl esters (FAME) as fuel in diesel engines and heating boilers were conducted in France by the French Petroleum Institute (IFP) and the Agency for Environment and Energy Management (ADEME). At about the same time, PROLEA, one of the major French agricultural associations, began a study that focused on the diversification of outlets for agricultural products. Oilseed crops were included in the study. Although these early French efforts eventually resulted in some of the earliest biodiesel legislation in Europe, the use of biofuels in general, and biodiesel in particular, did not start to

develop on a commercial basis until after the European Union set-aside rules for agricultural lands were implemented in 1992. In the same year, concerns about the potential for pollution caused by large-scale farming of rapeseed in France helped spur the adoption of the so-called Agro-Environmental Charter.

A major series of advanced laboratory and field tests was conducted in the early 1990s on the use of rapeseed methyl ester (RME) in diesel engines by the ADEME and IFP. Most biodiesel in France now is produced from RME (about 80 percent), although sunflower ester accounts for another 10 percent. Imported soybean oil (5 percent) and palm oil (2 percent) make up most of the remainder. In France, biodiesel generally is referred to by the name *diester*, a contraction of *diesel* and *ester*. (*Diester* is also the trademarked name of the rapeseed methyl ester produced by Diester Industrie, the nation's major biodiesel supplier.)

The first biodiesel in France was produced in an existing methyl ester plant in Péronne, owned by Castrol, and another in Boussens, owned by Sidobre-Sinova. Following the success of a small-scale pilot plant in Compiègne, the first dedicated biodiesel plant was built in the same location by the Robbe Company in 1993. Additional expansion of the French industry occurred in 1995, when Italian-based Novaol added 70,000 tons of capacity to an adapted chemical plant in Verdun. That was followed by a 120,000-ton-capacity dedicated biodiesel plant (the largest in the world at the time) in Rouen, a joint venture between Sofiproteol and VaMo Mills.[44]

Government Policies

Most biodiesel in France is made from rapeseed oil grown specifically for that purpose, mainly on set-aside land. The rapeseed pulp produced after the oil has been extracted is used for animal feed. As in the rest of Europe, the EU's Common Agricultural Policy with its set-aside lands provisions has had a major impact on French biodiesel production. The reformed EU Common Agricultural Policy adopted in June 2003 set a "carbon credit" payment of 45 euros per hectare for growing nonfood crops. But French oilseed farmers generally have felt that this is not enough support for them to grow industrial crops and believe that 90 to 100 euros per hectare would be necessary to maintain stability in nonfood production in their country. Around 2005, the EU Parliament proposed that this support should be raised to 90 euros. However, in October 2007, the EU Commission announced that it would scale back the program because EU farmers already had exceeded the 2007 target of 2 million hectares by planting 2.84 million hectares for energy crops (mostly rapeseed).[45]

Beginning in 1993, the French government has supported biodiesel by offering full excise-tax exemption, but for limited state-set quotas. (This strategy has been followed in Italy as well.) These quotas, however, have acted as a damper on the expansion of biodiesel production capacity. As of 2003, the quota for tax-exempt biodiesel was 317,500 metric tons. However, there was an overproduction of 47,500 tons that did not qualify for the exemption, and this amount was either exported or used in the chemical industry for other purposes.[46] Since then, biodiesel production quotas have been raised by the French government in a series of annual steps to 417,500 in 2005, and to 1,342,500 tons in 2007, with an eventual target of 3,232,500 tons by 2010 to help the nation reach its new EU-mandated biofuels percentage levels.[47] There are also penalty taxes on companies that do not market biofuel blends.

In 2006, the French government reduced the tax break for biodiesel from 33 cents to 25 cents per liter.[48] What's more, there have been some recent government discussions about possible additional reduction of the tax advantage for biofuels by one-third. This move undoubtedly would cause the biodiesel industry in France some significant challenges.[49]

The Main Players

The top biodiesel producer in France is Diester Industrie, which controls more than 80 percent of the French biodiesel industry. Diester owns one of the larger European biodiesel plants at Grand-Couronne, Normandy; this plant now has an annual production capacity of 260,000 metric tons (78 million gallons). Diester Industrie also owns five other plants in various locations with a combined production capacity of 990,000 metric tons.[50] In addition to Diester, the Novaol plant in Verdun mentioned earlier, as well as a 150,000-ton facility in Dunkerque owned by Daudruy, an

80,000-ton plant in Limay owned by SARP, and a 55-ton plant in Liseux owned by Bionerval Saria round out the major biodiesel plants in the country as of 2007. Additional plants owned by various companies are scheduled to open in the next few years to try to help the country meet its new EU biofuel targets.

The popularity of diesel vehicles in France has created a somewhat unusual situation for French petroleum oil refiners, who now suffer from surpluses of gasoline and shortages of diesel fuel. The surplus gasoline must be exported (primarily to the United States) at low prices, while petrodiesel must be imported, sometimes at higher cost. This helps explain why the French refiners generally are happy to blend as much biodiesel into their petrodiesel as the French government and EU regulations will allow, since it reduces the need for expensive imports.[51] Some large oil companies like Total are among the biggest consumers of biodiesel, and virtually all French biodiesel is sold to petroleum companies for blending purposes.

In a significant new initiative, the French National Railways (SNCF) began an extensive 42-month biodiesel trial in June 2006 with B30 rapeseed biodiesel as well as a more spectacular B100 trial on two high-speed express trains. Three to four freight locomotives were also part of the initial trial. This was the first step in the railway's "zéro pétrole" program, which aims at replacing 6 percent of petrodiesel fuel in its diesel locomotive fleet by 2010. However, if the trials are successful, the target could be raised to 30 percent. In July 2007, the biodiesel was further tested on the TER (transport express régional) Poitou-Charentes region, followed by tests in several other regions by the end of the year. So far, the trials look promising.[52]

The arrival of McDonald's fast-food restaurants in 1979 was viewed by many French citizens as a foreign invasion (bordering on gastronomic sacrilege). Today, however, these same restaurants have begun to play a productive role in the biodiesel industry. In July 2003, McDonald's France, along with its partners Ecogras and Sud Récupération, which collect used frying oils from McDonald's, signed an agreement with the Italian biodiesel producer Novaol for the reprocessing of McDonald's frying oils into biodiesel. Ecogras and Sud Récupération are supplying Novaol with approximately 1,200 tons a year of oils collected in French McDonald's restaurants, which use partially hydrogenated rapeseed oil. The biodiesel made from the used frying oil has generally been exported to Italy.[53]

The French strategy of blending biodiesel seamlessly into the diesel-fuel sector has been a success and has created a mutually beneficial partner-ship between biodiesel producers and the traditional petroleum industry. Although the state-set limits on biodiesel tax exemption initially were viewed by many in the industry as a constraint, to some extent this strategy has had the effect of partially insulating French biodiesel pro-ducers from some of the worst effects of the recent downturn in Germany and other EU member nations. There is no question that the export market for French biodiesel has been reduced by the sharp drop in German demand, but overall the industry remains on relatively solid footing.

The French biodiesel industry is well established and has matured over the years to the point where it now has carved out an important niche in the nation's energy portfolio. The industry's proven track record should allow it to continue to grow to help meet the new biofuels targets set by the EU for 2020. However, if the French government decides to cut or eliminate the remaining biofuels tax credits, this optimistic outlook could change dramatically.

Other European Countries

Although Germany and France are the two largest players in the European biodiesel industry, a number of other nations, such as Italy, the UK, Austria, Poland, the Czech Republic, and Slovakia, also have played significant roles in the development and growth of the industry, although they frequently have followed slightly different paths. Here are some of the highlights of those developments.

ITALY

Italy ranks third in EU biodiesel production with about 447,000 metric tons (134 million gallons) produced annually in 2006. Italy traditionally has been highly dependent on energy imports, which amount to about 80 percent of total energy consumption. This makes Italy vulnerable in the energy sector, but it also offers biodiesel producers considerable long-term prospects for continued growth. At the moment, however, liquid biofuels (mostly biodiesel) represent less than 1 percent of Italian liquid fuels. In the past decade or so, Italy has promoted the increased use of all types of bioenergy, including biodiesel, and that emphasis is expected to grow as the nation attempts to meet the goals of the Kyoto Protocol and the EU.

The earliest biodiesel production in Italy started with some small, existing, nondedicated methyl-ester plants in the northern part of the country. In 1992, two Italian companies, Novaol (then named Novamont) and Estereco, took part in a joint venture to develop a biodiesel industry, funded in part by the European Commission. A large-scale, dedicated plant with a capacity of 80,000 tons per year was built by Novaol in the port city of Livorno.[1] The Livorno facility, which processed both rapeseed and sunflower oil, was one of the first large-scale commercial biodiesel plants in the world. Initially the entire output of the plant was directed to the heating oil market, but some of it subsequently was marketed as transport fuel. Unlike the case in most other European nations, in Italy both transport fuels and heating oil have been taxed at fairly high levels. Since heating oil provided biodiesel producers the same tax advantages as vehicle fuel, but at slightly lower quality levels, most Italian producers initially opted to focus primarily on the heating oil market.

Until very recently, biodiesel blends sold for use in Italy have been exempted from fuel taxes, but only up to certain production quotas established annually by the government. In prior years the limit had been set at 125,000 metric tons per year, but in 2001 the limit was raised to 300,000 tons, where it remained through 2004. But in 2005, the quota was reduced to 200,000 tons because of budget considerations. The quota was again increased from 220,000 tons in 2006 to 250,000 tons for 2007.

In April 2001, a National Voluntary Agreement for the use of biofuel in national transport was implemented. The agreement anticipated the introduction of up to 5 percent biodiesel in Italian public transport.[2] In 2007, about 70 percent of Italy's biodiesel was produced from (mainly imported) rapeseed oil, while about 20 percent was from soybean oil, with sunflower oil and palm oil making up the remainder. Due to limited domestic demand, about two-thirds of Italy's total biodiesel production in 2006 was exported to Germany, France, Austria, and Spain. It's interesting to note that Italy imports large quantities of rapeseed and soy oil, which are then processed into biodiesel and then reexported to the EU, sometimes to the same countries from which the original feedstock oil came.[3]

Like many other EU member nations, Italy failed to meet its initial EU biofuel target in 2005. In January 2007, the Italian government signed a new national agreement on biofuels aimed at reducing the country's reliance on fossil fuels and encouraging the production of domestic biofuels. The 2007 national budget provided substantial financial support for these initiatives and also called for 70,000 hectares to be used for biofuels production by the end of the year. The government further hopes that 240,000 hectares will be used for biofuel crop production by 2010. In addition, gasoline and petrodiesel suppliers are now required to blend at least 1 percent biofuel as of 2007 and 2 percent in 2008.[4] But since Italy does not have enough land to grow all of the biodiesel feedstock it will need, it has been looking overseas for additional supplies. In March 2007, a number of Italian companies signed agreements to build four new biodiesel refineries in Brazil for a total investment of $480 million, to take advantage of Brazil's vast feedstock potential.

There are currently around 10 major producers of biodiesel in Italy with combined production capacities of about 1.1 million tons. In addition, there are six other new plants under construction that should open between 2007 and 2009 that could add another 600,000 tons of capacity. The vast majority (about 90 percent) of biodiesel produced in Italy traditionally has been used for heating purposes. However, this percentage has been declining dramatically in recent years to only about 3 to 5 percent due to changes in heating oil taxes and strong increases in demand for biodiesel in transport.[5] That growing share of biodiesel for transport fuel is generally at a B5 blend, with a B30 blend used for some fleets, particularly in public transport. Public transit buses in Florence, Gorgonzola, Padua, and Perugia now run on biodiesel. As a heating fuel in Italy, biodiesel generally has been used in its pure B100 form (but sometimes as B20), and even the Vatican is reported to have biodiesel-fueled boilers.

In early 2007, the Italian biodiesel industry began to experience some problems of its own with a significant drop in production caused by uncertainty about regulatory changes introduced on January 1. The 250,000-ton tax exempt quota mentioned previously was replaced with an excise duty equal to 20 percent of that levied on petrodiesel sales. Not surprisingly, the new tax immediately put the brakes on domestic biodiesel sales. As a result, the 1 percent EU target for Italian biofuels almost certainly will not be met. To make matters worse, later in the year the turmoil in the German biodiesel industry began to spill over into the

Italian biodiesel export market. Overall, 2007 Italian biodiesel production could fall by as much as 40 percent, according to one industry
observer.[6] All of this is taking place while the EU is encouraging member
states to use more, not less, biofuel.

UNITED KINGDOM

The United Kingdom's oil production from the North Sea peaked in 1999
and has been in decline ever since. The UK became a net oil importer in
2004. Despite the fact that the country consumes more than 15 million
tons (4.2 billion gallons) of petrodiesel fuels a year, it was one of the
slowest adopters of biodiesel in Europe. Perhaps it's the isolating influence of the English Channel. In 2002, the UK was barely on the list of
EU biodiesel producers. But this situation has changed dramatically. As
of 2006, the UK has catapulted itself to number four on the producers list.

The foundation for this dramatic growth was laid in August 2002, when
a reduction of 20 pence per liter in the fuel tax was implemented by the
government to help level the playing field for biodiesel producers in
England. And in May 2004, the government launched a 12-week consul-
tation aimed at setting a minimum sales target for biofuel beginning in
2005 and deciding whether a "renewable transport fuels obligation" should
be imposed on fuel producers.[7] Unfortunately, like many other EU member
states, the UK failed to meet its EU biofuels target in 2005. The long-
awaited Renewable Transport Fuels Obligation (RTFO) was finally
announced in November 2005. The RTFO calls for 2.5 percent biofuels to
be sold at the nation's forecourts (filling stations) by April 2008, and 5 percent by 2010. In addition, tax rates for biodiesel recently have been cut in
half and there is also a new penalty for noncompliance in blending targets.
On the other hand, in 2007 the government announced new "sustainability standards" for biofuels that were scheduled to come into effect in
2008. The standards will require manufacturers to demonstrate how much
CO_2 has been emitted during the biofuel production process and that the
biomass used for feedstock has come from a sustainable source.[8]

Initially, there was quite a lot of activity at the local level, where rapidly increasing numbers of diesel-powered cars offered a significant opportunity for biodiesel producers. A group of entrepreneurs set up a number of small-scale biodiesel plants that relied mainly on used frying oils for their feedstock. These producers developed a modest but growing customer base, including municipal governments, local truckers, supermarkets, and breweries, as well as individual drivers. Ironically, finding enough used frying oil, rather than customers, was the main problem for these producers.

But biodiesel activity has expanded well beyond the small local producer. The Green Shop garage, located near Stroud, Gloucestershire, became the first garage in the United Kingdom to sell biodiesel in July 2002.[9] As of mid-2007, there were approximately 500 biodiesel forecourt locations in England that sell biodiesel blends to the general public (up from just a handful a few years earlier). London-based Greenergy has been marketing a B5 blend it calls GlobalDiesel in southwestern England and more recently at 150 Tesco forecourts. An independent biodiesel producer and reseller, Rix, as well as the Asda supermarket chain, now offer biodiesel to their customers. However, with 331 service stations that now offer biodiesel blends, Total has the largest number of retail outlets in the nation. And in July 2007, McDonald's Corp. announced that its British chain had started converting its used cooking oil into biodiesel for use in its fleet of 155 trucks.[10] Despite all of this activity, though, biofuels still represent only about 0.2 percent of total transport fuel in the UK.[11]

There is also quite a lot of biodiesel activity in the northeast of England, where area farmers have been encouraged to create "onshore oil fields" by growing biodiesel feedstock crops to help boost jobs and income in their rural economy. Similar initiatives have been promoted in other rural parts of England as well. What's more, Petroplus, the owner of a crude-oil refinery in North Tees, has been processing approximately 20,000 tons of biodiesel (called Bio-Plus) a month for use in commercial and other vehicles. Until fairly recently, Petroplus reportedly accounted for about 60 percent of the UK's biodiesel market share.[12] In April 2004, Greenergy unveiled plans for a new 100,000-ton-capacity, multifeedstock plant. The £13.5-million ($23.4-million) continuous-process facility located in Immingham was finally completed in March 2007.[13]

Another new player in the UK biodiesel sector is D1 Oils, which opened a 42,000-ton plant in Middlesbrough in 2006. The company also planned to open a 100,000-ton-capacity plant in Bromborough, but scaled back the initial opening in October 2007 to 50,000 tons due to "market conditions." Like many other EU biodiesel producers, D1 has called for government action to be taken against the recent flood of B99 from the United States. The company had plans to increase its total refining capacity in the UK to 320,000 tons by the end of 2008, but has decided to take a wait-and-see approach. D1 recently has formed a joint venture with BP to plant jatropha bushes in countries such as India and South Africa, as well as in Asia and Central and South America, to provide an alternate biodiesel feedstock to edible oils like soy or palm.[14] The new company, D1-BP Fuel Crops, will invest approximately $160 million in the project. The company hopes to import the first commercial quantities of jatropha oil in 2008.[15] In April 2008, D1 Oils, blaming cheap U.S. imports, decided to close its UK refineries and focus instead on its jatropha joint venture with BP.

But the really big news in the British biodiesel industry was Biofuels Corporation's huge 250,000-ton-capacity (75-million-gallon) plant at Seal Sands, Middlesbrough, on the northeastern coast. The facility, one of the largest and most advanced in the world, cost £45 million (considerably more than the original £25 million estimate) and finally opened in June 2006, well over a year behind schedule. A second 250,000-ton-capacity unit tentatively was planned at the same site, although its future is now unclear. Only seven months after the plant opened, however, Biofuels Corporation announced that production had been reduced to just 25 percent of capacity due to "unfavorable market conditions." The company claimed that subsidized U.S. imports were "severely damaging" the industry. Then, the company restructured its £100-million debt, delisted from the stock market, and reorganized as a private company under the name of Earls Nook Limited. The company's troubles, at least in part, were unquestionably a matter of bad timing, but the current turbulence in the EU biodiesel sector continues to spread. Many observers, however, hope that the situation is only temporary.

Scotland

In Scotland, there was not much biodiesel activity until the United Kingdom biodiesel tax reduction of 2002. However, since then there have been some significant developments. In November 2002, Rix Petroleum introduced Rix Biodiesel (a B5 blend with petrodiesel) to the Scottish market. The fuel was an instant hit at the company's 25 Scottish outlets, and sales of the biodiesel blend now account for a significant percentage of the company's total diesel sales. The biodiesel blend is doing so well, in fact, that some of the Rix stations have eliminated standard diesel fuel entirely.

But the most significant activity in Scotland has taken place in Newarthill near the city of Motherwell in Lanarkshire, where a 45,000-ton-capacity biodiesel facility that uses tallow as its main feedstock was built next to an existing animal-rendering plant by Argent Energy. The £15 million (22.5-million-euro) facility, which opened in March 2005, produces enough biodiesel to supply about 5 percent of Scotland's requirements for diesel.

AUSTRIA

Austria is the fifth-largest EU producer of biodiesel, with about 123,000 tons (37 million gallons) produced annually in 2006. (production was projected to double by the end of 2007.) Despite the country's relatively modest production figures, it would be hard to overemphasize the role that Austria has played in the biodiesel industry. The first research and development projects in the biofuels sector in Austria began as early as 1973. As noted in Chapter 2, much of the most important and sustained early biodiesel research was conducted at the Austrian Federal Institute of Agricultural Engineering (BLT) in Wieselburg and at the University of Graz in the early 1980s. As an outcome of that research, one of the first industrial-scale biodiesel plants in the world—with a production capacity of 10,000 metric tons per year—was started up in Aschach, in the state of Upper Austria, in 1991. "Unfortunately, it was also the first one which was closed because of oil prices that were lower than $10 per barrel at the time," notes Manfred Wörgetter of BLT.[16]

One of the key private-sector collaborators with some of this early work was Vogel & Noot GmbH of Graz, which, beginning around 1982, was active in developing technologies to produce biodiesel from plant and animal oils and fats. In 1988, the company received a patent for a transesterification process that used waste cooking oils and waste fats for the production of biodiesel. In 1991, Vogel & Noot provided the rapeseed oil process technology for a 1,000-ton biodiesel plant in Mureck, and the following year the company was involved in the development of a larger-scale biodiesel facility in Bruck with a capacity of 15,000 tons (more recently expanded to 25,000 tons), now operated by Novaol Austria. Vogel & Noot also were connected with the construction of a number of small farmers' cooperative biodiesel plants in various Austrian communities that process different types and qualities of oils and fats.

In 1994, the plant in Mureck was altered and expanded by Vogel & Noot to enable the facility to produce biodiesel from used frying oil collected from the surrounding region. The 9,000-ton biodiesel output from the multifeedstock plant now is used for high-quality vehicle fuel. In 1996, the biodiesel division of Vogel & Noot was sold and a new company, BioDiesel International (BDI) Anlagenbau GmbH, was created. BDI, with headquarters in Graz, subsequently has been involved in a number of multifeedstock plant projects in Germany, Spain, the United Kingdom, and the United States.

A new biodiesel plant located in Zistersdorf, in the state of Lower Austria, was opened in April 2002 by Biodiesel Raffinerie GmbH, a subsidiary of DonauWind KEG. DonauWind, which began operations in 1997, has a four-turbine wind farm in Zistersdorf as well as another turbine near Vienna. Several additional wind farms in other locations are planned. DonauWind is a good example of an increasing number of renewable energy companies that are expanding the scope of their activities to include a wide range of renewables.

A more recent addition to the list was Biodiesel Vienna GmbH's 95,000-ton plant in the Ölhafen Lobau district of Vienna. This state-of-the-art facility cost 30 million euro, and there are plans to expand the plant to a 400,000-ton capacity in 2008.[17] The most recent major addition of all was the new 110,000-ton capacity plant in Enns (near Vienna), which opened in April 2007. The plant, owned by Global Alternativ Energy International (Gate), is expected to provide for about a third of Austria's demand for blended biodiesel.[18] Altogether, there are 12 commercial biodiesel plants in Austria, with their production capacities ranging from 2,000 to 110,000 tons per year. Several additional plants are in the planning or early construction stages. The majority of Austrian biodiesel is produced from rapeseed oil, with sunflower oil, used frying oil, and animal fats playing lesser roles. As much as 90 percent of total biodiesel production in Austria has been exported to neighboring countries for many years.

Early on, Austria adopted a 95 percent tax reduction for biodiesel, as long as the fuel was used full-strength (B100) in vehicles. As of January 1, 2000, all fuels from renewable sources were free of petroleum fuel taxes. Since October 2005, there has been a mandatory blending target for biofuels, and refiners now must meet a 2.5 percent market share. The target rises to 5.75 percent by October 2008. Biodiesel has been widely marketed in Austria to large municipal bus fleets, transport and taxi companies, and other diesel users, especially in environmentally sensitive locations. What's more, there are currently more than 100 filling stations in Austria that offer biodiesel to the general public. The large oil company OMV has been selling a B5 blend at all of its filling stations in Austria since October 2005.

One of the earliest and most successful biodiesel experiments in an Austrian city transit fleet took place in Graz. Beginning in 1994, the Grazer Verkehrsbetriebe (GVB) began field tests with two of its city buses running on biodiesel made from recycled frying oil. The three-year test was considered a success, and in 1997 eight additional buses were switched to biodiesel, followed in 1999 by 10 more. By 2002, GVB had 55 of its 140 buses running on biodiesel, and the entire fleet was switched over by 2005.[19]

In 1991, the Austrian Standardization Institute developed and implemented a biodiesel standard (ON C 1190) for rapeseed methyl ester (RME), becoming the first country to do so, and in 1997 the institute published a more sophisticated standard (ON C 1191) for fatty acid methyl ester (FAME). The former helped encourage biodiesel warranties for diesel tractor engines from numerous manufacturers, while the latter opened the way for use of a wider range of biodiesel feedstocks, including used frying oil. The emphasis in the latter standard was in defining the quality of the fuel that ends up in the fuel tank rather than focusing on the feedstock from which it was made. One of the indirect results of that second standard is that about 1,400 tons of used frying oil is collected every year from 135 McDonald's restaurants in Austria and then turned into biodiesel for the city buses in Graz. More recently, some of that fuel has been used in McDonald's own Austrian delivery trucks as well.

Although the biodiesel industry in Austria is fairly mature, there is still room for additional growth if it can weather the current problems with high feedstock prices and cheap foreign imports.

POLAND

The first research work and testing of rapeseed methyl ester in Poland was conducted between 1989 and 1991 at Wyzsza Szkola Inynierska (WSI–College of Engineering) in Radom. Preliminary road trials on a Tarpan car also were conducted. Between 1991 and 1994, this work was continued and expanded at the Institute of Aviation in Warsaw, where the Austrian standard ON C 1191 was used as a reference for testing. In 1996, a small biodiesel pilot plant was built in Mochelek by the IBMER Institute of Warsaw, financed by the Polish Committee for Scientific Research.[20]

More recently, Mosso, a food-processing company with a new oilseed-pressing plant located near Warsaw opened a small plant in 2004. Since then, three commercial-scale plants with a combined production capacity of 350,000 tons have been opened by different companies in various locations.[21] In 2006, total Polish biodiesel output was about 116,000 tons (up from 70,000 tons the previous year).[22] Most of this output was exported to Germany. At least half a dozen companies have plans to build new biodiesel plants in the next few years.

The Polish government established basic biofuels legislation in August 2006 in response to long-standing complaints from the biofuels industry that government support was insufficient. Recent legislation has given the green light to the sale of B100 and B20 (previously sales were limited to B5), and this is expected to have a dramatic effect on domestic sales of biodiesel.[23] In addition, there has been a modest 5 percent increase in the tax break for biofuels, while the excise tax for B100 has been reduced to almost zero.[24] Since Poland has generally been a major producer of rapeseed, there is good potential for further growth of the Polish biodiesel industry.

CZECH REPUBLIC

Until fairly recently the Czech Republic had the largest number of biodiesel production plants in the world, totaling 16, with a combined production of about 50,000 tons annually. However, with the exception of the 30,000-ton Milo plant in Olomouc, the 12,000-ton Bionafta plant in Mydlovary, and a 3,000-ton Agropodnik facility in Jihlava, the plants were small-scale farmers' cooperatives with less than 1,000 tons capacity. The developing Czechoslovakian (and later Czech) biodiesel industry was highly influenced by the early research and development work in Austria, and consequently the Czechs followed the Austrian model of small farmers' cooperatives. The large-scale plant in Olomouc (later expanded to 50,000 tons) was completed in 1994 with the technical assistance of the Austrian biodiesel plant designer Vogel & Noot. The Agropodnik plant in Jihlava was later expanded to 68,000 tons capacity. As of 2007, there are still about 16 plants that produce biodiesel in the

Czech Republic, but their combined production has increased dramatically in recent years.

Initially, biodiesel was used in the Czech Republic at full strength (B100) as a vehicle fuel, but later on a variety of blends—especially B31—also became popular. In 1994, the Czech standard (CSN 65 6507), developed with the assistance of Austria, was published for rape methyl esters. A subsequent standard for biodiesel blends in the B30 to B36 range was published in 1998. Engine warranties for biodiesel have been issued on a wide range of Czech cars, trucks, and tractors.

Biodiesel now receives a 31 percent tax reduction in the Czech Republic for blended fuels, and B100 recently has been reintroduced to the market. Motorists can refuel their biodiesel-powered vehicles at more than 176 fueling stations across the country. Agricultural supports for rapeseed production in particular have played a major role in encouraging the domestic biodiesel industry. With the acceptance of the new European standard EN 14214 in 2004, the previous Czech standards were replaced. As of September 1, 2007, a 2 percent biofuel blend for diesel fuel is now mandated throughout the nation, and this will rise to 4.5 percent by 2009. This has spurred a lot of renewed interest in biodiesel (and ethanol) in the industrial and agricultural sectors. Recent efforts to simplify the bureaucratic procedures for biofuels supports should improve the future prospects for domestic biodiesel production and consumption. Based on a solid foundation of experience, the Czech biodiesel industry should continue to grow as the EU-27 nations attempt to meet the new biofuels targets in the years to come.

SPAIN

Although many European countries have been developing their biodiesel industries for well over a decade, Spain was initially slow to jump on the bandwagon. In 2003, Spain didn't even show up on European biodiesel statistical charts. This is partly due to the fact that, until fairly recently, Spain has focused much of its attention on bioethanol, and as a result it is now the largest EU producer of that fuel. Another reason was that the Spanish climate is not well suited for the growing of rapeseed. But in the past few years, in response to the new EU targets for biofuels percentages, Spain has been making up for lost time in the biodiesel sector. Spain's late entry into the biodiesel market may actually work to its advantage, since it should be able to avoid many of the early challenges experienced by its pioneering neighbors and make use of the latest technological advances, especially multifeedstock processes. What this means is that much of Spain's biodiesel can be made from relatively inexpensive used frying oils, especially olive oil, which the country has in great abundance. More than 1 million tons of plant oils are consumed in Spain every year, about half of which end up as used frying oil.[25]

One major incentive for the rapid growth of the biofuels industry in Spain is the complete tax exemption for pure biofuels as well as blends through December 2012. More recently, on June 15, 2007, the Spanish Congress approved new mandatory biofuels blending requirements that will take effect in 2009. This legislation also includes a voluntary 1.9 percent blend level in 2008.[26]

By way of background, in 1999 the company Grupo Ecológico Nacional (GEN) in Barcelona participated in a pilot project (sponsored by the Austrian Biofuels Institute) to recycle waste fats under the EU's ALTENER-2 renewable energy initiative. The goal of the project was to establish a low-cost system for recycling used frying oils from restaurants, hotels, and households and then using them as biodiesel feedstock.[27] Following up on that program, biodiesel was first introduced to the Spanish market for tests, mainly in pilot projects with public transport buses.

In October 2002, Spain's first dedicated biodiesel production plant was opened by Stocks del Vallès in Montmeló. The 6,000-ton-capacity, multifeedstock plant, which cost 5 million euros to construct, utilizes used cooking oil as its primary feedstock and was later expanded to 31,000 tons. That was followed in 2003 by a 20,000-ton-capacity plant in Álava built by Bionor, and a 50,000-ton-capacity plant in Reus constructed by Bionet Europa.

In February 2003, the first public biodiesel pump in Spain opened in the Catalan town of Tárrega. Located in a Petromiralles service station, the pump dispenses B30 manufactured by Stocks del Vallès. Petromiralles installed additional biodiesel pumps in a number of other towns as well. The public response was enthusiastic and Petromiralles soon added many more locations. Since then, biodiesel pumps have spread across the nation, and it is now possible to purchase biodiesel blends at more than 250 locations.

In recent years biodiesel refineries have multiplied dramatically across the country. Altogether, there are approximately 19 operational biodiesel plants in Spain, with a combined capacity of 760,000 tons. Up until mid-2007, there were also about 23 additional plants under construction nationwide, with a combined production capacity of 2,443,000 tons. What's more, there were an additional 19 plants on the drawing board for a staggering future total capacity of 5.5 *million* tons. If this total is reached in the next few years, Spain could be challenging the top EU producers for bragging rights.

However, this prospect is now in doubt. The recent troubles in the biodiesel industry in the EU due to high seed oil prices and cheap foreign imports have been causing serious problems for the industry in Spain as well. As of January 2008, about 80 to 90 percent of the nation's biodiesel production capacity had been shut down. Many of the planned new construction projects have been delayed or even cancelled, and the industry seemed to be teetering on the edge of collapse, according to some industry observers.[28] This has become a distressingly familiar story in some parts of the EU, and there are no easy solutions in sight, at least in the short term.

SLOVAKIA

Agriculture is a very important part of the economy in this mostly rural nation. Slovakia also imports most of its energy supply. Consequently, Slovakia is determined to increase its use of domestic renewable energy sources, and it has had a long history of biodiesel activity. Like the Czech Republic, Slovakia has a large number of smaller, mostly farmers' co-op plants (500 to 1,500 metric tons per year) and a few larger-scale facilities. The first small-scale plant in Slovakia was built by a group of technicians from the ZTS-Martin company (formerly known as a manufacturer of heavy weapons and tanks) in collaboration with the University of Bratislava in 1992. Most of the small-scale plants that followed were supplied with process equipment from the Slovak company MTD.[29]

The larger Slovak producers are Palma-Tumys a.s. in Bratislava (60,000 tons), AgroDiesel in Revúca (1,500 tons), and BIO BHMG in Spišský Hrušov (1,500 tons). There are also about six smaller plants with production capacities of about 500 tons scattered throughout the country. Total annual biodiesel production capacity in Slovakia is approximately 99,000 tons.[30]

The Slovakian biodiesel standard (STN 656530), which went into effect in 2001, was subsequently replaced by the new EU standard, EN 14214. Biodiesel has been used both as a vehicle and heating fuel in Slovakia. B30, sometimes referred to by the name *bio-naptha*, is the most popular blend of biodiesel in the country. In the late 1990s, biodiesel enjoyed full detaxation in Slovakia, but that changed in 2002 when a new national excise-tax law came into effect that caused considerable confusion and removed some of the previous incentive for the use of biodiesel blends, resulting in a dramatic drop in local demand.[31] The biodiesel that was produced after that date was primarily exported. More recently, production has rebounded, and biodiesel has been introduced to the market in a B5 blend with a reduced excise tax for use in agricultural and forestry production and public transport. Some higher blends have been utilized in various fleets.

One sign that the regulatory landscape has changed in Slovakia was the 2006 introduction of a B5 blend by Slovnaft at all of its fuel stations across the country. (Slovnaft is a subsidiary of the Hungarian oil giant MOL.) About 65 percent of the biodiesel required was imported from other EU states, while the remainder was produced in Slovakia. Slovnaft, along with a consortium of other partners (known as Meroco), is building a new 100,000-ton biodiesel plant in Loepoldov. When the plant is opened in early 2008, it will provide Slovnaft with the ability to meet all of its domestic needs from a local source.[32]

The Slovakian biodiesel industry, although relatively small, has demonstrated that it knows how to produce biodiesel and has the growing capacity to do it. The government, however, has been slow to support the industry with financial and other incentives, but is making progress in that direction. The structural changes in the Slovak economy in recent years and the transition to EU membership have proved challenging and often difficult for the country. However, if these transitional problems can be overcome, the long-term outlook for the Slovak biodiesel industry is fairly good.

BELGIUM

Belgium has had a long but uneven history of biodiesel involvement. Unlike many of its neighbors, it has not followed through with widespread national implementation until fairly recently because of lingering reservations about the environmental and economic performance of the fuel. A 1937 Belgian patent for the use of ethyl esters of palm oil and an experiment with a commercial bus operated between Brussels and Louvain in 1938 were two of the earliest biodiesel-related activities anywhere in Europe. In addition, the article "The Use of Vegetable Oils as Fuel in Engines" was published in Belgium in 1952.[33] More recently, between 1991 and 1992 an experiment was conducted in Mons with 15 buses belonging to Transport En Commun (TEC)-Hainaut running on B20 (the most common biodiesel blend in Belgium). The drivers were satisfied with the performance of the buses, and the only negative comments received were from a few people who were bothered by the "barbecue" odor of the exhaust.[34]

In 1992, a biodiesel pilot plant was built by the De Smet Company in Feschaux, and for the next three years a number of additional tests with buses and other vehicles running on the fuel produced by the plant were conducted, with favorable results. In 1996, a two-year test was conducted by the Flemish Institute for Technological Research (VITO), in cooperation with the University of Graz in Austria, on used vegetable oil as a biodiesel feedstock. Despite all of this activity, the majority of Belgian biodiesel output (which amounted to 5 percent of the European total in 1998) was exported to France, Germany, and Italy. Then, lacking a domestic market, Belgian biodiesel production ceased.

But in May 2005, with crude-oil prices on the rise and the EU pushing for higher biofuels usage, the Belgian government had a change of heart and decided to support biodiesel with a tax reduction of 36 euro cents per liter for biodiesel blends used as a transport fuel. Oleon then announced plans to build a new biodiesel production unit at Ertvelde with a capacity of 95,000 tons. The facility opened in early 2007. In addition, Cargill and two Belgian partners announced plans in January 2006 to build a new 25-million-euro rapeseed-oil-based biodiesel plant at Cargill's existing location in Ghent with an annual capacity of 200,000 tons. The companies also announced that they would convert an existing soybean-processing facility at Ghent into a 1-million-ton rapeseed facility.[35] In September 2006, Neochim SA announced that it would begin production of biodiesel at a renovated BASF complex at Feluy.[36] And in late 2007, Proviron, a specialty chemical and manufacturing company, jumped into the fray when it opened its new 100,000-ton-capacity biodiesel plant in Oostende at its existing chemical complex.[37] What a difference government support makes.

With all of this increased interest and activity, biodiesel production in Belgium grew to 25,000 tons in 2006, according to the EBB. In November 2006, oil giant Total announced that it would offer a biodiesel blend at its filling stations throughout Belgium. (The biodiesel for the blend was produced by Oleon.) A number of even larger biodiesel plants are in the planning or early construction stages, but in mid-2007, the turmoil in the EU biodiesel market delayed most biodiesel projects in Belgium by about 12 months as the various companies involved tried to assess their prospects for profitable conditions.

PORTUGAL

Until very recently, Portugal wasn't even on the EU biodiesel production chart. This has changed rather dramatically in the past few years. In 2004, in response to the EU biofuels directive of 2003, the Portuguese government released a report, "Biofuels: Which Strategy for Portugal?" The report highlighted the nation's various potential response choices to the EU directive. The government ultimately decided to reduce taxes on biofuels blended with petroleum and to establish tax benefits for biofuels producers, among other actions. The government also set a maximum ceiling of 100,000 tons on which the tax exemption can apply. These government decrees have had the desired effect. In 2005, about 160 tons of biodiesel were produced in Portugal. In 2006, that figure had vaulted to 91,000 tons. Another 500,000 tons of new biodiesel plant capacity is slated to open between 2007 and 2010 in various locations around the country. Prior to 2006, what little biodiesel production there was in Portugal came primarily from used cooking oils.[38]

One constraint on the growth of the domestic biofuels sector in Portugal is the limited agricultural land base available for biofuels feedstock production, and it almost certainly will be necessary to import substantial amounts of feedstock oils in order for the nation to meet its EU-mandated biofuels targets. This already has led to a number of joint ventures between Portugal and the Brazilian government. (Brazil was a Portuguese colony for more than 300 years.)

DENMARK

Although Denmark was involved in some successful biodiesel tests with buses in Copenhagen in the mid-1990s, only limited progress has been made in the country's biodiesel sector due to a long-standing lack of political support for the use of biofuels in transportation. Most efforts to promote the use of biofuels in transportation have been in the form of research. One possible explanation for this apparent lack of urgency may be Denmark's North Sea oil reserves, although crude-oil production from that resource is now in decline. Another reason is the government's strong emphasis on the use of biomass for the production of electricity and heat, which it claims is far more cost-effective than liquid biofuels.

Nevertheless, between 2001 and 2006 Danish biodiesel production increased from 10,000 tons to 80,000 tons at the Emmelev A/S plant in Otterup, the nation's only large industrial-scale biodiesel producer for many years. All of that production, however, has been exported to Germany and Sweden, where government tax exemptions have (until very recently) created better market conditions. In order to feed that demand, Danish farmers have been growing more and more rapeseed, and production of the yellow-flowered crop increased by 54,000 hectares (133,000 acres) in 2007, up 44 percent over the previous year.[39] Daka Biodiesel, a cooperative formed in 2006, is building a new 50,000-ton facility at Løsning that will use animal fat (and possibly used cooking oil and fish oils) as feedstocks. The plant, which is designed for future expansion to 100,000 tons, opened in December 2007.[40]

Denmark would seem to be well suited for additional future growth in biodiesel production and use, since both its climate and agricultural sector are ideal for high yields of rapeseed oil. What's more, most Danes are environmentally conscious and most likely would embrace greater use of biodiesel as well as recycling programs for used frying oil, which has been an underutilized resource.

THE NETHERLANDS

The Netherlands has been slow to adopt biodiesel. However, beginning in 2005 with the rise in crude-oil prices, that situation changed dramatically as Dutch consumers began to demand alternatives. The nation's first new biodiesel production plant opened in July 2005 in the northern Dutch port city of Delfzijl. The relatively small facility is owned by Solar Oil Systems and a group of local farmers.[41]

In 2006, the government announced a series of tax incentives to spur the development of the biofuels sector. The first industrial-scale biodiesel plant owned by Sunoil Biodiesel B.V. opened in August of that year. Originally planned for 35,000 tons, the new 4-million-euro refinery in Emmen now produces 70,000 tons of biodiesel from rapeseed oil but also uses small quantities of waste vegetable oil from restaurants in the region. Sunoil also has installed a retail biodiesel pump for local B100 customers and hopes to set up a network of retail B100 pumps throughout the nation.[42] Another Dutch producer that relies on used frying oil is Biodiesel Kampen BV, located in Haatlandhaven.

On January 1, 2007, the Dutch government abolished the tax breaks, but finally initiated a long-awaited 2-percent biofuels mandate for gasoline and diesel, spurring additional biodiesel activity. As a result, four large new biodiesel refineries with combined production capacities of about 900,000 tons have opened or are under construction in or near Rotterdam. Rotterdam is Europe's largest refinery center as well as the major hub of the vegetable-oil market, making it an ideal site for these large new biodiesel projects.

Reflecting the sudden increase of activity in the industry, eight biodiesel producers in the Netherlands recently joined together to form the Netherlands Biodiesel Industry Association (Vereniging Nederlandse Biodiesel Industrie—VNBI) with the goal of making the nation one of the top producers in the EU within two years. The new group hopes to produce a combined total of 1.5 million tons of biodiesel annually by 2009.[43] In December 2007, however, in a sign of the ongoing shift in public opinion, the Dutch government announced that it would no longer subsidize the importation of palm oil, its main source of green electricity generation. This will probably also affect any biodiesel plants that planned to use palm oil as their main feedstock.

SWEDEN

The southern part of Sweden offers excellent potential for the production of rapeseed, but for many years most of the emphasis on Swedish biofuels has been directed toward ethanol made from wheat and other biomass. Stockholm, in fact, has Europe's largest fleet of ethanol-fueled buses. As a result, the use of biodiesel in Sweden is not as widespread as in many other European countries. Nevertheless, in 1992 the first Swedish biodiesel production plant was opened by Svenska Ecobränsle AB (Swedish Eco Fuel) in Klippan; it produced about 1,000 tons annually. Rapeseed was the primary feedstock, but some tests were conducted with linseed, lard, and used frying oil. Around 2001, the plant's capacity was expanded to 8,000 tons annually.

In 2005, the Swedish government announced plans to break the nation's dependence on fossil fuels by 2020. The country has embarked on a wide-ranging series of initiatives in order to achieve that ambitious goal. In 2006, Swedish regulations were changed to allow a 5 percent biodiesel blend (up from a previous 2 percent). This shift greatly accelerated the growth of the domestic biodiesel industry. Following the announcement, Statoil ASA (a major Norwegian oil company) announced that it would offer a 5 percent biodiesel blend at all of its service stations in Sweden.

In April 2006, Svenska Ecobränsle AB opened a new 40,000-ton biodiesel plant in Karlshamn on the country's south coast. Beginning in 2008, the company hopes to be able to produce a total of 100,000 tons annually. Up until the construction of this plant, approximately 80 percent of the biodiesel consumed in Sweden was imported from Germany and Denmark. Another new biodiesel refinery was reportedly opened in late 2007 by Preem Petroleum AB (Sweden's largest oil company) and Pestorp AB (a specialty chemical company), who collaborated on a 160,000-ton facility in Stenungsund.[44]

Biodiesel is fully tax-exempt in Sweden. For many years, much of Sweden's biodiesel production was used as B100 for vehicle fuels, while most of the remainder was blended at a B2 level with petrodiesel. With the new regulation allowing B5, the amount of blended biodiesel is expected to rise dramatically. The government also promotes biofuels by offering lower taxes as well as free parking for "clean" vehicles and by mandating at least one biofuel pump at every large filling station in the nation. There are numerous diesel engine warranties for biodiesel use in Sweden, based on the Swedish biodiesel standard (SS 1554 36). The Federation of Swedish Farmers (LRF) supports increasing biodiesel production, and rapeseed farmers in particular see this as a great opportunity.

LITHUANIA

Rapeseed is the largest oilseed crop in Lithuania, with an area of 60,000 hectares (148,000 acres) given over to its cultivation in 2002 (more recently 80,000 hectares) for domestic vegetable-oil production and export. However, at that time only a few hundred tons of the crop was being processed domestically for biodiesel. There is quite a lot of fallow land in Lithuania, some of which is well suited to the production of industrial energy crops.[45] Consequently, in early 2003 the Kaunas Technical University in Kaunas, Lithuania, began working on a biodiesel research project related to testing a variety of known and new oilseeds, with assistance from the EU's ALTENER-2 project.

Interest in biodiesel clearly extends beyond the laboratory. The Rapsoila company completed a 100,000-ton-capacity biofuels plant in the Mazeikiai region of Lithuania in 2004.[46] Approximately 1,500 small farms in the surrounding region are reportedly supplying the rapeseed for the plant. In 2006, Lithuania actually produced 10,000 tons of biodiesel. More recently, a new 100,000-ton biodiesel refinery was completed in late 2007 by Mestilla in an economic free zone near the town of Klaipeda.[47]

On January 18, 2007, the government of Lithuania published an Updated National Energy Plan that emphasized its strong commitment to renewable energy, including both biomass and biofuels (ethanol and biodiesel). Lithuania now permits a 5 percent biodiesel blend, and the government believes that it can meet or surpass the EU targets for biofuel use. The government also subsidizes the production of biodiesel feedstock (primarily rapeseed) with a 160 lita (about $53) payment per metric ton.[48] Prospects for future growth of the biodiesel industry in Lithuania seem good.

LATVIA

Delta Riga Ltd. was the first company to open a biodiesel refinery in Latvia in November 2001. The 2.5-ton facility (later expanded to 12,000 tons) is located in Nauksene (Valmiera district) in the northern part of the country. The plant uses rapeseed oil as its primary feedstock and also burns its glycerin byproduct in a boiler to produce process heat for the refinery and domestic heat and hot water for the surrounding community.[49] In 2004, another biodiesel plant was opened in the Jelgava district by Mezrozite Ltd. with an annual capacity of 5,000 tons. And at the end of 2005, another company, Mamas-D Ltd., opened a 3,000-ton facility.

One reason for the relatively slow development of the biodiesel sector in Latvia is that until fairly recently rapeseed was not widely grown. Consequently, there is an absence of an established infrastructure for cleaning and drying rapeseed, a situation that should be resolved in the not too distant future. There is also a large amount of idle agricultural land in Latvia that offers substantial opportunities for additional biofuel crop expansion. An excise-tax exemption for biofuels by the government has offered additional incentive for the development of the biofuels industry.

In response to EU emphasis on biofuels and increasing interest in domestic biodiesel, three more biodiesel refineries with combined production capacities of about 185,000 tons have recently opened in various locations. Considering the amount of idle agricultural land in Latvia, there seem to be good prospects for future expansion of the biodiesel industry there.

BULGARIA

Bulgaria's first commercial biodiesel plant is located in Brousartsi, near the border with Romania, and is operated by Sampo AD, headquartered in Sofia. Originally a vegetable-oil-processing facility, the 3,500-ton-capacity plant was converted to biodiesel production in 2001 with a simplified transesterification process and now relies primarily on used frying oil (UFO) as a feedstock. The biodiesel produced by Sampo has been used mainly at B50 and B100 concentrations in trucks, buses, and automobiles in the region. Some of the glycerin by-product from the plant has been refined for use in pharmaceuticals, while some has been used as a liquid fertilizer and also as an ingredient in hydraulic fluid. Sampo has spearheaded the formation of the National Biofuels and Renewable Energy Sources Association.[50]

One of the main constraints on the biodiesel industry in Bulgaria is the lack of local raw materials. Rapeseed has not traditionally been produced in large quantities due to unfavorable growing conditions. In late 2006, as part of the EU accession process, the Bulgarian cabinet approved a long-term program to encourage the use of renewable energy sources, including biofuels, which was seen as an important step toward harmonizing local legislation with EU targets. Biodiesel is currently exempt from excise duties, but only for B100.

In addition to the Sampo plant mentioned above, there are about 4 major producers of biodiesel operating in Bulgaria with combined production capacities of about 125,000 tons. Two more plants with combined capacities of 130,000 tons were scheduled to come online in late 2007, and another even larger 175,000-ton refinery was scheduled to open in 2008.[51] There are also about 15 much smaller biodiesel operations scattered around the country, mostly located at farms, which produce biodiesel for their own use.

The rapidly growing biodiesel industry still faces substantial regulatory, legislative, and tax issues, all of which need to be addressed if it is going to reach its full potential in Bulgaria.

GREECE

Greece has been a very late entrant into the biodiesel sector, but the country is beginning to make up for lost time. Like many other countries around the world, Greece sees the production of biofuels as a way to increase demand for some of its agricultural products while providing income for its producers. An increasing number of Greek farmers have become enthusiastic about growing rapeseed as a biodiesel feedstock. The Greek government has adopted the EU targets for biofuels and has also detaxed biodiesel for the amount necessary to reach the targets.

Almost overnight, biodiesel refineries have sprouted up across the country, and there are about five currently in operation, with around six more that have plans to open in the next few years. Most of these facilities are supported with funds from the EU and the Greek government. It has been estimated, however, that only about a third of the feedstock materials needed for all of these plants can be met with locally grown feedstocks, requiring substantial imports.[52]

SLOVENIA

Slovenia is another recent entrant into the biodiesel sector. Nevertheless, the country is making measurable progress. The government now exempts pure biodiesel from excise taxation and blended fuels receive up to a 5 percent exemption. The percentage of blended biofuels is now also mandated at 2 percent for 2007, 3 percent in 2008, and up to 5 percent by 2010. Slovenia's largest fuel retailer, Petrol, now offers a B5 blend at more than 20 of its service stations, and the Ljubljana public transport company tested the use of a biodiesel mixture on 2 of its buses and now runs 20 of them on B100. EU payments for the production of energy crops also have been incorporated into Slovenian agricultural policies. In 2005, about 2,500 hectares of land were sown to rapeseed, which yielded about 7,500 tons of oilseed. The country has approximately 6,000 to 7,000 hectares of land available for rapeseed production, according to agricultural officials.

In 2005, biodiesel was being produced by three major biodiesel companies in Slovenia. Two of the operations were relatively small-scale, at about 2,000 tons each, while the third refinery had a larger output.[53] A number of additional facilities are planned to open in the next few years, including a 60,000-ton refinery in Lendava scheduled to begin operations in early 2008.[54] The feedstock for the new plants will come primarily from imported oils, waste vegetable oil, and animal fats.

IRELAND

Ireland is capable of producing significant quantities of rapeseed, providing the potential basis for a domestic biodiesel industry. However, the present high cost of producing that crop is a limiting factor. The main strategy of the Irish government to promote greater biofuels usage has been excise-tax relief. The Biofuels Mineral Oil Tax Relief scheme took effect in 2005 and was initially limited to a group of small allocations: 1 million liters (733 tons) each for biodiesel and ethanol. (The following year, the allocations were raised to about 53,000 tons and 62,000 tons, respectively.) There has been quite a bit of interest in using waste cooking oil for biodiesel production in recent years, leading to some small-scale production in various locations. More ambitious production targets subsequently have been set by the government to try to meet the new EU biofuels mandates. And in February 2007, the government announced that the country would shift to a biofuels obligation, requiring specific percentages of biofuels beginning in 2009. One of the first large-scale responses came when Green Biofuels Ireland began construction on its first biodiesel refinery in New Ross in April 2007. The 30,000-ton plant is scheduled to come online in 2008.[55]

ROMANIA

Romania is yet another country that has recently jump-started its biodiesel industry, with a new 25,000-ton refinery in Vaslui owned by Romanian vegetable-oil producer Ulerom Vaslui that began production in July 2007. The new computer-controlled, continuous-flow facility is seen as a key part of Romania's attempt to meet its new EU biofuel targets. Several other large projects are also in the works. Rompetrol is investing about 19 million euro in a new biodiesel refinery, and Portuguese biodiesel developer Martifer is building a new 45-million-euro refinery in Lehliu Gara. Some observers believe that Romania has the capability to export biodiesel to help other EU member nations meet their biofuel quotas, which partly explains the recent flurry of foreign investment.[56] Rompetrol Downstream, the retail division of Rompetrol Group, began to offer B3 at many of its 350 service stations around the nation in January 2008.

SERBIA

Serbia has had an on-again, off-again biodiesel production history related to trade embargoes. Biodiesel production peaked in 1995 at 10,000 tons in the former Yugoslavia, but then returned to zero in 1996 after the embargo was lifted. Today, Serbia is experiencing a significant biodiesel renaissance. In August 2006, a series of tests were conducted with biodiesel blended fuel on several bus lines in Belgrade without any adverse effects on the buses. In addition, a new 100,000-ton biodiesel refinery owned by the Victoria Group was opened at the end of June 2007 in the town of Sid, located about 100 kilometers northwest of Belgrade.[57] However, so far the Serbian government has not taken a major role in promoting biodiesel.

HUNGARY

Hungary's first large biodiesel plant opened in May 2007. The 25,000-ton facility, located in Bábolna (northwest Hungary) is owned by Öko-Line Hungary Kft. The company bought the facility in 2003 from previous investors who had gone bankrupt before the plant could produce any biodiesel. There are plans to double the capacity of the new facility in 2008. Öko-Line also has been producing biodiesel in two small batch-process plants in Kunhegyes and Mátészalka. The Hungarian oil and gas group MOL Nyrt will be the primary customer, but some of the new plant's output may be exported to Austria and Germany.[58] In January 2008, MOL opened a new 150,000-ton biodiesel plant in Komárom.

NORWAY

In Norway, biodiesel is used for vehicle fuel, but most of it is imported. There are currently more than 18 public fuel pumps serving Norwegian drivers with biodiesel blends. Estra AS, located in Trondheim, is the nation's largest producer of biodiesel at around 8,500 tons per year. For many years, the company used fish oil from processing waste as the main feedstock, but switched to imported rapeseed oil, mainly from Denmark, because the fish oil biodiesel was too high in iodine to meet the EU Standard.[59] Recently, several companies have been exploring the possible production of biodiesel from wood. And, in June 2007, the Norwegian government announced plans to build a biodiesel plant in Fredrikstad as part of its plan to try to meet the new EU targets.

FINLAND

Finland has been consuming modest amounts of biodiesel for a number of years. But in a dramatic move, Neste Oil opened a new 100-million-euro biodiesel plant in Porvoo at the end of May 2007. The new 170,000-ton facility uses Neste's proprietary process that can use a flexible mix of vegetable oil, animal fat, and other raw materials as feedstock.[60] And in December 2007, Neste Oil announced plans to open a mammoth 2.6-million-ton (800-million-gallon) biodiesel refinery in Singapore in late 2010 that would use palm oil as its main feedstock. The 440-million-euro plant would be the largest in the world. Neste Oil has committed itself to using only palm oil certified by the Roundtable on Sustainable Palm Oil when sufficient supplies become available, probably by 2008.

MORE EUROPEAN ACTIVITY

Before moving on to the rest of the world, it should be noted that there also has been some biodiesel activity in a number of other European countries including: Estonia, Croatia, Luxumburg, Switzerland, Cyprus, Malta, and Macedonia.

RUSSIA

It's difficult to decide whether Russia and some of the former Soviet republics (now more or less independent nations) belong in a discussion of Europe or Asia, since the territories involved tend to straddle the boundaries and are so vast. Due to Russia's huge oil and natural gas resources, there has been very little domestic demand for biofuels. Nevertheless, there is growing interest in biofuels in some locations. In Russia, there does not appear to be any large-scale biodiesel production yet, although modest initiatives are taking place at the regional and local levels, especially on farms. However, a number of relatively large biodiesel refineries are planned or under construction, including the 100,000-ton Rusbiodiesel refinery in Krasnodar Territory that is scheduled to open in 2008. A number of other companies reportedly have plans for commercial-scale plants as well.[61] One of the main barriers for domestic consumption, however, is that there is no national standard for the production of biodiesel and virtually no coordinated national support.[62]

There are, however, very large quantities of potential biodiesel feedstocks being grown in Russia and some of its former republics. In 2006, for example, Russia produced 525,000 tons of rapeseed, and about 20 percent was exported to Europe.[63] Russian officials note that the nation has approximately 20 million hectares (49.4 million acres) of arable land not currently in production that potentially could be used for biofuels feedstock production. These same officials suggest that Russia will be a major biodiesel producer in the not-too-distant future, although there are numerous technical issues that need to be resolved before that can happen. In any case, most of the current focus on biofuels is aimed at the export market.

A number of former Soviet republics are moving ahead on biodiesel initiatives of their own. The government of Ukraine recently endorsed a Program of Diesel Biofuel Production Development aimed at promoting the construction of biodiesel facilities and increasing rapeseed production on farms.[64] And Belarus has announced plans to cooperate with Ukraine for a planned biodiesel production facility in the Ukrainian oblast (province) of Ternopol. The equipment for the project would come from Belarus while the feedstock for the plant would come from Ukraine.[65]

A relatively new organization, the Russian Biofuels Association, is active in promoting ethanol and biodiesel and has already hosted three international conferences in Moscow. Expect a lot of biodiesel activity from Russia and its former republics in the years ahead at both the local and national levels.

TURKEY

Turkey presents a similar problem. It's not (yet) technically part of Europe, but its proximity to the EU market offers substantial opportunities for both Turkey and the EU. There is a good deal of interest and activity in the biodiesel sector in Turkey, although its exact status is somewhat difficult to assess. Nevertheless, there is already significant production from hundreds of mostly small producers, including a number of cooperatives. However, only about 50 of those producers are expected to be able to ultimately comply with new government registration requirements. Nevertheless, these small producers can help meet the needs for fuel in their local communities.

In 2005, there was somewhere between 450,000 and 878,000 tons of biodiesel produced nationwide. (Available statistics do not agree.) About 70 percent of the biodiesel is produced from imported palm oil, while safflower and canola make up much of the balance. Waste vegetable oil is rarely used, except by the smallest of producers. Biodiesel made from local feedstocks is tax-exempt, and a biodiesel standard, TSEN 14214, recently has been established.[66]

Biodiesel could play a significant and positive role in Turkey's future agricultural and energy sectors, as long as the industry follows a careful path that avoids the excesses that have developed in some other countries.

7

Africa and Asia

Although Europe has been the main focus of most biodiesel activity for many years, there have been a lot of interesting developments in numerous countries elsewhere more recently. When the official survey of the international status of the industry was published in 1997 by the Austrian Biofuels Institute for the International Energy Agency, 24 countries were included. Since then, that number has increased to about 90. Especially in the past few years, biodiesel activity has begun to expand exponentially as more and more governments, concerned about the growing volatility and dramatically increasing prices in the international oil market, discover the potential benefits of biodiesel.

In the absence of strong leadership on renewable energy issues at the national level in some of these countries, individuals or small groups of citizens as well as nongovernmental organizations have taken matters into their own hands to initiate small local biodiesel projects. However, because many of these grassroots activities have tended to take place "under the radar" of the international press, it is often difficult to assess just how much of this activity is going on. In many cases, however, the efforts of these early adopters have been overshadowed by more recent, large-scale commercial projects. This does not mean that the smaller-scale efforts are unimportant; they are important, especially for the local communities that rely on them. In fact, in developing countries with limited food supplies and struggling economies, these smaller projects can be far more beneficial to the local population than the large multinational corporate projects that have been sprouting up across the globe.

The survey that follows is an attempt to fill in at least some of the larger blank spaces on the global biodiesel map. In this chapter, the journey begins in Africa and follows a meandering but generally eastward course through Asia.

SOUTH AFRICA

South Africa has had a long but inconsistent history of biodiesel activity. As mentioned in Chapter 2, South Africa was involved in early biodiesel experiments prior to World War II, which were subsequently abandoned in favor of coal and synthetic fuels. A second round of research and testing took place between 1980 and 1984 and made some significant advances. But, again, the work was abandoned. In recent years, though, biodiesel has made yet another comeback.

A June 2001 report published jointly by the Council for Scientific and Industrial Research (CSIR) and the Agricultural Research Council investigated technologies that could provide alternatives to crude-oil-based fuels. Biofuels were highlighted in the report as offering a good deal of potential. A follow-up report published in March 2003 recommended placing a greater emphasis on biodiesel in South Africa. In addition to decreasing the country's dependency on imported fossil fuels, the report said that greater biodiesel production would strengthen the agricultural sector and create new jobs in rural areas, a key national priority.[1]

The South African potential for biodiesel production is enormous, according to a 2003 United Nations study. With the use of 2.3 million hectares of land (5.7 million acres), it is estimated that 1.4 billion liters (517 million gallons) of biodiesel could be produced, without having an adverse impact on food supplies. Even half of this projected amount could supply about 17 percent of current road and rail use of petrodiesel in the country. The study suggested a more modest (but still ambitious) target of 10 percent within 10 years.[2] Another study, commissioned by Earthlife

Africa, Johannesburg, and the World Wide Fund for Nature, Denmark, noted that much of this production could come from a large number of relatively small biodiesel facilities spread across the country, offering significant employment opportunities for farmers and other local entrepreneurs.[3]

In order to supply enough biodiesel to create a B5 blend for the entire nation, 280 million liters (approximately 260,000 metric tons) of biodiesel would have to be produced in South Africa annually. Current production doesn't even begin to approach that figure. In Merrivale, located in the province of KwaZulu-Natal, Biodiesel SA, a company that was established in January 2001, produced up to 600 liters of biodiesel a day (about 100 tons per year) from used vegetable oil. The biodiesel produced in Merrivale is used mainly by companies that operate heavy earth-moving equipment, according to the plant's owner, Darryl Melrose. The plant was upgraded to a capacity of 2,000 liters a day, but Melrose could not find enough used cooking oil to meet all the demand for his biodiesel. As a consequence, Melrose has spent a good deal of time researching alternate feedstocks. He finally decided that the hardy *Jatropha curcas* (physic nut) would best suit his (and the industry's) needs. Melrose is now part of a private/public partnership to cultivate this prolific oilseed-bearing plant.[4]

"We're trying to empower poor farmers on the east coast of South Africa between Durban and Mozambique where the climate conditions are suitable," Melrose explains. "*Jatropha curcas* can really benefit the farmers we are working with because the seed crop is toxic, so it can't be stolen to eat; animals won't even feed on it because it's inedible. So, I think, given our particular situation in this country, it's the perfect crop for this purpose."[5] About 300 hectares are scheduled to be planted with jatropha. Field tests to date have been extremely positive, and jatropha is reportedly taking off like wildfire in various parts of KwaZulu-Natal. About one thousand growers are already involved, and their number is increasing.

Another, somewhat larger biodiesel production facility with a 5,000-liter-per-day (about 1,000 tons per year) capacity began operating in May 2002 in Wesselsbron, in the province of Free State. The plant, operated by the sunflower farmer Johan Minnaar, made use of locally grown sunflower seeds as well as used cooking oil. Minnaar was the first to produce a vegetable-oil-based fuel on a commercial basis in South Africa. Since then, a number of other farmers in various locations also have begun to produce small quantities of biodiesel for their own (and their neighbors') use. In 2005, another new biodiesel project was launched with assistance from the North West provincial government near Mafikeng when the first phase of a jatropha plantation was started. The jatropha was intended for use as feedstock for biodiesel, and if the 45,000-hectare plantation is completed it could provide jobs for as many as 2,000 local residents. The cultivation of jatropha, however, will require a lengthy environmental impact study, and some sectors of the South African government do not favor its use.[6]

Initially, South Africa adopted the European standard for rape methyl ester. However, due to the different biodiversity and climate conditions in the country, a series of tests were conducted on soy, sunflower, canola, groundnut (peanut), cotton, and jatropha in order to establish an appropriate South African standard. The standard was eventually developed by the government's joint implementation committee for the biodiesel industry, with the assistance of the South African Bureau of Standards.

On January 25, 2007, the South African cabinet finally gave its approval to the long-awaited biofuels industrial strategy that should help the nation meet about 75 percent of its renewable energy target under the Kyoto Protocol. The strategy called for an overall 4.5 percent biofuels target by 2013, including an 8 percent ethanol (E8) and 2 percent biodiesel (B2) blending target. The strategy, among other provisions, also set a 40 percent tax rebate on ethanol, the same as for biodiesel.[7] Although the biofuels industry generally greeted the strategy with enthusiasm, some social protection groups were worried that the shift to greater biofuels feedstock production might cause food prices to rise beyond the means of South Africa's poor. On the other hand, the strategy was projected to create around 55,000 new jobs in rural areas. In any case, the strategy helped to revive the prospects for the biodiesel industry in South Africa. However, in early December, responding to concerns about rising food prices, the government lowered its 4.5 percent biofuels target to 2 percent by 2013 and specifically excluded maize (corn) from the strategy. South Africa is the largest producer of corn in Africa.[8] It is not entirely clear how this move will affect the biodiesel industry, since corn is not a major feedstock.

After years of deliberation, the large oil and chemicals group Sasol has finally completed a study on the feasibility of producing biodiesel in South Africa using soybeans as the main feedstock. The proposed 100,000-ton biodiesel plant most likely will be located at Sasolburg. However, the company maintains that the plant will not be viable without government incentives.[9] Another company with big plans is D1 Oils, Africa, a subsidiary of the UK-based global producer of biodiesel. However, the company has a rather different, decentralized plan for local production from local resources based on modular refineries. The D1 20 refinery is about the size of a large shipping container and can be set up in remote locations. The units can be used alone or in combination to create a refinery with a capacity from 8,000 tons (2.4 million gallons) per year and up. The modular refineries, coupled with D1's extensive new jatropha plantations in various African nations, could add up to a lot of biodiesel production for local transport or electricity generation in the next few years.[10] And, since many of these projects are based on jatropha rather than food crops, it potentially eliminates the food-versus-fuel issue.

Another company in the race to be among the first to produce biodiesel in significant quantities in South Africa is BioFuel Africa, which is designing a 60,000-ton plant in Alrode, Gauteng. The company is the first in South Africa to have a carbon credit trading agreement with a foreign (Danish) government. Another contender is Evergreen Biofuels, which is planning an 8,800-ton refinery in Bethal, Mpumalanga. The land and buildings for the facility have been purchased already and only await the equipment from Europe and the United States. The company intends to use locally produced soybean oil as its primary feedstock.[11]

Interest in biodiesel in South Africa is growing, and the long-term prospects for the industry appear to be good.

MALI

The nation of Mali is a landlocked former French colony located in West Africa. A Dutch-backed startup there, Mali Biocarburant, has decided to bypass the wait for new jatropha plantations to reach productive age, and plans to rely on jatropha nuts that are already available from the approximately 22,000 kilometers (more than 13,000 miles) of existing jatropha hedgerows used by local farmers to protect other crops and stop soil erosion. The jatropha nuts harvested from these living fences will be processed into biodiesel at small, local facilities. Admittedly, the logistics of the Mali Biocarburant plan are going to be a challenge. In order to encourage local farmers to participate, Mali Biocarburant is giving farmers jatropha seeds to increase future production and also is offering them an ownership stake in the business. The main idea is to keep the profits (and the biodiesel) from the venture circulating in the local economy, rather than sending them out of country to foreign corporate owners. This strategy differs substantially from many of the large-scale new plantations of jatropha that are being planted or promoted by other large biodiesel companies in other countries. "I believe in small-scale, decentralized biodiesel plants, where there is local production of jatropha nuts to minimize transport," says Hugo Verkuijl, the company's chief executive. "By making [the farmers] shareholders in the business and showing them that it pays dividends, we can make it work," he adds optimistically. The first test runs are scheduled for early 2008. If the project succeeds, it may very well become a model for other biodiesel ventures in Africa.[12]

Quite a few other jatropha-based initiatives also are active in Mali, and some 700 communities have already installed biodiesel-fueled electric generators for lighting, and to power water pumps and grain mills. The government of Mali hopes eventually to power all of its 12,000 villages with renewable energy sources, and biodiesel unquestionably is going to play a major role in that initiative. A number of private foreign companies have made offers to assist the development of the jatropha industry in Mali, but they have been told that biodiesel would not be available for export until the nation's domestic needs were met first; a wise decision.[13] The future of biodiesel development in Mali for local needs at least seems to be good.

MOZAMBIQUE

In Mozambique, the former Portuguese colony in southeastern Africa, millions of hectares of unused land have recently been identified as suitable for the production of biofuels crops. And since the relatively small size of the nation's economy means that domestic energy needs could be met with biofuels fairly easily, that leaves quite a lot of extra potential capacity for the production of biofuels for export. The government of Mozambique has been actively promoting its biofuels potential to large investors around the world. Not surprisingly, there have been numerous requests to the government to open up about 5 million acres of land for biodiesel production from jatropha, sunflowers, and coconuts.

Sweden has been supporting the production of biofuels in Mozambique for some time. A Norwegian group also has been involved with several local companies that use sunflower oil for biodiesel production. Brazil's Petrobras and Italy's ENI both have announced plans to begin biofuels projects in Mozambique. A Canadian company, Energem Resources, has recently bought a 70 percent controlling interest in a jatropha-based biodiesel project in Mozambique that involves the planting of 1,000 hectares of land for a jatropha seedling nursery, as well as the clearing and planting of an additional 5,000 hectares in the immediate future. The project could involve as much as 60,000 hectares for jatropha plantations.[14] China and Portugal also figure prominently in various biofuel-related initiatives. In other words, the international rush is on to get a piece of the biofuels action, and $700 million has already been committed to biofuel production in the nation. Expect a lot more biodiesel activity in Mozambique in the years to come.

OTHER AFRICAN COUNTRIES

For a number of years, there has been interest in developing biodiesel initiatives in various other African countries where the cultivation of a variety of oil-bearing crops has been seen as a way to strengthen the agricultural sector, generate jobs, and reduce expensive imports of petrodiesel. Quite a few studies have been conducted by universities and other organizations on various aspects of biodiesel production, but aside from a few small projects, this work has not generally been translated into large-scale commercial biodiesel development until fairly recently. But with oil prices topping $135 a barrel in mid-2008, most African countries have suddenly begun to scramble to find alternate fuels for their own use and also to take advantage of the biofuels potential that their agricultural sectors offer for the export market.

Although some of this activity has been relatively large in scale, some has been quite small, since biodiesel technology—unlike that of the petroleum industry—can be viable at almost any size or scale. Consequently, in many parts of Africa—especially rural areas—a different model from what has been taking place in Europe makes a lot of sense. This alternate strategy involves local, community, or regionally based projects that rely less on sophisticated technology and more on hard work and determination. There is one project that has planted the seeds of this version of the industry—literally. Jatropha, moringa, and neem seeds have been used by the Africa Eco Foundation, a nonprofit organization based in South Africa, to implement a biodiesel initiative in southern Africa, where the best growing conditions for these crops exist. The seeds have been used to create plantations, which can be harvested for oils and other biomass by-products. To date, approximately 17 countries in Africa have established pilot programs.[15]

The largest of these programs, composed of 220,000 hectares (543,620 acres) is in Ghana. A very small-scale biodiesel initiative also has been set up in the country's eastern region by the Dumpong Pineapple Growers Cooperatives. This local, community-based pilot project assisted by U.S. partner Dumpong Biofuels has been using palm-oil–based biodiesel to power a generator and to fuel their vehicles.[16] In addition, Anuanom Industries has installed a large 500-ton capacity processor for turning jatropha seeds into biodiesel in the farming community of Gomoa Pomadze in Ghana's central region. The company also has installed equipment for producing organic fertilizer from the glycerin by-product of the biodiesel production process. This is part of a much larger jatropha initiative that has been active in the country for a number of years. Tests of the jatropha-based biodiesel have been made and approved by the Ghana Standards Board, the Tema Oil Refinery, and the Environmental Protection Agency.[17]

In Kenya, located in eastern Africa, another biodiesel initiative based on jatropha has been under way since March 2006. The Trees for Clean Energy project involves 950 farmers who are learning how to cultivate jatropha that is native to the area. The project was launched by a Kenyan agricultural scientist who was looking for a way to increase the incomes of local farmers while addressing greenhouse gas emissions. The project is located in Kibwezi, a semi-arid district adjacent to the Tsavo National Park, and it is designed to help the farmers grow and process the jatropha nuts into biodiesel. This project is intended as a pilot program for a much larger initiative in the Kisumu, Kajiado, and Kitui districts.[18] A Japanese company, Biwako Bio-Laboratories Limited, has recently announced its plans for a $20-million jatropha plantation in Kenya that it expects will produce 200,000 tons of jatropha oil per year. The company hopes to expand the operation to 100,000 acres in the near future.[19]

In Rwanda, another landlocked nation located in east-central Africa, a Brazilian/African agricultural machinery company, BrazAfric, has recently decided to begin commercial production of biodiesel from native jatropha trees. This initiative came after researchers, including those at the Institute for Scientific and Technological Research in Butare, Rwanda, discovered that the nation is home to millions of the trees, and the government has launched a campaign to promote the expanded cultivation and harvesting of jatropha.[20]

In Nigeria, where large petroleum resources have tended to act as a disincentive for large-scale biofuels initiatives, the international telecommunications company, Ericsson, in collaboration with several other organizations, has launched a number of biodiesel-fueled mobile base stations in early 2007. The first pilot station was set up in Lagos, and a biodiesel supply initiative has been set up to benefit local farmers who will grow peanuts, pumpkin seeds, jatropha, and palm oil as feedstocks.[21]

In Zimbabwe (formerly Southern Rhodesia), a large new biodiesel refinery opened in November 2007 in Mount Hampden, about 15 kilometers northwest of Harare, the nation's capital. The Transload biodiesel plant, a joint venture between a Zimbabwean and a South Korean company, will use cottonseeds, soybeans, jatropha, and sunflower seeds as feedstock. The plant reportedly has an annual production capacity of around 88,000 tons.[22] Questions have been raised, however, about the availability of these feedstocks from the struggling local agricultural sector.

Additional biodiesel projects are either under way or in the planning stages in Angola, Senegal, Madagascar, the Republic of Congo—in fact, in virtually every African nation with idle arable land. Much of southern Africa clearly offers huge potential for massive biofuels development, and that development is now escalating at a dramatic pace. The people of Africa could benefit enormously from the widespread implementation of local biodiesel initiatives that would offer more jobs, higher incomes, and greater self-reliance. Whether that will actually occur, or whether the continent will simply be used to satisfy the seemingly insatiable demand for liquid fuels for the developed nations remains an open question. A number of international organizations, including Oxfam, have warned about the possibility of arable land being used up for biofuel plantations at the expense of local populations.

As we move on to the Indian subcontinent, along the way it should be noted that there also has been some recent biodiesel activity in Middle Eastern countries like Saudi Arabia, where UK-based D1 Oils has signed a 50/50 joint venture with a Saudi partner, Jazeera for Modern Technology. The venture will develop jatropha plantations and then process the harvested oil into biodiesel for local use and possible export. Jazeera provided 5,000 hectares of land, with an additional 100,000 earmarked for development if the project goes well. Most of the initial investment of up to $10 million came from Jazeera.[23] There is also at least one small biodiesel plant in Pakistan, and there are probably others in the region as well, especially at the local level.

INDIA

India imports 75 percent of its oil (60 percent from the Middle East), at an annual cost of about $30 billion. In recent years, with those costs continuing to rise, the country has been investing in biodiesel as a way of stabilizing energy prices and increasing its energy security. Since petrodiesel represents about 82 percent of Indian transport fuels, the potential market is enormous. But India is following a strategy uniquely different from that of many other nations. Much of the nation's future biodiesel industry will be powered by oilseed-bearing trees (OSBT), rather than the field crops preferred by the EU and the United States. And unlike countries that rely on palm oil, India will focus primarily on inedible oils. The country plans to use native plants such as jatropha, mahua, and karanji to produce the feedstock oil for its biodiesel. What's more, plans call for growing feedstock crops in many areas that are not presently in agricultural production. And, in another departure from standard biodiesel wisdom, straight vegetable oils from OSBT will be used to fuel many rural electrical generators.[24]

The use of vegetable oil as a fuel is not a new idea for India. During the 1930s, the British Institute of Standards in Calcutta examined a group of nonedible oils from indigenous plants as potential diesel fuels. In 1940, a textile mill in Warangal, in the state of Andhra Pradesh, powered its entire operation (and the surrounding community as well) with nonedible oils. These facts were generally forgotten until 1999, when the residents of the village of Kagganahalli, in the state of Karnataka, told Dr. Udupi Srinivasa, a mechanical engineering professor at the Indian Institute of Science and head of a rural development agency, about an inedible oil from the seeds of the honge tree (also known as the pongam tree or Indian beech) that their grandparents had used for lamp oil. Dr.

Srinivasa immediately recognized the oil's potential as a biodiesel fuel. "Here we were—all scientists—looking at technical solutions like windmills, gasifiers, solar panels, and methane generators for rural India, and we had not made the obvious connection with the potential of nonedible oils known from Vedic times as fuels," Dr. Srinivasa said later.[25]

The rediscovery of this local biodiesel feedstock led one company, Dandeli Ferroalloys, to convert all five of its 1-megawatt diesel generators to run on honge oil, resulting in significant savings on their energy expenses. And thanks to a separate rural development project, the villagers of Kagganahalli now have water to irrigate their crops from pumps powered by honge-oil-fueled electrical generators. The previously dry and desolate village now produces watermelons, mulberry bushes, sugarcane, and grains.[26] The Kagganahalli project has been the catalyst for numerous other recent Indian biodiesel initiatives.

Deep in the tropical forests of southern India, the Kolam people of the isolated village of Kammeguda, in Andhra Pradesh, also have had their lives changed by biodiesel. Until recently the Kolam had no telephones, televisions, running water, or electricity. Then, in 2002, Dr. Srinivasa walked into the village and literally lit up their lives. He showed the Kolam how the seeds of the karanji trees in the nearby forest could be turned into biodiesel fuel to power an electric generator. The generator has brought electric lights as well as running water to the village. "With lights, we can chase away snakes and animals that stray into our village in the night. We can catch the occasional thief also," says Lakshmi Bai, who manages the tiny power station. "Earlier, we used to put our children to sleep early, but now we make them study under the lights."[27]

Back in the mainstream of national commerce, Indian Railways has also been experimenting with biodiesel. With approximately 4,000 locomotives that burn approximately 1.7 million tons (510 million gallons) of diesel fuel per year, there is ample opportunity for a shift to a greater reliance on biodiesel. On December 31, 2002, the first successful trial run of a high-speed passenger train was conducted when the Delhi-Amritsar Shatabdi Express used a B5 blend as fuel. The test was successful, and India Railways has now expanded the use of biodiesel with higher-percentage blends in more locomotives on other lines. The feedstock for the biodiesel was the jatropha bush, which can be grown on either side of India Railways' tracks.[28] This strategy is being implemented with the recent planting of about 1 million seedlings, and some India Railways trains are now running beside rows of the same plants that will eventually fuel its locomotives. There are plans by India Railways to set up a number of biodiesel production facilities to produce some of the biodiesel for their trains.[29]

Indian biodiesel activity really began to heat up in 2003 when India's Planning Commission recommended an ambitious two-phase national plan for biofuels. The first phase involved the planting of jatropha on 500,000 hectares of government land that would be administered mainly by local governments. The second phase would involve the planting of 12 million hectares and then privatizing the production of the jatropha biodiesel. But, due to lack of a strong national emphasis on the plan, only the local part of phase one has made much progress so far.

In November 2003, DaimlerChrysler announced a new public/private partnership in India for the production of biodiesel from jatropha on eroded ground. In addition to yielding fuel, the five-year plantation project also is expected to create new jobs and reduce CO_2 emissions. It is hoped that the plantations created by the test will subsequently be operated by municipal authorities and can serve as models for other parts of India as well as for other countries. After several years of extensive vehicle tests, the ongoing project entered its second phase in December 2006 to study rural income and employment, indigenous biodiesel crops, and the reclamation of wasteland. Partners in the project include the University of Hohenheim (in Germany) and the Indian Central Salt & Marine Chemicals Research Institute. DaimlerChrysler is providing some of the money, test vehicles, and engineers for the project.[30]

In December 2003, the Indian Oil Corporation announced that it would begin field trials of buses belonging to Haryana Roadways, which would run on a B5 biodiesel blend with petrodiesel. About 450,000 liters of biodiesel were used in the pilot project.[31] Since then, the trials have been expanded to include B10 and participants have included other bus companies, as well as India Railways, Escorts Tractors, and Tata Motors. Automobile manufacturers such as Mahindra & Mahindra Ltd. as well as Ashok Leyland also have endorsed biodiesel as fuel for their vehicles. And, in December 2003, a state (as opposed to national) government initiative for the construction of five biodiesel pilot plants was announced. Each plant project cost about 16,850,000 rupees ($372,000), and was a partnership with a different regional organization. This initiative was in addition to a national government proposal to set up similar plants in cooperation with the Indian Oil Corporation.[32]

The state of Chhattisgarh, which has the most advanced biodiesel program in the nation, has distributed 380 million free jatropha seedlings to farmers, enough to cover 150,000 hectares, as of late 2007. The state has also provided 80 oilseed presses to various village authorities, and guarantees to buy back the jatropha seeds. UK-based D1 Oils has played a major role in the seed-buyback program. In addition, several local businesses have started up across the state and are now operating small-scale refineries. These refineries are providing biodiesel for local transport, water pumps, and electric generators.[33]

In July 2007, India's first commercial-scale biodiesel plant opened in the southern state of Andhra Pradesh. The Southern Online Bio Technologies Ltd. plant can produce 40,000 liters per day (about 12,500 tons per year), and hopes to supply biodiesel to the state Road Transport Corp. as well as India Railways.[34] And in October 2007, India's second commercial-scale biodiesel plant also started up in Andhra Pradesh. The 100,000-ton (30-million-gallon) facility, owned by Naturol Bioenergy Limited, is located in Kankinada, and will sell its products in India as well as in U.S. and European markets.[35]

In August 2007, the Indian Ministry of Petroleum and Natural Gas announced a biodiesel purchase policy to replace fossil fuels with nonconventional types of energy. The main features of the policy are that the public-sector oil-marketing companies will purchase biodiesel that meets the Bureau of Indian Standards specifications through selected purchase centers. Biodiesel producers who want to sell their fuel will have to register with state-level coordinators, and, in an effort to address food-versus-fuel concerns, priority will be given to producers who use nonedible oils as their feedstocks. The ministry identifed 20 biodiesel procurement centers in nine states.[36]

According to a recent estimate, meeting a 5 percent biodiesel target for India could generate 2.4 million new jobs and bring about 2.8 million hectares of wasteland back into productive use. What's more, an additional, combined yearly farm income of 42 billion rupees could be generated in the fourth year of the initiative.[37]

India appears to be well on its way to establishing a widespread and diverse biodiesel industry that should benefit a broad spectrum of its citizens in many different ways, as long as the primary emphasis remains on local production for local use.

THAILAND

Diesel consumption far exceeds the use of gasoline in Thailand. Since 2001, Thailand has had an active biodiesel program, officially inaugurated and sanctioned by the nation's king. Experiments have been conducted with various plant oils that could replace petrodiesel, such as palm, coconut, soybean, peanut, sesame, and castor. For a number of years there were numerous requests for national standards due to the uneven quality and different biodiesel formulations found in various parts of the country. Consequently, the government empowered the Department of Energy to set biodiesel standards for the entire nation, and they were finally announced in 2005. After several years of study, the government also announced tax incentives for biodiesel.

Biodiesel is used for many different purposes in Thailand, powering everything from cars, trucks, and small farm tractors to fishing boats and stationary engines. In May 2003, the Royal Thai Navy completed a series of biodiesel tests and reported that it was satisfied with the results. The navy subsequently set up two biodiesel production plants that were later used as models for local community biodiesel production facilities. More extensive biodiesel testing was later conducted with palm oil and used cooking oil. Once the best formulation was determined, the biodiesel was used in vehicles owned by the armed forces and the Energy Ministry, and later in city buses in Bangkok.[38]

In June 2005, the nation's first community-based biodiesel refinery was opened amid much fanfare. The small plant, capable of producing 2,000 liters (about 525 gallons) a day from used cooking oil, is located in the Sansai district of Thailand's northern province of Chiang Mai. Officials hoped that the project would act as a model for similar local initiatives in other parts of the country.[39] In September 2006, a major Thai oil refiner, Bangchak Petroleum Plc, officially launched a program to make biodiesel out of used cooking oil, and a few months later, completed work on a 6,000-ton biodiesel refinery. The biodiesel from the plant was to be sold at the company's filling stations.[40] The initiative was successful, and the company subsequently built several additional biodiesel refineries.

In June 2007, the Thai minister of energy announced that, as of April 1, 2008, a B2 blend of biodiesel and petrodiesel would be mandatory at all 10,000 service stations across the nation. The nation now has embarked on an ambitious plan to replace not less than 20 percent of its liquid vehicle fuels with renewables by 2012. More than 60 percent of the nation's vehicles use diesel fuel, and by the April 1 deadline, palm oil will account for 1 million of the 50 million liters of diesel fuel consumed. Energy ministry officials are convinced that the nation's palm-oil industry can meet the demand without serious environmental or food price repercussions. Unlike some other nations, most oil-palm plantations in Thailand already are well established and are not generally tied to massive deforestation. Improved output is mainly due to the replanting of older plantations with higher-yielding varieties of palm.[41]

In response to the new B2 mandate, Bangchak Petroleum announced in August 2007 that it will invest approximately $30 million in a new 100,000-ton plant to be located near the company's existing oil storage facilities in Bang Pa-in, Ayutthaya. The new facility is scheduled for completion in 2009, and the company is now the largest producer of biodiesel in the nation.[42] Thailand appears to be following a steady path toward greater energy independence, and biodiesel will definitely play a role.

CHINA

The Liaoning Province Research Institute of Energy Resources in China reported trials with pure vegetable oil as well as transesterified oils between 1991 and 1994. The Chinese researchers focused on guang-pi, an oil-bearing tree that can be grown on marginal land.[43] In 1998, a pilot project partly financed by the European Commission was conducted in cooperation with the Austrian Biofuels Institute, the Center for Renewable Energy Development in Beijing, and the Scottish Agricultural College in Aberdeen, Scotland, as partners. The project evaluated the overall potential of feedstock available for possible future biodiesel production. The study looked at all known oilseeds, animal fats, and recycled cooking oils in China.[44]

The initial research in the 1990s has been followed by a lot of dramatic activity. The nation is now the second-largest importer of petroleum on the planet, and relies on this oil for about 50 percent of its total needs. Consequently, China has been actively promoting renewable energy in an effort to reduce its dependence on those foreign sources, and biodiesel is playing a major role in this effort. One of the early companies to produce biodiesel was Gushan Environmental Energy Limited in Fujian Province, which started making biodiesel from used cooking oil in 2001. The company now has three plants in three provinces with a total capacity of 170,000 tons, and hopes to increase that capacity to 400,000 tons by the end of 2008. As early as 2003, several companies were also producing large quantities of biodiesel from domestic rapeseed oil. China passed its renewable energy law in 2005, and since then there has been a virtual explosion of biodiesel activity across the nation as part of an effort to try to meet a minimum of 15 percent of the nation's transport fuels with biofuels by 2020.[45]

In 2004, China produced only 60,000 tons (16 million gallons) of biodiesel. But by 2006, total production had taken a great leap forward to approximately 300,000 tons. By 2010, demand is expected to increase to about 1 million tons, exceeding probable supply by about 20 percent, according to the National Grain and Oil Information Center.[46] In response, there has been a flood of new biodiesel plant construction. All of this biodiesel refinery construction has created a huge demand for feedstocks, and China is now the top buyer of palm oil from Malaysia at 3 million tons per year. That amount is expected to climb to 4 million tons by 2009–10.[47] However, the high cost of palm oil has prompted many Chinese biodiesel producers to switch to used cooking oil as a feedstock, dramatically increasing the price of used cooking oil as well.

But China has been doing more than building new biodiesel refineries. In an exciting development, a new kind of hybrid rapeseed has been developed by the Chinese Academy of Agricultural Sciences, one that has a record oil content of 54.7 percent. And China has the perfect place to grow it: The Yangtze River Valley is the world's largest rapeseed-producing region, and accounts for nearly one-third of total rape yield. In addition, a four-year initiative to encourage local farmers to grow jatropha in three southwestern provinces has been launched by the United Nations Development Programme and China's Ministry of Science and Technology.[48] In early 2007, China claimed to have 2 million hectares (4.9 million acres) of jatropha already under cultivation, and announced plans to add an additional 11 million hectares in its southern provinces by 2010.[49]

The industry has grown so fast, in fact, that no one really knows for sure how many biodiesel refineries there currently are in China. Many smaller companies aren't even on the national radar screen. All of this sudden growth has resulted in a good deal of confusion due to the lack of close coordination at the national level, and there have been calls for more coherent biofuel standards and regulations. There is also growing concern about the expansion of biofuels feedstock production coming at the expense of food production, since much of the biodiesel produced in China is being exported to Europe.[50] Consequently, the government has recently drawn up policies to encourage the production of nonfood biodiesel feedstocks. The recent escalation in vegetable-oil prices has begun to put the squeeze on profits for many Chinese biodiesel plants, although with domestic demand so high, there does not seem to be any likelihood of a serious overcapacity developing any time soon.

Nevertheless, the building boom goes on. State-funded biodiesel plants with capacities varying from 300 to 600,000 tons have reportedly been constructed recently in Guizhou, Guangxi, Shandong, and Anhui. The feedstocks for these facilities range from used cooking oil to cottonseed, tung oil tree, and organic wastes. As of 2007, the total production capacity of biodiesel projects under construction in China exceeded 3 million tons, according to a biofuels expert in the Ministry of Science and Technology.[51] The government is reportedly in the process of drafting a biodiesel blend standard for a 5 percent mix with petrodiesel, and may announce it in early 2008. If China's great leap into renewables continues at its current pace, the nation stands to be the world's largest consumer of renewable energy in the near future, and biodiesel will unquestionably play a significant role.

JAPAN

Japan, the world's third-largest oil consumer (behind the United States and China), is almost totally dependent on imported, mainly Middle Eastern, oil. But, due in part to a lack of substantial capacity for growing feedstock crops, and partly because the nation uses almost twice as much gasoline as diesel fuel, most of Japan's biofuel focus has been on (mostly imported) ethanol. As a consequence, Japan has been slow to jump on the biodiesel bandwagon.

In 1995, a three-year study was instituted to look into the feasibility of biodiesel production and use in Japan. A small pilot plant was constructed that relied on used frying oil collected from the Tokyo area as feedstock. In 1997, Japanese businessman Soichiro "Sol" Yoshida contracted with Pacific Biodiesel in Hawaii to design and build a biodiesel plant that would use waste cooking oil from his Kentucky Fried Chicken franchise in Nagano, Japan. The plant, which also processes used cooking oil from 60 other restaurants, produces 600,000 liters (132,000 gallons) of biodiesel per year and was designed so its capacity could be doubled.[52] The fuel produced is used in numerous cars, trucks, and industrial engines.

Biodiesel also was used in buses during the 1998 Winter Olympics in Nagano.

Another series of biodiesel pilot projects have been conducted in Kyoto, where more than 200 city-owned trucks have been fueled with B100 and 81 municipal buses have been operated successfully on B20. In 2001, the Shizuoka Trucking Association, comprising about 1,441 trucking companies, began buying farmers' entire canola (rapeseed) harvests in Shizuoka prefecture's Daitö and Iwata. The canola, which had been grown on idle plots of land, was then processed into biodiesel and mixed with petrodiesel for use in trucks. The association has been collecting performance data and has received numerous inquiries from other trucking companies that are also interested in using biodiesel.[53]

In December 2002, the Japanese government unveiled a comprehensive plan to encourage the greater use of biomass fuel, which included the production of biodiesel from rapeseed and used frying oil. The plan also called for the creation of a biomass energy promotion council that would, among other things, lay out specific proposals for the construction of biomass energy production facilities.[54] More and more companies, municipalities, groups, and individuals across Japan are collecting used cooking oil and converting it into biodiesel for their own vehicle use or for sale to others. At the time, it was estimated that Japanese restaurants and households generated about 400,000 tons of used cooking oil every year.

As biodiesel use has increased, additional sources of feedstock have been investigated, and some researchers have turned to sunflower oil. Tsukuba University, in cooperation with the town of Hikawa, conducted sunflower-oil field tests, and in August 2003 a tractor fueled with sunflower biodiesel was demonstrated at the Hikawa Sunflower Festival.[55] Others have turned to canola (rapeseed) and are growing and processing their own canola oil in "nanohana projects" (*nanohana* means "canola blossoms" in Japanese) in what has to be the ultimate grassroots recycling program. The program, which has now been promoted nationwide, was started in Aito in 1998. This is how it works. Local, small-scale community cultivation of rapeseed in crop rotation with rice is encouraged. Cooking oil is pressed from the harvested rapeseed, and, after it is used for food preparation, it is collected by local housewives and converted into biodiesel. The pressed seedcake is used for fertilizer or animal feed. The biodiesel is consumed within the community. All of this happens at the local level in a closed loop. More than 100 groups across Japan now collect and recycle canola oil as part of this program. A national organization, the Nanohana Network, promotes the initiative and even holds an annual "Nanohana Summit."[56]

Even with these grassroots initiatives, overall use of biodiesel in Japan has been limited to relatively small programs at the local level totaling less than 800,000 gallons per year, and it wasn't until August 2006 that the first large-scale importation of biodiesel arrived from Malaysia. Japan also imports biodiesel (and ethanol) from Brazil.

In March 2007, Japan's quality standard for a B5 blend of biodiesel finally took effect. However, in response to the nation's powerful refining sector, the government will not force the refiners to sell it. Nevertheless, in early 2007, Toyota Motor Corporation, Hino Motors, the Tokyo Metropolitan Government, and Nippon Oil Corporation initiated a joint project aimed at commercializing what they describe as bio-hydrofined diesel. Hydrofined diesel is a so-called second-generation biodiesel produced by hydrogenating the vegetable-oil feedstock. Nippon Oil and Toyota have worked jointly on developing this technology since 2005. The project was developed as part of the Tokyo Metropolitan Government's 10-Year Project for Carbon Reduction in the city. The new fuel will be tested in a small number of city buses at a B10 blend level, which is 5 percent above the current official national blending level. This is in addition to the Tokyo Metropolitan Government's prior decision to use a B5 blend of conventional biodiesel in its city buses in 2007.[57]

Although Japan has limited domestic sources of biodiesel feedstock, they will undoubtedly be utilized in a very efficient manner in the years to come. Many Japanese have embraced biodiesel enthusiastically, and future prospects for additional growth of the biodiesel industry, especially in terms of collaborative feedstock projects overseas, are fairly good.

SOUTH KOREA

South Korea is entirely dependent on imported oil; the nation is the fifth-largest importer in the world. With oil prices rising and supplies tightening, South Korea has been looking at alternatives. In 2002, the South Korean Ministry of Environment as well as the Ministry of Commerce, Industry, and Energy (MOCIE) started looking at biodiesel as one solution to environmental concerns, global warming, and energy security issues. In February 2002, MOCIE conducted a series of biodiesel emissions tests, followed by a B20 pilot project in several locations, and then built a full-scale, 100,000-ton-per-year biodiesel refinery at the end of the year. In September 2003, work began on creating a national biodiesel standard, which was completed the following year. The South Korean standard is close, but not identical to, the European EN14214 standard.

In July 2006, the nation launched a biodiesel mandate with a 0.5 percent initial requirement, to increase by half a percentage point every year, up to 3 percent by 2012. The original plan was far more ambitious, but was eventually scaled back due to strong resistance on the part of the nation's powerful oil-refining lobby as well as South Korean automakers. Another issue was the lack of sufficient local feedstocks, since most feedstock is imported from South America and Southeast Asia.[58] In addition to the mandate, the government also initiated a tax break for biodiesel that lowers the price of blended biodiesel slightly below the price for straight petrodiesel at the pump.

For many years, most automobiles in South Korea have used gasoline as fuel, but in early 2006 the sale of diesel-powered cars was finally approved, boosting demand for biodiesel blends, which are now available at virtually every filling station in the country. Some locations offer B20, intended mainly for commercial transport and heavy construction equipment use. The recent growth of the Korean biodiesel industry has been dramatic. There are approximately 20 commercial-scale biodiesel companies registered with the government that have a combined production capacity of about 800,000 tons per year. Considering that domestic consumption in

2007 was only around 90,000 tons, there is now a huge imbalance between supply and demand, leading to cutthroat competition.[59]

Despite these current challenges, if the government raises the level of its biodiesel mandate in the years ahead, the South Korean biodiesel industry should have fairly good future prospects.

TAIWAN

Taiwan is almost entirely dependent on imported oil, which represents approximately 46 percent of the nation's total energy mix. In the road transport sector, diesel fuel is used mainly for heavy and light trucking, while about 11 percent is used by transit buses. There are very few diesel-powered cars on Taiwan.

In 2004, the Taiwan NJC Corporation began producing biodiesel at its 3,000-ton-per-year facility located in Chiayi County in the southern part of the country. This plant, the first commercial-scale operation in the nation, uses recycled cooking oil and animal fats as its primary feedstocks. In 2006, several hundred kiloliters of biodiesel were used to fuel commercial vehicles, mainly garbage removal trucks. In July 2007, the government announced ambitious plans for a 25-fold increase in the use of biodiesel, from 4,000 kiloliters to 100,000 kiloliters (26 million gallons) by 2010. As part of this initiative, a B1 blend of biodiesel was introduced on a trial basis at a limited number of filling stations in the counties of Taoyuan and Chiayi. The blended biodiesel is expected to be available across the island by the end of 2008, when the mandate officially kicks in, and the percentage is scheduled to increase to 2 percent by 2010. The initiative was supported by the state-run CPC Corporation, which offered the B1 blend at 82 of its filling stations. The private petroleum company Formosa Petrochemical Corp. also offers the B1 blend at its stations. In addition, buses in Kaohsiung have been using B2 since early 2007, and 35 bus lines in Chiayi County have been using B5 as well.[60]

There are presently four suppliers of biodiesel in Taiwan, and all of them are getting ready to meet the expected increase in demand once the mandate is implemented nationwide in 2008. Although Taiwan does not have a lot of land for feedstock production, the biodiesel industry would seem to have fairly good prospects.

SINGAPORE

Singapore is an island nation city-state at the southern tip of the Malay Peninsula. Located at the entrance to the Strait of Malacca, Singapore is strategically positioned as one of the most important transshipping centers in the world. Singapore is a major hub for the international petrochemical sector, the products of which have recently been the nation's major exports, second only to electronics. Its key location, adjacent to huge supplies of palm oil in Indonesia and Malaysia, coupled with its advanced petrochemical infrastructure, has made Singapore attractive for biodiesel development, and the government has been eager to add biofuels to its long-term energy mix.

In 2003, a local entrepreneur founded Biofuel Research Pty Ltd, the first biodiesel company in Singapore. Biofuel Research then built an 18,000-ton-per-year biodiesel plant at Tuas that relies on used cooking oil as its main feedstock. The company only sells B100.[61] On a much larger scale, in October 2005, two different groups announced plans to construct biodiesel refineries on Jurong Island, Singapore's petrochemical hub. The first project, proposed by a German company, was for a 200,000-ton-capacity refinery, that was scaled back to 100,000 tons, and eventually opened at reduced capacity due to the downturn in the biodiesel market in 2007. That was followed by plans for a 150,000-ton plant in a joint venture between Wilmar Holdings and Archer Daniels Midland Company. However, this plan was put on hold in mid-2006 as the biodiesel market began to suffer from high feedstock costs.[62]

In November 2006, construction began on a huge 600,000-ton refinery owned by Australian renewable energy firm Natural Fuel. Plans called for a three-phase construction schedule that would eventually result in an annual production capacity of 1.8 million tons. However, in November

2007, in response to the downturn in the international biodiesel sector, Natural Fuels announced that it would delay the latter phases of the project until the second half of 2008 or perhaps even until 2009. In January 2008, the existing plant was reportedly operating at just 10 percent of its capacity.

With its relatively small population of 4.6 million, Singapore's main opportunities in the biodiesel market appear to be as a refiner and transshipper, and the prospects for those activities are fairly good in the long term.

PHILIPPINES

Coconut oil was reportedly used as an engine fuel in the Philippines during World War II. Research into the use of biofuels was later conducted in the 1970s as a result of the OPEC oil crisis. A program for biodiesel and one for bioethanol emerged from the research, but both programs were abandoned in the mid-1980s due in part to low oil prices. However, the Philippines imports virtually all of its crude oil from overseas, leaving the nation extremely vulnerable to price increases or supply shortages. About 7 billion liters of diesel (1.8 billion gallons) are consumed in the Philippines every year.

The Philippines claims to be the world leader in coconut oil production (second only to Indonesia in terms of total coconut production). Every year the country exports about $760 million worth of coconut products and consequently has easy access to an enormous supply of biodiesel feedstock. Despite this enormous potential and the early research programs, it wasn't until November 2002 that the Philippines Department of Environment and Natural Resources (DENR) finally launched Esterol, a biodiesel made from coconut oil known locally as coco-biodiesel. In a practice-what-you-preach move, the department also initiated a policy of using biodiesel to fuel its own fleet of vehicles. Senbel Fine Chemicals Co. Inc. supplied the biodiesel, and the oil company Flying-V agreed to sell the fuel in all of its filling stations throughout the country. Since

2002, biodiesel blends also have been used extensively in a fleet of buses in Manila.[63] Biodiesel pilot testing programs for government vehicles also have taken place in at least four other cities.

In February 2004, a presidential order was signed directing all government agencies, including government-owned and government-controlled corporations, to switch to coconut methyl ester in their diesel-powered motor vehicles. In May 2006, anticipating the passage of a national biofuels act, Chemrez Incorporated of Manila opened what was then the largest biodiesel plant in Asia with a new $12.6-million facility located in Libis, Quezon City. The plant, with an annual capacity of 16 million gallons (53,000 tons), initially expected to export up to 80 percent of its output to Europe, Japan, and Australia. As of 2007, there were three commercial-scale coco-biodiesel producers in the Philippines with a combined production capacity of 160 million liters (42 million gallons) annually.

The Philippine Biofuels Act of 2006, which was finally signed into law in January 2007, called for the establishment of a National Biofuels Board spearheaded by the Department of Energy and also representing the other concerned governmental departments. The act also mandated a 1 percent blend of biodiesel with petrodiesel and a 5 percent blend of ethanol with gasoline, stepping up to B2 and E10 respectively by 2010. The Philippines' Department of Agriculture and the Philippine Coconut Authority have plans to develop 740,000 acres to meet the expected demand for coco-biodiesel. Coconut Authority officials estimate that the Philippines would need at least 70 million liters (18.5 million gallons) of coco-biodiesel to meet the targets of the biofuels act. This has some observers, even including some coconut farmers, concerned about the potential environmental damage this could cause to forests as other farmers rush to meet the demand.[64]

But coconuts are not the only biodiesel feedstock in the Philippines receiving attention these days. Tian Biogreen Energy Limited, based in Singapore, recently announced plans to plant 2 million hectares (4.9 million acres) with jatropha. The project is expected to produce about 1.2 million tons of jatropha oil.[65] Even the Philippines National Oil Company is getting into the jatropha business through its subsidiary, PNOC Alternative Fuels Corp., which has plans to plant up to 700,000 hectares of jatropha, mainly on Mindanao. The company will encourage farmers to plant jatropha on unused land to avoid competition with food crops. In a related move, recent tests conducted on samples of jatropha biodiesel by the Technical University of the Philippines in cooperation with Chemrez Technologies, confirmed that the locally sourced and produced biodiesel met European (EN 14214) and U.S. (ASTM 6751) standards.[66]

There are now about 300 coco-biodiesel distributors in the Philippines. Despite all of this activity, public awareness of the benefits of biodiesel is low. Moreover, the vast majority of the general public, especially in the transportation sector, still has not embraced biodiesel in everyday use since it is generally more expensive than petrodiesel.

The biodiesel industry in the Philippines appears to be well positioned for considerable future growth, but there are growing concerns about the potential environmental consequences of some of the large-scale feed-stock projects currently being planned or implemented.

MALAYSIA

As mentioned in Chapter 3, oil palm is the king of biodiesel feedstocks in terms of oil yield. Together, Malaysia and Indonesia account for about 87 percent of global palm-oil production. Malaysia's first oil-palm plantation was set up in 1917, and the country claims to be the world's largest producer and exporter of palm-oil products. (Malaysia was actually surpassed by Indonesia in crude palm-oil production in 2007.) The total crude palm-oil production for Malaysia in 2007 was 15.8 million tons.[67]

It should come as no surprise, then, that the palm-oil industry in Malaysia has experimented with biodiesel for some time. The Palm Oil Research Institute Malaysia (PORIM) has been the prime mover in this initiative, and their 10,000-ton-capacity biodiesel pilot plant has been the main fuel source for research and field tests. Between 1987 and 1990 PORIM, together with Mercedes-Benz, conducted palm ester tests with

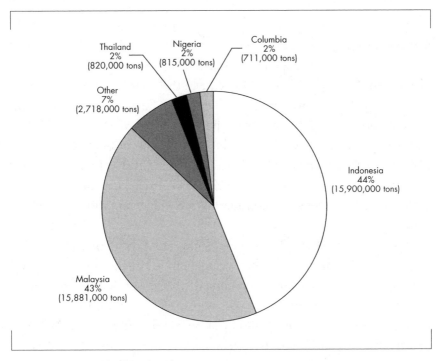

Figure 6. 2006 World Palm Oil Production U.S. Department of Agriculture

city buses in Kuala Lumpur. Additional tests also were run, with hundreds of trucks participating. Although palm-oil biodiesel does not perform well in cold climates because of its high cloud point, there were no problems under the relatively hot local temperatures.

In early 2005, the Malaysian Palm Oil Board announced that it had entered a partnership agreement with Golden Hope Plantations to build a biodiesel plant in Labu in Negeri Sembilan. The 60,000-ton facility was completed in late 2006 and, of course, used palm oil as its primary feed-stock. Plans for a number of additional plants were announced a short time later. In August 2005, the Malaysian government published a new National Biofuel Policy. The primary goal of the policy was to reduce the consumption of imported petroleum. In order to accomplish that, the policy promoted greater production and use of B5 palm-oil biodiesel, set up palm-oil biodiesel standards, and provided price support for palm oil at times of low demand. Those supports have not been needed recently, since the demand for palm oil has soared and its price has doubled between 2006 and 2007, causing a lot of problems for many biodiesel producers.

In 2007, two new commercial-scale biodiesel refineries were completed. But due to the dramatic increase in the price of palm oil, the new facilities opened at only partial production.[68] The record prices for palm oil have slowed or delayed some other biodiesel projects as well. High palm-oil prices also have delayed the introduction of biodiesel into the domestic Malaysian market, as the government felt that higher biodiesel prices would be a burden on consumers.[69] Nevertheless, in early 2008, the Malaysian government was considering the implementation a B2 blend mandate.

At the end of 2007, 91 biodiesel projects, with a total production capacity of 10.1 million tons, had been approved by the government. Of that number, 4 plants with a combined capacity of 300,000 tons were operating, while another 6 with a capacity of 471,000 tons had been completed and were undergoing initial trial runs.[70] But, in response to record high prices for palm oil, only 7 plants across the country were actually operating, most well below capacity as of early 2008. The growth of biodiesel capacity in Malaysia recently has been truly remarkable. The growth of environmental concerns surrounding oil palm plantations also has been significant, although the amount of land available for large new plantations in Malaysia is somewhat limited. This has led some Malaysian companies to look at Indonesia as a better place to grow more oil palm, which has raised even more environmental concerns in Indonesia. There also have been some concerns raised about the treatment of indigenous peoples whose lands have been appropriated by oil palm plantations, especially on Borneo. The growing controversy over palm oil's environmental and social impacts, along with its soaring price, had caused Malaysian palm-oil exports to Europe to drop about 20 percent in 2007.[71]

In partial response to the growing controversy over palm oil, the government announced in November 2007 that it was planning on planting "jarak" (jatropha) trees as possible future feedstock for biodiesel. The Malaysian Palm Oil Board is carrying out the research to try to find suitable climate and land conditions in the country for a 300-acre test plantation.[72] If Malaysia hopes to maintain its strong growth in the biodiesel sector it needs to find ways to respond to the growing international concerns surrounding oil-palm plantations and rain forest destruction.

INDONESIA

Indonesia's petroleum industry is one of the oldest in the world. However, the nation's oil production peaked in 1977 and has been in decline ever since. In fact, Indonesia became a net oil importer in 2005, as its own supplies are unable to keep up with domestic demand. The government's own estimates project that Indonesia's remaining oil reserves will be depleted in less than 20 years. That fact, coupled with dramatic recent price increases in oil, has prompted growing Indonesian interest in biofuels.

In an effort to reduce air pollution in Jakarta, the city's administration announced plans in January 2004 to develop biodiesel fuel as an alternative to petrodiesel. The city collaborated with the Riau provincial administration, which had a small biodiesel plant that provided fuel for the test. The Jakarta Environmental Management Agency (BPLHD) and PT Energy Alternatif Indonesia, a biodiesel supplier, had previously conducted a joint experiment on 10 public buses in the Indonesian capital using biodiesel blends ranging between B5 and B10. The initial tests were successful, and the subsequent larger, more sustained road tests were the result.

In early 2006, a presidential decree was issued to implement the government's long-term energy strategy aimed at reducing the consumption of oil to 30 percent of the nation's energy mix by 2025 from the current level of 60 percent. This was soon followed by government plans to build four pilot biodiesel plants on the islands of Kalimantan and Sumatra, with a combined total production of 6,000 tons per year.[73] By the end of 2007, it was estimated that there were about 9 biodiesel plants with a total production capacity well over 675,000 tons in the nation. There are many other plants under construction or in the planning stages, and if they all are completed, Indonesia's total biodiesel production capacity could well reach 1 million tons by 2009.[74] By early 2008, however, many of these plans had been cut back or shelved due to the record high price of palm oil.

The dramatic recent growth in Indonesian biodiesel production capacity has not been matched by sufficient increases in palm-oil supply, and the strong demand for that feedstock has raised the price substantially, putting upward pressure on food prices as well. Nevertheless, palm-oil production has increased in recent years, and in 2007 Indonesia surpassed Malaysia as the world's top producer of palm oil with a total of 15.9 million tons (versus Malaysia's 15.8 million tons).[75] The government also is looking at jatropha and cassava as additional biodiesel feedstocks that offer the advantage of reducing the competition between biofuel and food. Jatropha is still at an early stage of development in the nation, and requires more labor than palm oil. There are also some concerns that jatropha may not be suitable for large-scale production.[76] In addition, there is growing alarm, both in Indonesia and abroad, about the destruction of forests and peatlands for oil-palm plantations.

The government-owned oil and gas company, Pertamina, now sells a biodiesel blend at more than 220 fuel stations, mainly in Jakarta, home to some of the world's worst air pollution. That blend had been set at B5, but due to the strong increase in the price of palm oil and reduction in profit margin, the company temporarily cut the blend back to B2.5 and more recently to B1. It is expected, however, that the blend will be restored gradually to B5 by 2010. In 2007, biodiesel consumption averaged about 82,000 tons per month for the nation as a whole.[77] As a result of Pertamina's decision to cut back to B1, along with what many biodiesel producers see as "inconsistent government policy," there has been a dramatic drop in demand and a sudden crash in domestic production. In only one month, between December 2007 and January 2008, biodiesel production plummeted 85 percent. As many as 17 proposed biodiesel projects have now been put on hold, according to industry sources.[78]

Despite these difficulties, there is still a good deal of long-term potential for the Indonesian biodiesel industry to expand. However, that expansion needs to be conducted in an environmentally responsible manner if the industry wants to avoid the growing anti-palm-oil-biodiesel sentiment around the world.

PAPUA NEW GUINEA

Papua New Guinea, a group of islands including the eastern half of the island of New Guinea, is dominated by rugged, mountainous terrain largely covered by tropical rain forest. During a lengthy period of civil unrest in the 1990s, diesel fuel became scarce in some locations, and coco-biodiesel was used as an emergency substitute by some residents. In November 2003, it was reported that Papua New Guinea intended to develop coconut oil as a biodiesel feedstock. The development of coconut-oil biodiesel was viewed as a high priority by the PNG government and its various departments. The University of Technology in Lae made a good start with the initiative. The Biofuel Research Group at the university had carried out a successful test of its coconut oil methyl esters the previous November. The initial results were promising and demonstrated that the fuel performed well.[79] Despite this initial activity, the government has not yet implemented a comprehensive biodiesel initiative nationwide.

More recently, some of the residents of the island of Bougainville have responded to increasing prices and shortages of diesel fuel by taking matters into their own hands. A number of people have started producing coco-biodiesel in their own backyard mini-processors. One resident who lives in Buka Town, Mathias Horn, has a processor of his own, and he sells small quantities of coco-biodiesel to others. "The coconut tree is a beautiful tree," he says. "Doesn't it sound good if you really [can] run your car on something which falls off a tree?"[80] Although Papua New Guinea has been described by some observers as one of the "biofuel superpowers of the twenty-first century," it still has a long way to go before it achieves that status.

OTHER PACIFIC ISLANDS

There also has been some recent small-scale biodiesel activity, mostly at the local level, in a number of other Pacific Island nations. Fiji, the Solomon Islands, Samoa, Vanuatu, the Federated States of Micronesia, Tonga, Kiribati, the Marshall Islands, the Cook Islands, and Palau (in addition to Papua New Guinea) collectively spend more than $800 million a year on fuel imports, according to the 20-member South Pacific Applied Geoscience Commission (SOPAC). In about half of these countries, imported fuel accounts for more than a quarter of their total imports. These generally impoverished nations could benefit from the greater use of locally produced coconut oil biodiesel. Electric companies in Vanuatu, Fiji, and Samoa have been testing blends of coconut oil and diesel to fuel their power generators. The Australian Biodiesel Group has reportedly been considering a number of biodiesel projects in Papua New Guinea, the Solomon Islands, and Vanuatu.[81]

8

Australia, New Zealand, and the Americas

Our biodiesel survey now arrives in Australia. The nation's widely dispersed population, in part, accounts for Australia's dependency on petroleum. Despite being a net energy exporter (mostly uranium and coal), Australia is a net importer of liquid petroleum fuels. Nevertheless, the country has considerable potential to grow industrial oilseed crops on unused agricultural lands. Following the Gulf War in 1991, there was a brief flurry of biodiesel activity in Australia, but it soon waned.

In January 2002, the Australian Biodiesel Consultancy, in collaboration with Collex, one of Australia's leading environmental services companies, began operation of a pilot plant in Wyong, New South Wales. The plant used recycled cooking oil and tallow as feedstock and supplied fuel for trucks, buses, cars, earth-moving equipment, and boats. On March 12, 2003, Australia's first commercial-scale biodiesel production facility, operated by Biodiesel Industries Australia, was opened in Rutherford, New South Wales. The multimillion-dollar plant, which has an annual capacity of 40 million liters (10.5 million gallons, or 37,000 metric tons), can use virgin vegetable oil, animal fats, or used frying oils for feedstock. The unique modular design of the production system allowed the plant to be set up in less than three months.[1]

In September 2003, an Australian biodiesel standard was finally implemented, and additional provisions were gradually introduced over a number of years to allow existing producers time to get up to speed with testing and accreditation. What's more, the government set a voluntary biodiesel and bioethanol target of at least 350 million liters (93 million gallons) by 2010. The government also has initiated tax credits and various types of industry assistance to help achieve those goals. Biodiesel has begun to spread to different parts of the country, and blends have been used by various industries, a number of commercial fleets in Perth, and, beginning in early 2005, nearly all of the trains and buses in Adelaide.

In 2006, there was a flurry of large plant openings across the nation, and as of 2007, there were about 10 commercial biodiesel production facilities in Australia with a combined production capacity of approximately 600 million liters (158 million gallons). However, at least some industry observers were concerned that production capacity had expanded beyond demand.

In 2007, in a dramatic reversal of fortune, the Australian biodiesel industry was suddenly plunged into crisis. Although this reflected a global trend, the Australian government's decision to revise its previous biodiesel tax credit of 38 cents per liter was the first step on a slippery downward slope for the industry in Australia. In addition, the federal government has so far ruled out mandates to meet its voluntary 2010 biofuels target. In the biofuels sector, voluntary targets almost never work.

But that wasn't all. With the doubling of the price of palm oil in just 18 months, it became virtually impossible for Australian biodiesel producers who relied on that feedstock to make a profit, and a wave of plant closings swept across the nation. Even biodiesel plants that relied on tallow experienced problems as global demand for this feedstock climbed and prices soared. By November, with the closure of the Australian Renewable Fuels plant in Picton, all of the large Australian biodiesel production facilities had shut down.[2] When, or even whether, they will reopen remains unclear, although the ouster of longtime prime minister John Howard in the November 2007 elections and his replacement with the Labor Party's Kevin Rudd may signal a change in biofuels policy.

NEW ZEALAND

In 2003, oil represented about 48 percent of New Zealand's total energy consumption. With dwindling reserves, the country imports more than five times the amount of oil it is able to produce domestically. Consequently, the New Zealand government has been looking at ways to encourage the greater production of biofuels.

Various energy crops were investigated in New Zealand in the late 1970s and early 1980s, but all were found to be noncompetitive with fossil fuels at the time. However, there are *a lot* of sheep and cattle in New Zealand, and the meat industry produces about 130,000 tons of tallow every year. A number of years ago, studies were conducted into the production of biodiesel from tallow. It was estimated that about 5 percent of the nation's diesel-fuel needs could be met from this source.[3] Other studies concluded that rapeseed would probably be the other best potential biodiesel feedstock for the country.

In a dramatic development in May 2006, Aquaflow Bionomic Corporation, located in Marlborough, produced biodiesel from algae in sewage ponds. It is thought that this was the first production of biodiesel from "wild" algae outside of a laboratory. The algae for this project came from the Marlborough District Council's sewage treatment plant. Although algae play a useful role in helping to purify the water, too much can taint the water and cause an unpleasant odor. Harvesting the excess algae for biodiesel feedstock removes this problem and simultaneously creates clean water.[4] In December, the company demonstrated a vehicle powered by some of its algae-biodiesel outside of the parliament buildings in Wellington for lawmakers and the media. And, in 2007, the company moved into new, larger facilities where it continues its groundbreaking work.

In February 2007, after a good deal of debate, the New Zealand government released mandatory sales targets for biofuels that should become effective in 2008, with the eventual target of 3.4 percent of total fuel sold by 2012. This proposal was subsequently introduced into legislation for a mandatory biofuels sales obligation (BSO). Part of the legislation included a sustainability standard sponsored by the Green Party to ensure that the production of the fuel did not impact on food supply or the environment. The government has also recently introduced biofuel standards that will help ensure the quality of the fuel produced. Most major vehicle and machinery manufacturers have been producing new models with biodiesel specifications, and several fishing companies are looking into producing biodiesel from fish oil. As of 2007, however, there were only a few pilot projects and relatively small local producers of biodiesel in New Zealand.

Nevertheless, one recent project with considerable potential is a six-month trial of a B5 blend in two locomotives belonging to Toll Rail, New Zealand's rail service. One freight and one passenger locomotive, dubbed the "bioloco" engines, will use the biodiesel blend to give the company a chance to evaluate how the fuel performs and to consider whether to fuel all of its 165 locomotives with biodiesel in the future. With annual fuel consumption of 50 million liters (13 million gallons), the switch to B5 could potentially result in the use of 2.5 million liters (660,000 gallons) of biodiesel.[5]

Things are beginning to heat up in other parts of the biodiesel sector as well. Recently, Shell New Zealand signed a letter of intent with Argent Energy to help Shell meet the new biofuels sales targets. Argent will proceed with a feasibility study for an 85-million-liter (22.5-million-gallon) biodiesel plant that would use tallow as its feedstock.[6] Another major energy player, coal company Solid Energy, announced plans in May 2007 to invest $20 million over three years into the production of biodiesel using canola grown on the South Island, along with used cooking oil. Finally, the most recent entry into the biodiesel race is Auckland-based Ecodiesel Ltd., which plans to set up a 20-million-liter (5.3-million-gallon) plant that will rely on tallow from the meatpacking industry for feedstock. The company hopes to open the facility in 2008 and to ramp up production to 40 million liters by 2009.[7]

As long as it is not derailed by high feedstock costs, New Zealand, with its mandatory biofuels sales obligation, appears to be on its way toward meeting its biofuels targets.

BRAZIL

Our biodiesel tour now moves on to South America. Brazil is the pioneer in promoting the use of biodiesel in South America. During the oil shocks of the 1970s, the rising price of oil contributed to Brazil's foreign debt crisis and led to the stagnation of the economy in the 1980s. In response to the oil crisis, research conducted at the Ceará Federal University in Fortaleza by Professor Expedito Parente resulted in what is claimed to be the world's first patent for the production of biodiesel as an industrial process in 1980. Dr. Parente also conducted substantial research into bio-jet fuel.[8] There were also reportedly some early engine tests with biodiesel conducted in the 1980s at Motores MWM Brazil in Sao Paulo, as well as a few road tests, but this research was subsequently discontinued.

Nevertheless, Brazil is a huge consumer of biofuels, but the big seller has been ethanol rather than biodiesel. Roughly 40 percent of Brazilian automobiles operate on ethanol, either E100 or in various blends. In 2005, Brazil consumed about 10.5 billion liters (2.3 billion gallons) of ethanol, which was produced primarily from sugarcane. Despite the focus on ethanol, there is a substantial opportunity for biodiesel to become a significant part of Brazil's renewable energy mix, since trucks and long-distance buses consume 80 percent of the petrodiesel sold in the country, amounting to 48 percent of all petroleum fuels.[9]

The Brazilian government, which had previously focused much of its attention on ethanol, now considers biodiesel to be a top priority. Beginning in 2002, the government initiated a new biodiesel-fuel research program and also developed plans to offer financing and other fiscal incentives to help jump-start the industry. In 2004, the government launched the National Biodiesel Production Program (PNPB), and in January 2005 a government mandate for biodiesel use was initiated. The law authorized the voluntary use of B2 until January 1, 2008, when the mandatory requirement took effect. The percentage is scheduled to rise to B5 in 2013. This ultimately would require about 2.4 billion liters (635 million gallons, or 2.1 million tons) of biodiesel to meet the mandate.

This should not be a major problem if soybeans are used as a feedstock, since the country produces about 57 million tons of soybeans annually and is second only to the United States in soybean production. What's more, Brazil's huge ethanol industry can easily supply the alcohol needed for the transesterification process to make the biodiesel. A new biodiesel formula developed by researchers at the University of Sao Paulo (USP) in Ribeirão Prêto uses ethanol instead of methanol as the alcohol component, making the resulting biodiesel completely renewable.[10] Brazil's first commercial-scale biodiesel plant, Ecomat PR & Partners located in Cuiabá, Mato Grosso, has an installed capacity of about 14,000 liters a year.

There have been a number of urban biodiesel road tests in Brazil. In 1998, and again beginning in 2001, the world-famous bus fleet in Curitiba, in the southern Brazilian state of Paraná, was involved in tests of B20 and other blends. The tests not only have worked well in the buses but also resulted in substantial reductions in the city's overall air pollution. In the first half of 2002, Rio de Janeiro's military police and electric utility vehicles were fueled with a mixture of biodiesel, petrodiesel, and alcohol. Additional trials with six city garbage trucks were conducted in 2003. The biodiesel feedstock for the tests was used frying oil from local McDonald's fast-food restaurants, prompting some locals to refer to it as "McDiesel." Based on this and other tests, biodiesel is now being considered for a wide range of uses.

In 2003, Eletrobras (the main Brazilian electric utility) announced a program that will promote the use of biodiesel as an alternative to petrodiesel at diesel-powered electrical-generating plants in the Amazon. The project calls for the local utility to gradually replace the petrodiesel used in 91 isolated systems with biodiesel made from mamona (castor oil plant) and palm oil. The government of the state of Rio Grande do Sul is particularly interested in using mamona as a biodiesel feedstock. The Farming Cooperative Mourãoense Ltd., one of the largest farming enterprises in Brazil, with 17,000 rural producers, planned to install a biodiesel production facility close to its headquarters in Paraná. The pilot unit will produce 2,500 liters (1,650 gallons) of biodiesel a day from soy oil. (Soy accounts for between 60 to 70 percent of biodiesel feedstock in Brazil.) The fuel initially will be used in the cooperative's vehicles.[11]

As mentioned previously, some Brazilian trains in the southern part of the nation plan to expand biodiesel use even further. In December 2003, the transportation company América Latina Logística (ALL), with 15,000 kilometers (9,321 miles) of railroad lines in southern Brazil and Argentina, decided to replace a quarter of the petrodiesel it consumes (150,000 tons per year) with biodiesel. Preliminary tests were conducted on two trains in early 2004, and there are plans to expand biodiesel use to the entire system. More recently, in May 2007, BR Distribuidora signed an agreement with the iron ore mining company CVRD to supply B20 for its locomotives on the Carajás (EFC) and Vitória-Minas (EFVM) railroads. CVRD, the world's largest iron producer, had previously been experimenting with B2 in its locomotives, off-road trucks, and power generators.[12]

In March of 2005, a new 12,000-liter biodiesel plant was opened by Brazil's flamboyant president, Luiz Inacio Lula da Silva, in Cassia, located in the state of Minas Gerais in the southeastern part of the nation. The feedstock for the plant was sunflower, turnips, and castor beans and other crops grown by local farmers in the semiarid northeastern part of the state.[13] Two months later, President da Silva, a strong biodiesel advocate, attended the opening of an 8-million-liter (2.1-million-gallon) plant, this time in the northern region of Brazil in the state of Pará. In August of 2005, the president traveled to the city of Floriano in the northeastern state of Piaui, where he dedicated yet another biodiesel plant and officially kicked off a multipronged national biodiesel initiative aimed at reducing dependence on imported fossil fuels and also benefitting the inhabitants of the nation's poorest region.

Virtually all of the biodiesel projects in the north and northeast of Brazil have a strong emphasis on creating jobs for impoverished local farmers.[14] The National Biodiesel Production Program includes tax incentives for biodiesel producers that source their feedstock from local communities, and many small farmers have enthusiastically taken part in the program. Biodiesel producers that meet certain social criteria are awarded the Social Fuel Stamp from the Ministry for Agrarian

Development. By the end of 2005, there were approximately 7 commercial producers of biodiesel in the nation, with a combined output of about 176,000 tons (52.8 million gallons) that generated revenues of US$54.5 million.

The biodiesel construction boom continued throughout 2006, and included an unusual combined biodiesel and ethanol facility owned by Barralcool in Barra do Bugres, state of Mato Grosso, located in the heart of Brazil's center-west soybean belt. The plant had been producing ethanol for about 20 years, and a new biodiesel plant was integrated into the existing operation, creating the world's first integrated ethanol, biodiesel, and sugar refinery. The opening of the new soy-biodiesel facility also marked the beginning of the turnaround of the region's soy industry, which has been suffering from declining exports for a number of years.[15] Some environmentalists, however, have complained that the recent expansion of industrial soy farming in the region has caused the destruction of tropical rain forest in the Amazon. In response, Brazilian soy interests have imposed a two-year moratorium on the trading of soy grown on newly deforested lands in the Amazon basin. The moratorium went into effect in October 2006.[16]

The initial success of Brazil's biodiesel mandate was dramatically displayed when it was announced in May 2007 that all of the diesel sold at the 5,000 Petrobras BR Distribuidora service stations across the country contained 2 percent biodiesel. In addition, 3,350 large business clients in all states in the nation were receiving B2 as well. The company had invested Brazilian R$35 million (US $17.3 million) in improvements to infrastructure in order to make the rollout as smooth as possible.[17] Two months earlier, Shell Brazil had announced that it had added two more distribution centers and 65 Shell stations to its B2 network. This brought the company's total to 200 service stations and 188 business clients.[18] As of January 2008, all of the nation's 35,000 filling stations were required to offer B2.

In August 2007, Brazil's first commercial jatropha biodiesel project got off the ground when Compahnhia Productora de Biodiesel de Tocantins set up its new biodiesel processor in the state of Tocantins. The project involves a collaboration with local cooperatives and small farmers in the region to supply jatropha feedstock to the 40,000-ton facility. The project has spawned 48,000 hectares of jatropha plantations, providing an important boost to the local agricultural sector.[19]

As of late 2007, the number of registered commercial biodiesel plants across the nation had mushroomed to 42. These plants are capable of producing 1.6 billion liters (423 million gallons)—almost double the amount needed to meet the 2 percent national requirement. Unlike many European countries, which have missed their biofuel targets, Brazil is actually ahead of schedule. This is a remarkable achievement. Even more remarkable, the production for 2008 is projected to be 2 billion liters (528 million gallons), enough to allow for higher biodiesel blends in certain circumstances as well as an acceleration of the dates by which the country can achieve its B5 mandate.[20] The only clouds on the horizon for the industry are the record-high prices for vegetable oils, including soy, the main biodiesel feedstock in Brazil.

If all of this biodiesel activity continues at its present rate, by the end of 2008 Brazil might overtake Italy in actual production and move the nation into fourth place after Germany, the United States, and France (in descending order). This would be a truly remarkable achievement. (Whether this will actually happen is now uncertain, since much of Brazil's biodiesel production capacity was idle in early 2008 due to high feedstock prices.) Although biodiesel production still lags well behind ethanol, the national biofuels mandate has given biodiesel in Brazil a major boost, and the future prospects for the industry appear to be excellent.

PARAGUAY

Paraguay, a landlocked nation in the southern part of South America, imports all of its petroleum and petroleum products, and these imports are a heavy burden on this relatively poor nation. Disruptions to international oil supplies as well as the upward trend in oil prices in recent years have given the Paraguayan government a strong incentive to investigate the use of biodiesel. In October 2000, the vice-minister of Mines and Energy announced a biodiesel initiative and the government officially launched studies to verify the technical and economic feasibility of the fuel. As part of the initiative, Hardy S.A., an oleochemical company based in Asunción, began to produce small quantities of biodiesel. The government was particularly impressed with the fact that biodiesel technology is relatively simple and can be scaled to meet the needs of local communities. Subsequently, a number of government agencies began to travel throughout the country to explain the initiative and set up pilot projects and trials. Soybeans are the principal biodiesel feedstock in Paraguay, and a national B5 biodiesel blend would require the production of about 50,000 tons of soybean oil every year. Since Paraguay is the fourth-largest exporter of soybeans in the world, this would not be a problem. Ethanol produced from sugarcane is the principal source of alcohol for the transesterification process.

In October 2001, the first collaboration with the municipal government of the capital city of Asunción began, and a number of road tests were conducted. In September 2002, B30 was introduced at local filling stations at a price lower than that of petrodiesel. In the same month, a bus company, Linea 25, began using a biodiesel blend in all of its buses, and production at the Hardy S.A. plant reached 5,000 liters (1,320 gallons) per month.[21]

Recent uncertainties in the international oil market along with dramatic price increases have bolstered the Paraguayan government's resolve to place even more emphasis on biodiesel to strengthen its energy security. Paraguay has recently launched a major biofuels plan aimed at cutting its dependence on imported oil, and the government has now implemented a mandate to include a 3 percent mix of biodiesel in petrodiesel in 2008 and a 5 percent blend by 2009. The mandate is expected to give the nation's small biodiesel industry a huge boost.

Paraguay has substantial biofuels production potential. By 2011, the nation hopes to export $50 million worth of biofuels. In the same time frame, Paraguay hopes to replace $150 million worth of petroleum fuels with domestically produced biofuels. In order to meet those goals, the plan calls for eventual production of 300 million liters (79 million gallons) of biofuels by 2011.[22] In a related event, in May 2007, the government announced plans to produce and export biofuels in a joint venture with Brazil, in which Paraguay would provide the feedstock and Brazil would provide the technology for the production of both biodiesel and ethanol.[23]

In 2007, biodiesel production in Paraguay was estimated to be about 2.5 million gallons. That figure is expected to climb to 8 million gallons by the end of 2008, according to the Ministry of Industry and Commerce. However, it's important to note that all of this biofuels activity is not without its downside. Paraguay has some of the most unequal land distribution in South America, with 95 percent of land under private ownership in large estates. What's more, there are reports that an estimated 90,000 rural families in Paraguay have been pushed off their land by the expansion of large corporate soy farms, turning what had once been small subsistence farming communities into vast monocultural landscapes dependent on industrial fertilizers and pesticides. It should come as no surprise, then, that there is a good deal of resistance at the grassroots level to many of the large agribusiness projects in the country. Opponents of these projects often refer to "agrofuels" rather than "biofuels" to highlight the food-versus-fuel issues that are a major part of their protests. Similar problems have also been identified in other South American nations. These types of issues need to be properly addressed if biofuels are to be considered truly "sustainable."[24]

ARGENTINA

Argentina is the world's third-largest soybean producer and also grows substantial amounts of sunflowers, offering opportunities for substantial biodiesel production. Although Argentina has put most of its recent alternate-energy emphasis on compressed natural gas, there has been increasing biodiesel activity in the country recently. As of 2001, there were at least 14 small-scale biodiesel production projects in operation in

various locations. Most of these biodiesel production facilities are reportedly still operating, and a national biodiesel-fuel standard has now been
adopted.

In 2005, a young Argentine engineer, Edmundo Defferrari, developed a
small-scale biodiesel processor and set it up it in Chacabuco, Buenos
Aires province. The prototype cost only 450,000 pesos ($152,000).
When fed with 12 tons of soybeans, the processor can produce 1,400 liters
(364 gallons) of biodiesel in a day, enough to fill the fuel tanks of five
large farm trucks. The cost of the fuel was about 70 centavos per liter
(about 95 U.S. cents per gallon), roughly half of what petrodiesel then
cost at the pump. Defferrari plans to sell similar processors to farmers so
they can make their own fuel. "The farm sector consumes 76 percent of
diesel in Argentina and farming is done throughout the nation,"
Defferrari said. "So we have an enormous logistical advantage because we
consume fuel in the same place where we could produce it."[25] This is an
idea that makes sense almost anywhere.

Yet another Argentine biodiesel entrepreneur, Ricardo Carlstein, has
been thinking along similar lines. "The key is small-scale, decentralized
processing, based on individually owned and operated small-scale plants,"
he says. Carlstein is highly critical of the large-scale, multinational,
agribusiness model of most biodiesel projects and thinks that a "democratic" biofuels revolution is what is needed. Carlstein's company,
Biofuels SA, has designed small biodiesel processors with capacities from
45 to 45,000 tons to help that revolution along. More than 200 of the
units he designed operate successfully worldwide.[26]

Following Brazil's lead, a new Biofuels Law and a Biofuels Decree have
been implemented by the Argentine government. Under the law,
biodiesel will account for 5 percent of all diesel sold in the nation by
2010. Promotional tax incentives are also part of the new rules and apply
to biodiesel produced for the domestic market. However, various regulatory and other issues for the local market remain to be clarified,
prompting many larger companies to export most of their output.

In January 2006, Repsol YPF, a Spanish-Argentinean petroleum company, set up a new Argentinean Bio Fuel Investigation Center near its La

Plata Refinery in Ensenada, a distant suburb of Buenos Aires with a large port. The plan called for the construction of a 31.7-million-gallon biodiesel plant that was scheduled to open in late 2007.[27] And, in December 2006, a small Argentinean company, Oilfox, signed the nation's first biodiesel export contract. The agreement was for 12,000 tons of biodiesel a year. The company owns two biodiesel plants located in San Luis and in Chabas, Santa Fe province.[28]

At the end of 2007, Argentina's total biodiesel output exceeded 319,000 tons (96 million gallons), and is expected to expand to about 1.5 million tons by the end of 2008, according to the nation's agricultural secretariat. About 76 percent of exports in 2007 went to the United States, while about 24 percent was shipped to Europe.[29] There are an estimated 20 additional biodiesel projects under way, and if they are completed, total production could easily exceed 1.7 million tons (528 million gallons) by 2010, making Argentina one of the largest producers and exporters of biodiesel in the world. It should be noted, however, that there is growing local opposition to many of these large-scale multinational corporate projects with their vast agricultural monocultures. The protesters say that small farmers are being evicted from their land, and that the large-scale agribusiness model being used by most corporate biodiesel projects in Argentina is simply not sustainable in the long run.

COLOMBIA

In October 2003, a group of residents from the internationally famous sustainable-living community of Gaviotas and six volunteers from Boulder Biodiesel and the University of Colorado, Boulder, built the first biodiesel plant in Colombia. The 60,000-gallon (210-ton) capacity plant, located in Bogota, uses palm oil gathered from local farmers as its feedstock.[30] The team also converted one Gaviotas truck to run on crude palm oil.

Since then, the Colombian government has become actively engaged in encouraging biofuels development. In 2004, a group of biofuel laws were enacted, calling for a 10 percent mandatory blend of ethanol in gasoline by 2009 and 2 percent biodiesel blended with petrodiesel by 2008. Tax cuts for ethanol and biodiesel also have been implemented. Colombia is now the second-largest producer of ethanol in Latin America after Brazil. All of this emphasis on biofuels has led to dramatic increases in the size of African oil palm plantations. (Palm oil is the main feedstock for biodiesel in Colombia.) Since 2001, Colombia has been the largest producer of palm oil in the Americas. In 2003, there were 118,000 hectares of these plantations, but as of 2007 that number had more than doubled, to 285,000 hectares.

The palm-oil producers and Ecopetrol, the state oil company, have recently become involved in a number of joint ventures to produce biodiesel. Many other companies have jumped into the biodiesel rush in Colombia as well. Total biodiesel production capacity in 2007 was an estimated 321,000 tons. Reflecting the dramatic growth of the industry, it is estimated that a total of 645,000 tons of biodiesel will be produced by eight large producers by the end of 2008, according to the National Federation of Oil Palm Growers.[31]

Unfortunately, much of this large-scale, monoculture, corporate agricultural activity in Colombia has tended to come at the expense of some black and indigenous populations. This has created intense competition among these populations, private companies, the state, and right-wing paramilitary groups for control of valuable land in some locations. Intimidation of small farmers is widespread, and some who have refused to sell or give up their land have been murdered. An investigation conducted by the Agriculture Ministry in 2006 into reported abuses concluded that some of the land acquired for oil palm plantations had, indeed, been illegally acquired. However, the abuses continue. The ongoing destruction of tropical rain forest for the expansion of oil palm plantations in Colombia has drawn increasing criticism from both within the country and abroad.[32]

But palm oil isn't the only feedstock attracting interest in Colombia. In October 2007, the Oilsource Holding Group formed a joint venture with Abundant Biofuels Columbia SRL to produce jatropha-based biodiesel in Colombia. Oilsource Holding Group is a consortium of public companies and private entities engaged in the biofuels industry that are focused on the production of biodiesel and biomass for electricity needs in rural areas. Abundant Biofuels has had previous jatropha experience in Africa. The companies plan to plant 100,000 hectares with jatropha, with an estimated investment of $45 million.[33]

The recent growth of the biodiesel industry in Colombia has been dramatic, but the growing social and environmental concerns raised by opponents to these large-scale corporate agribusiness projects need to be addressed.

PERU

Most of the biofuels activity in Peru has been in the ethanol sector. Nevertheless, Peru's first large-scale producer of biodiesel, Interpacific Oil SAC in Lima, began turning out commercial quantities back in 2002. The plant currently produces about 7.2 million gallons a year. In August 2003, the Peruvian government enacted a new biofuels law that has assisted the development of the biofuels industry. Among its many provisions, the law called for the establishment of voluntary 7.8 percent blends for ethanol and 5 percent for biodiesel. Although the targets are voluntary, there has been some discussion about possible mandatory requirements. There are no other laws or fiscal policies specific to the biofuels industries, leaving a good deal of uncertainty about specific details for investors, producers, and consumers alike.

Despite the lack of a comprehensive set of rules and regulations, Peru has recently attracted the interest of a number of foreign biodiesel investors. In May 2007, Pure Biofuels Corporation announced the purchase of Interpacific Oil for $6.3 million. Pure Biofuels plans to expand the capacity at the Interpacific facility to 10 million gallons. Pure Biofuels also is constructing a new 180,000-ton (54-million-gallon) biodiesel refinery in the Port of Callao near Lima. The output from the company's two refineries will be sold almost exclusively within Peru.[34] The prospects for additional biodiesel projects in Peru appear to be fairly good, especially if the gaps in the current laws concerning biofuels can be filled.

CHILE

Chile imports most of the fuel it consumes. The Chilean government has recently signed a pact with Brazil to help it identify areas of potential biofuel development and cooperation. The German firm Südzucker also has expressed interest in making heavy investments in the Chilean biofuels sector, and the Chilean government has been working on a new biofuels regulatory package. In May 2007, the government announced the elimination of taxes on bioethanol and biodiesel in a bid to increase demand and help jump-start the industry. A commission representing a broad cross-section of stakeholders was set up and has generally been supportive of biofuels for local use as long as they are produced in a sustainable manner that does not harm the nation's 178,000 farm families. The emphasis on local use was intended to avoid some of the worst aspects of large-scale production for export seen in some other Latin American nations. The government then published a comprehensive technical set of specifications for biofuels, apparently patterned after similar regulations in Germany. Also, in the northern part of the country, the government is working with the University of Tarapaca to develop a test plantation of jatropha on 1,500 hectares.[35] Chile appears to be making careful progress toward greater biodiesel use in its future energy mix.

BOLIVIA

Although Bolivia has an existing biodiesel target of 5 percent, with an eventual goal of reaching 20 percent by 2015, at the present time it appears unlikely that these goals will be achieved. The new government of Evo Morales is not currently pursuing large-scale biodiesel projects due to its strong political support from the low-income majority of the population. The government has promised not to divert land presently in food production for biofuel feedstocks.

GUATEMALA

Our biodiesel tour now arrives in Central America. Guatemala is home to a successful local biodiesel initiative based on jatropha. The project was initiated by Ricardo Austurias, an entrepreneur with a background in the agricultural industry, who saw an opportunity with rising petroleum prices in recent years. The vegetable oil he had in mind for his biodiesel venture was to come from the native piñon tree, widely used in Guatemala for livestock fencing. He discovered that the tree was known elsewhere as jatropha, and that it had been the focus of a lot of attention around the world. "We were surprised to find out that this was the same plant as the piñon, which we had known about all our lives in Guatemala," he says. After securing a number of grants, he proceeded to acquire land for jatropha testing, and with a group of other investors formed a company, Octagón, and built a $1.5-million biodiesel refinery that produces 600 gallons a day (about 215,000 gallons per year). In another location, he is testing about 50 varieties of jatropha to see which ones do the best in local growing conditions. Although his company needs to increase its production in order to make a profit, Austurias is convinced that he is heading in the right direction.[36]

In another small-scale biodiesel initiative, the Toledo sausage factory (in the Campero Group), located in Guatemala City, recycles the grease from their plant to make biodiesel to run their delivery fleet. Also in Guatemala City, Guatemala Bio-Diesel reportedly uses all the waste oil from Frito-Lay as feedstock for the production of substantial amounts of biodiesel. In addition, in 2005, a new organization, Combustibles Ecológicos S.A. was formed to study and produce biofuels in Guatemala. The organization, with headquarters in Quetzaltenango and Guatemala City, has grown to include biofuels work throughout the nation. The group currently is working on a medium-sized biodiesel facility in

Quetzaltenango and plans to fuel public transportation mini-buses there. The organization also continues to work with rural campesino communities to build and operate small-scale biodiesel reactors to power generators and farm equipment in rural communities.[37] There appears to be a lot of small-scale biodiesel activity across much of Guatemala, which is an excellent model for other Central American nations to follow.

COSTA RICA

Costa Rica, which spends about $500 million on petroleum fuels every year, appears to be on the verge of embarking on a national biodiesel initiative. At the end of March 2004, the Costa Rican Center for Cleaner Production announced that it would present a proposal for the wide-scale production and use of biodiesel created from natural oil sources such as the African palm.[38] Since then, a number of small biodiesel plants have been set up. Although the production of biodiesel in Costa Rica is relatively small in scale, it is large enough to supply about 1 percent of total diesel consumption in the nation. Also, in the small-scale sector, CentralBiodiesel HTP S.A. began operations in their new biodiesel reactor factory in San Jose in April 2006. Under a licensing agreement with Ricardo Carlstein (see Argentina, p. 180), CentralBioDiesel now manufactures one of Carlstein's new reactors, which can produce 800 liters of biodiesel per day.[39] These units are sold throughout the region. On a larger scale, in October 2007, Amelot Holdings, Inc. of Massachusetts, a leading producer of renewable fuels, and Pan-Am Biofuels, Inc., a Utah-based company with biofuel feedstock plantations located in Costa Rica, announced details of their jatropha joint venture in Costa Rica. The 2,000-acre plantation, located in Guanacaste, is expected to yield about 3 million gallons of jatropha oil. Amelot will have exclusive rights to all the oil produced.[40]

NICARAGUA

A successful study on the use of jatropha oil as a biodiesel feedstock by the National Engineering University (UNI) in Managua, Nicaragua, supported by an Austrian development program, was completed in 1996. This was followed by the construction of the country's first biodiesel pilot plant as well as a quality-control laboratory. The facility, which opened in 1997, has an annual production capacity of 3,000 tons. In addition to the construction of this facility, 1,000 hectares (2,471 acres) of previously degraded and idle agricultural land was planted with *Jatropha curcas* and was expected to yield about 426,000 gallons of oil after the bushes matured in about five years. (The plantation has reportedly been expanded to 1,500 hectares.) This plantation was expected to supply enough oil to meet about 3 percent of the country's total annual petrodiesel consumption. The biodiesel initiative was part of the larger Biomass Project, a collaboration between UNI, the Nicaraguan Ministry of Energy (INE), and the Austrian firm of Sucher & Holzer.[41] Also, a biodiesel filling station has reportedly opened in Managua. Since then, aside from a few small processors at the local level, the biodiesel sector does not appear to have made a lot of progress.

OTHER CENTRAL AMERICAN COUNTRIES

Panama imports all its oil from other countries. The nation recently signed an agreement with Brazil to cooperate on the development of the biofuels sector in Panama. In an unrelated initiative, in March 2007, construction began on a $65 million biodiesel refinery located in the Burica region near Port Armuelles, Panama. The facility, capable of producing 50 million gallons per year, is owned by Houston-based Texas Biodiesel Corp. and is expected to come online in 2008. The company also is building a $2.6 million oil-extraction facility nearby to process the palm, mustard, and other oilseeds provided by a large palm-oil cooperative as well as by other local farmers in the region.[42]

Honduras imports more than 1 million tons of petrodiesel fuel every year. At the same time, the nation has 70,000 hectares in existing oil-palm plantations that could produce enough palm-oil biodiesel to replace about 20 percent of that imported fuel. With this resource, it is no surprise that there have been a number of initiatives to use at least some of the palm oil for biodiesel. There are reportedly a number of public biodiesel pumps that sell biodiesel, and the fuel is even being used in part of the public transport sector. In addition, a commercial fish farm in El Borboton is using fish guts to produce 300,000 gallons of its own biodiesel every year. The biodiesel then is used to fuel the 10 trucks and 8 buses that bring company employees to work.[43]

Mexico's largest oil field, Cantarell, peaked in 2004 and is now in serious decline. The nation is expected to become a net oil importer in the next few years. Mexico recently passed a new biofuel law that should encourage farmers to produce biofuel feedstock crops such as yucca root, sorghum, and beets. The law, which comes into effect in 2008, offers as yet unspecified support for farmers who grow biofuel crops other than corn or sugarcane, which are already in relatively short supply in Mexico. It is thought that biodiesel feedstock production could help many of the nation's poorest farmers, and that none of the crops currently grown for food would be replaced by biofuel feedstock crops under this plan.[44]

THE CARIBBEAN

The Caribbean region also has seen some early biodiesel activity recently. In the Dominican Republic, a dozen government vehicles have recently been used in a B50 biodiesel trial. In Haiti, a small producer, Haiti Biodiesel Group, makes about 300 gallons of biodiesel per month from used cooking oil. Jamaica has entered into an agreement with Brazil on the possible construction of two biodiesel plants, and in the West Indies, the Petra Group has plans for a jatropha plantation and a biodiesel refinery. There are also a number of very small, local biodiesel producers who are making biodiesel for themselves, but their numbers are hard to quantify.

CANADA

Canada's Alternative Fuels Act was implemented on April 1, 1997. The main purpose of the act was to increase the use of alternate transport fuels (ATF) in Canada by gradually increasing the percentage of vehicles in government fleets capable of operating on alternate fuels to 75 percent by 2004. Although those targets have been met, Canada has been slow to adopt the use of biodiesel. This may, in part, be due to concerns about the fuel's performance in the country's notoriously cold winters. Nevertheless, Agriculture and Agri-food Canada reported that if the world's 30 major economies replaced just 8 percent of the fossil fuel they consumed with biofuels, commodity prices would rise enough to solve the farm income crisis that has bedeviled the agricultural sector for many years. Canadian farmers have been taking note, and in recent years biodiesel activity in Canada has been heating up.

In April 2001, the BIOX Corporation started biodiesel production in a demonstration plant in Hamilton, Ontario, capable of producing 1 million liters (927 tons) per year. The multifeedstock process for the plant had been developed earlier at the University of Toronto. BIOX hoped to be first in Canada to open a commercial-scale biodiesel plant based on this technology in 2005. However, due to a number of delays, the $30 million plant with a 56,000-ton capacity did not actually open until late 2006. Another company, Rothsay, was actually the first major plant to come online. This company had been producing biodiesel in their Montreal plant since September 2001, using restaurant oil and animal fat as feedstocks. The 4-million-liter-capacity, batch-process plant was designed by Rothsay, and the company subsequently upgraded and expanded the facility to produce 35 million liters (9.3 million gallons). A $14.5 million plant expansion was completed in November 2005. Rothsay is a wholly owned subsidiary of Maple Leaf Foods Inc. of Toronto, Canada's leading rendering company.

In March 2002, the city of Montreal began testing biodiesel in some of its municipal buses, and by midsummer 155 Montreal Transit Corp. buses on 19 routes serving the downtown area were running on B20. The biodiesel was made from used vegetable oil and animal fats. Although the program was a technical success, it was discontinued the following summer when the Quebec provincial government refused to extend a tax break of 16.4 cents per liter (62 cents per gallon). However, in March 2007, Montreal announced that it would again be using biodiesel in its bus fleet. In June 2002, the province of Ontario initiated a tax exemption of 14.3 cents per liter for biodiesel, and the city of Brampton, Ontario, became the first municipality in Canada to commit to the regular, ongoing use of biodiesel. That summer, the city began using biodiesel in its fleet vehicles, and by October all of its 137 buses began running on B20.[45] Today, all of Brampton's 190 buses still run on a biodiesel blend. Other bus systems in Canada that are using or testing biodiesel include those of Halifax, Nova Scotia; Guelph, Sudbury, and Toronto, Ontario; and Saskatoon, Saskatchewan. In 2002, electric utility company Toronto Hydro switched its entire 400-vehicle fleet to B20. To date, thousands of fleet vehicles have driven millions of kilometers using biodiesel in Canada.

In May 2003, a new organization composed mainly of oilseed growers and processors, Biodiesel Canada, met for the first time in Winnipeg, Manitoba. The group quickly began to work with the government on a variety of issues related to oilseed production and fuel taxes. And on March 2, 2004, Canada's first public biodiesel fuel pump finally opened in Unionville, near Toronto. The B20 blend for the pump was supplied by Ottawa-based Topia Energy Inc., which was the first company to open a Canadian biodiesel terminal (also in the Greater Toronto area).[46] In May 2004, a second public pump was opened by Topia, this time on Queen Street in Toronto. A third pump was subsequently opened by the company in Ottawa. And in July 2005, Canada's first biodiesel pump at a retail truck stop opened in Burnaby, British Columbia. Since then a number of additional biodiesel pumps have been opened across the nation.

More recently, in December 2006, the federal government called for a 2 percent biodiesel standard by 2012, and offered $200 million to encourage farmers to invest in biodiesel production. That was followed in March 2007 by an announcement that the government would invest $2 billion over seven years to help develop the nation's biofuels capacity.

"Ten years from now, we will look back on this day as the start of a national biodiesel industry in Canada," said Barb Isman, president of the Canola Council of Canada.[47] The following month, the government launched another part of this plan, known as the federal ecoAgriculture Biofuels Capital (ecoABC) initiative. The initiative was part of the government's comprehensive renewable-fuels strategy focused on increasing the demand for these fuels through regulation, assisting farmers in getting involved in the sector, and accelerating the commercialization of new, renewable technologies.

Alberta-based Western Biodiesel, Inc. was able to leverage some of that new ecoABC money for its new $8 million biodiesel refinery in Alsersyde. The federal initiative agreed to contribute $638,559 to help build the biodiesel plant. One condition of that support, however, was farmer participation in the project, and local farmers had already invested $275,000. The 19-million-liter (5-million-gallon) facility will use canola oil and beef tallow as feedstocks, and was expected to start up operation in early 2008.[48]

On a much larger scale, in October 2007, Canadian Bioenergy Corporation of Vancouver announced plans to build a 225-million-liter (59.4-million-gallon) biodiesel refinery in Fort Saskatchewan, Alberta, adjacent to a canola-crushing plant owned by Bunge Ltd., the world's largest oilseed processor. The $90 million plant was twice the size of the original proposal, but the company felt that new federal and provincial renewable-fuel standards would create a 750-million-liter market in Canada by 2010. When completed in 2009, the new plant will have the capacity to consume about 5 percent of the nation's entire canola crop.[49] Despite this, canola growers insist that producing biodiesel from canola will not raise food prices in Canada. This assertion is mainly based on the sheer size of the Canadian canola crop—14 million acres, with an estimated value of $13 billion.[50]

Nevertheless, there has been some recent controversy about the environmental benefits of canola as a biodiesel feedstock. A 2006 study by four Canadian researchers concluded that the use of canola-based biodiesel would reduce greenhouse gas emissions by as much as 85 percent and that the fuel's environmental benefits were "overwhelming." This was contradicted in four 2007 studies by other researchers who concluded that producing biodiesel from heavily fertilized crops such as canola could generate as much as 70 percent *more* greenhouse gas emissions than using petrodiesel. This finding relates to the generation of nitrous oxide by soil microbes reacting to the fertilizer. The main debate is centered on just how large that reaction is and on the quantity of gas emissions actually involved. Researchers on both sides of the argument stand by their findings.[51]

In addition to large-scale projects, there are also a number of small biodiesel companies, cooperatives, and individual producers (especially farmers) in various locations across Canada. These organizations definitely have a productive role to play in the economies of the local communities they serve.

Overall, the Canadian biodiesel industry has the potential for substantial growth over the next few years.

PART THREE

Biodiesel
in the
United States

9

A Brief History

With only 4.5 percent of the world's population, the United States consumes about 25 percent of global energy and produces roughly 25 percent of the planet's CO_2 emissions. Because of this dubious distinction, the opportunities for positive change in U.S. energy practices are enormous—and the need couldn't be more urgent. The United States presently imports about 60 percent of its oil, and that figure is going to rise in the years ahead. Americans collectively spend about $475,000 per minute to pay for that imported oil.[1] This enormous outflow of money not only contributes substantially to the U.S. balance-of-payments deficit but also unquestionably has helped fund at least some anti-American activities around the globe.

Unfortunately, many U.S. politicians—especially those who spend most of their time in Washington, D.C.—still don't get it. For many years they have repeatedly resisted simple but important steps such as improving federal fuel economy standards, failing to understand that the consequences of inaction are becoming increasingly costly and far outweigh the short-term political advantages of maintaining the status quo. The recent modest increases in fuel economy standards contained in the 2007 Energy Bill are a small first step in the right general direction, but fall far short of what is needed. If there was ever a place and time for courageous and visionary political leadership regarding renewable energy, the place is the United States, and the time is now. Because it plays such a key role in the energy dilemma currently facing the planet, anything the United States does to try to clean up its act will have a significant impact on the rest of the world. This is why the next four chapters are devoted entirely to the biodiesel industry in the United States. First, a little background.

EARLY BIOFUELS

Although biodiesel is a relatively recent development in the United States, the basic chemistry involved has its roots in the nineteenth century. As early as the mid-1800s, transesterification was used as a strategy for making soap. Early feedstocks were corn oil, peanut oil, hemp oil, and tallow. The alkyl esters (what are now called biodiesel) resulting from the process were originally considered just by-products. Ethanol also had a significant place in early U.S. biofuels history. Prior to the Civil War, an ethyl alcohol (ethanol) mixed with turpentine, known as camphene, was widely used as a lamp oil. A tax on alcohol enacted during the Civil War dramatically reduced the use of industrial alcohol, and it was not until 1906, when the tax was finally repealed by Congress, that ethanol began to make a comeback as a fuel, especially for internal combustion engines.[2]

A number of Europeans have been credited with inventing the internal combustion engine, especially Nikolaus August Otto, who produced an early version around 1866. However, Samuel Morey of Orford, New Hampshire, built the first prototype internal combustion engine in the United States in 1826. He used a biofuel, alcohol, as the main fuel in his experiments. Unfortunately Morey, who was many decades ahead of his time, was unable to attract financial backing for his invention, and he has been largely ignored by historians. As noted previously, Rudolf Diesel was also a firm believer in the potential of biofuels for powering his engine. But Diesel was not alone in his enthusiasm.

In 1896, Henry Ford built his first automobile, a quadricycle, to run on ethanol. In 1908, Ford's famous Model T was designed to run on ethanol, gasoline, or a combination of the two. Ford was so convinced that the success of his automobile was linked to the acceptance of "the fuel of the future" (as he described ethanol) that he built an ethanol production plant in the Midwest. He then entered into a partnership with Standard Oil Company to distribute and sell the corn-based fuel at its service stations. Most of the ethanol was blended with gasoline. Ford's biofuel turned out to be fairly popular, especially with farmers, and in the 1920s ethanol represented about 25 percent of Standard Oil's fuel sales in that part of the country. In retrospect, Ford's alliance with Standard Oil may not have been such a good idea. As Standard Oil tightened its grip on the industry, it focused its attention on exploiting its petroleum markets— and eliminating any competition. Nevertheless, Ford continued to promote ethanol through the 1930s. But finally, in 1940, he was forced to close the ethanol plant due to stiff competition from lower-priced petroleum-based fuels.[3]

During World War II, because of the disruptions to normal oil supplies, virtually all the participating nations made use of biofuels to power some of their war machinery. At the same time, transesterification used in the soapmaking process became the subject of great interest because of the by-product glycerin, which is a key ingredient in the manufacture of explosives. But after the war, the return of steady oil supplies and low gasoline prices brought an end to biofuels production in the United States. Some observers maintain that the demise of the biofuels industry in the 1930s and '40s was due to the deliberate actions of a small group of individuals such as William Randolph Hearst, Andrew Mellon, the Rockefellers, and a number of "oil barons." Others insist that the biofuels industry was simply the victim of larger market forces. In either case, for all practical purposes the industry ceased to exist after World War II. It was not until the oil shocks of the 1970s that biofuels began to experience a renaissance.

THE OPEC OIL CRISIS

The oil crisis of the 1970s was a rude awakening for most Americans, dramatically underscoring the nation's dependency on imported oil. On October 17, 1973, when the OPEC nations shut off the petroleum spigot to the West, oil imports from Arab countries to the United States quickly dropped from 1.2 million barrels a day to a mere trickle. As the price of oil increased dramatically and long lines at gasoline stations grew even longer, people across the country began to look for alternate sources of energy. In 1973, President Richard M. Nixon created a cabinet-level Department of Energy headed by William E. Simon, who became known as the nation's "energy czar." Many of the early federal renewable energy programs at agencies such as the National Renewable Energy Laboratory (formerly known as the Solar Energy Research Institute) were also initiated during the late 1970s.

The 1979 revolution in Iran that resulted in the ousting of the U.S.-backed shah precipitated yet another global energy crisis. Oil prices doubled, sending the industrial world into a recession. When the crisis finally subsided, oil prices fell, and the Reagan administration ended the tax incentives and other support for the renewable energy industry. The initial steps that had been taken toward a comprehensive national renewable energy initiative were largely abandoned.

EARLY EXPERIMENTS

Despite the loss of federal support, some of the enthusiasm for renewable energy lingered on and even began to grow. As mentioned in Chapter 2, some of the earliest experiments in the production and use of biodiesel in the United States took place at the University of Idaho, beginning in 1979 with the work of Dr. Charles Peterson. For the first few years, Peterson's experiments were relatively low-key and focused mainly on farm tractors. It wasn't until the late 1980s and early 1990s that the scope of his research began to expand exponentially. But Peterson was not the only one involved in biodiesel research.

Colorado
In the summer of 1989, Dr. Thomas Reed, who was on the faculty at the Colorado School of Mines, first learned about the conversion of animal

fats and vegetable oils into biodiesel. Reed, who had a longtime interest in biofuels, began to wonder whether used cooking oil would make a viable feedstock. In November, after researching the available literature (especially Dr. Peterson's), he decided to conduct his first transesterification experiment. "I got a gallon of used vegetable oil from the grease dumpster at a local McDonald's," he recalls. "Then, I brought it into my lab, made some minor adjustments in the methanol and lye recipe, and stirred it up well, and when I got back from lunch, there it sat—a gallon of beautiful fuel from that awful grease. I couldn't believe how easy it was."[4]

Reed was so enthusiastic that he continued his transesterification experiments with a variety of used oils, even making biodiesel from bacon grease on Christmas in his daughter-in-law's kitchen. In the spring of 1990, Reed approached the Denver Regional Transportation District (RTD) to see if they would be willing to try some biodiesel in one of their buses. They agreed, but first Reed had to make use of a larger laboratory at the School of Mines in order to produce the 100 gallons of biodiesel needed for the tests. After the fuel had been made, tests were conducted on an RTD bus, and they showed substantial reductions in emissions at various percentages of biodiesel. The remaining fuel was used to operate a bus in Denver for about a week. This may have been the first public test of biodiesel in a commercial vehicle in the United States.

Reed decided that "transesterified waste vegetable oil" was not a very catchy name, and, considering the source, he decided to call it McDiesel. "I applied for a copyright," he says. "I even approached McDonald's to see if they were interested. They were, but said they would sue me if I used that name. Later people came to call these fuels 'biodiesel,' and I now live with that. However, I would love to have had McDonald's sue me—what publicity!" With two other partners, Reed formed a company to make and promote his fuel. They published a number of papers, held numerous discussions with many people, and finally entered into an agreement with a chemical engineering company. But when the agreement with that company descended into protracted legal wrangling, Reed lost his enthusiasm for biodiesel and moved on to other renewable-fuel interests.[5]

Missouri

At the University of Missouri at Columbia, Leon Schumacher, an associate professor in the Agricultural Engineering Department, also became involved in the second wave of biodiesel research that began in the early 1990s. Schumacher was approached by members of the Missouri Soybean Merchandising Council in 1991 about conducting research on new uses for soybean oil. He was intrigued. Schumacher studied reports of earlier research, but he wanted to focus on a project that had not already been conducted in the United States. "An acquaintance of mine pointed out that most of the early testing had been done on farm tractors, but that no one had done much with the use of biodiesel as a fuel in regular diesel-powered vehicles like cars and trucks," Schumacher recalls. That was what he had been looking for. Schumacher applied for the Soybean Council project.[6]

With financial assistance from the Agricultural Experiment Station at the university, a new 1991 Dodge diesel pickup truck was purchased. "We put signs on the side of the truck that said 'Fueled by 100 percent SoyDiesel,' and it wasn't long before it began to catch the eye of a lot of people," Schumacher says. "I distinctly remember one day there was a guy riding a motorcycle who was passing me, and he looked at the sign on the side of the truck and took his hand off the throttle, made a fist, and raised his thumb as a sign of approval. There was a lot of that kind of public acceptance and support." The biodiesel used in the test came from Midwest Biofuels, a subsidiary of Interchem Environmental Inc. headquartered in Overland Park, Kansas. The company's small-scale production facility actually was located just across the state line in Kansas City, Missouri. "I think the first 100 gallons they produced for us were made in garbage cans in the backyard," Schumacher says, laughing.

Before long the truck was on the road almost constantly, and it was used as part of an education and promotion initiative for farmers, legislators, and the general public. The truck ran on B100 for 90,000 miles before the test was ended in 1996. During that time, fuel consumption was carefully monitored and dynamometer and other tests were conducted. "One thing we noticed was that there was virtually no soot coming out of the exhaust pipe, and that really caught our attention," Schumacher says. In 1992, a new Dodge pickup truck was purchased for a second test and it ran for 100,000 miles on B100. Then the engine was removed and sent to the manufacturer in Indiana, where it was torn down and carefully inspected. The engine was found to be in excellent condition.

Over the years Schumacher has been involved in an eclectic series of biodiesel experiments with garbage trucks in Columbia, buses in St. Louis, and even a diesel locomotive in St. Joseph, Missouri. For the locomotive test Schumacher collaborated with Jon Van Gerpen, at the time an associate professor from the Mechanical Engineering Department of Iowa State University in Ames, Iowa, who headed UIA's biodiesel program from its inception. The locomotive, a veteran 1945 GM diesel-electric switcher, was being used to move grain cars for the Bartlett Grain Company. The test was designed primarily to measure the amount of exhaust smoke produced with a B20 biodiesel blend. A light gray smoke was observed during engine acceleration, but no differences were noted by the locomotive operators concerning fuel economy, engine oil consumption, engine oil dilution, or fuel compatibility. "We had to install a special battery power unit for our testing equipment because there wasn't anything we could plug into on the locomotive," Schumacher recalls.

Schumacher later was involved in a biodiesel research project with the South Dakota Department of Transportation, which investigated a B5 blend for use in its vehicles. "I feel very good about it," Schumacher responds, when asked about his 15-plus years of biodiesel research work. "If I had it to do all over again, I might do a few things a little differently, but I guess that's normal. Each project has had its unique challenges, but they've all been interesting."[7]

EARLY PRODUCTION

Although there was unquestionably some biodiesel research taking place here and there across the country in the 1980s and early 1990s, most of the national biofuels emphasis at the time was focused on ethanol. But around 1990 that began to change. Some of the earliest U.S. production of biodiesel in commercial quantities took place in the early 1990s at the nondedicated plant of Procter & Gamble in Kansas City, Missouri.

Midwest Biofuels

In 1991, Midwest Biofuels, a subsidiary of Interchem Environmental Inc., set up a small-batch-process pilot plant in Kansas City, Missouri, to supply limited quantities of biodiesel for Leon Schumacher's SoyDiesel-powered Dodge pickup truck from the University of Missouri and similar demonstration projects. "I remember making the stuff by hand," recalls Bill Ayres, one of the company's founders. "Then Leon drove up in this brand new pickup truck with its twin fuel tanks. We filled up one of the tanks with B100, he flipped a switch, and from that point on the truck ran for many years on SoyDiesel."[8]

The following year, the pilot plant also supplied biodiesel for the first phase of a trial at Lambert Airport in St. Louis involving about 10 vehicles. "I didn't have the facilities to produce a lot of biodiesel at the time, and I was a little worried about the gel temperature in the winter, so, more or less out of the blue, I suggested a B20 blend," Ayres says. B20 subsequently became a standard for many other trials and is the most popular blend in use today across the United States. As demand for biodiesel for additional trials and tests grew, Midwest BioFuels set up a new, larger plant in Kansas City, Kansas, in 1993 with a production capacity of 100 gallons per batch (later expanded to 500 gallons). This was the first dedicated commercial biodiesel facility in the United States. But after about six months, the demand started to outstrip Ayres's ability to meet it, and he began to purchase biodiesel from the nearby Procter & Gamble plant.[9]

Ag Processing Inc.

In October 1994, Ayres and a business partner, Doug Pickering, left Interchem to become consultants for Ag Processing Inc. (AGP), the country's largest soybean-processing cooperative. AGP has about 200,000 members in 16 states and Canada. In the following year, Ayres and Pickering formed a new joint venture with Ag Processing called Ag

Environmental Products (AEP). In 1996, AGP opened a new batch-process biodiesel plant with a capacity of 5 million gallons (17,500 tons) in Sergeant's Bluff, Iowa, adjacent to an existing seed-crushing facility. (The plant's capacity was expanded to 30,000 million gallons in 2006.) In the spring of the following year, AEP provided biodiesel fueling stations at 10 farm co-op locations in six Midwestern states. Other stations were subsequently added. Over the years the $6 million soy methyl ester facility at Sergeant's Bluff has produced a wide range of products, including biodiesel, solvents, and agricultural chemical enhancers under the SoyGold brand name, which Ayres and Pickering promoted and marketed for AEP. Biodiesel produced by AGP has been used in a wide range of vehicles by customers across the country.

Twin Rivers Technology

In 1994, Twin Rivers Technology began business in a former Procter & Gamble oleochemical facility in Quincy, Massachusetts. The plant had a nominal production capacity of 30 million gallons, but the company never actually made any biodiesel at the Quincy facility, purchasing it instead from other plants. In 1996, Twin Rivers was the first company in the nation to receive U.S. Environmental Protection Agency certification of its Envirodiesel brand of biodiesel fuel for use in meeting Clean Air Act compliance standards under the agency's Urban Bus Retrofit Program. However, the company suffered from a long series of marketing and internal problems and closed its biodiesel division in 1998. (For more on Twin Rivers, see Chapter 10.)

NOPEC

In 1995, the NOPEC Corporation began initial production at its huge 18-million-gallon-capacity batch-process biodiesel plant in Lakeland, Florida, using roughly 50 percent virgin soybean oil and 50 percent used cooking oil as feedstocks. Although NOPEC (the company name poked fun at OPEC) landed a large contract to supply 23,000 gallons of biodiesel to the Iowa Department of Transportation, the company initially focused mainly on the marine market. NOPEC's main customers were boat owners in the environmentally sensitive waters of the Florida Keys and Maryland's Chesapeake Bay. The National Oceanic and Atmospheric Administration's patrol boats in the Florida Keys National Marine Sanctuary were early users of biodiesel from NOPEC.[10] Other customers included the U.S. Postal Service and numerous school districts across the state. In 1997 and 1998, the company ran into serious legal problems related to the sale of unregistered stock, and in December 2000, the NOPEC production facility was purchased by OceanAir Environmental, a California-based company. (For more on NOPEC, see Chapter 10.)

West Central Cooperative

In 1996, in response to increasing interest and demand, there was a sudden flurry of biodiesel plant construction across the country. West Central Cooperative, in the tiny community of Ralston, Iowa (population 119), built a new 2.5-million-gallon-capacity batch-process plant. The facility was a success and subsequently was expanded to increase its output. But there were some lingering inefficiencies with the expanded design, and in 2001 West Central decided to start with a completely new continuous-process facility. The new 12-million-gallon-capacity plant began production in 2002. "It's a state-of-the-art facility," says Gary Haer, who has been with West Central since 1998. (He was previously with Midwest Biofuels/Interchem.) "We recover all of the materials that are used to convert the soybean oil into biodiesel, so there's no waste. Everything is recaptured and reused."[11] With many years of experience in biodiesel plant design, construction, and operation under its belt, West Central launched Renewable Energy Group, Inc. LLC as a partnership with Todd & Sargent, Inc., an Ames, Iowa, engineering and construction company in 2003. (For more on Renewable Energy Group, see Chapter 10.)

Columbus Foods

The founding of West Central was followed in the fall of 1996 with the start-up of a biodiesel production facility in Chicago, Illinois, by Columbus Foods, a food-grade-oil producer and distributor for more than

60 years. Earlier that year, the company's owner, Michael Gagliardo, had read about biodiesel in a magazine article and had been inspired by Chicago's demonstration of biodiesel in buses and police patrol boats. He was intrigued by the possibilities of producing a new product from used oils, and after traveling to Austria to learn more about it, he had a 200,000-gallon-per-year plant constructed at a cost of $500,000. The City of Chicago donated a building through its brownfield program, which was designed to clean up environmentally degraded urban lands. Early demonstration users for a Columbus Foods B20 blend were the Chicago Transit Authority and the American Sightseeing Bus Company.[12] But after experiencing some cold-weather difficulties with biodiesel made from used oil, the company decided to switch to using virgin soybean oil as their primary feedstock. With some additional equipment and newer processing techniques, Columbus Foods now can produce up to 4.9 million gallons a year.[13]

Pacific Biodiesel

The last biodiesel production facility to come online in 1996 was built by Pacific Biodiesel in Kahului on the island of Maui in Hawaii. Bob King had been a diesel engine mechanic for more than 20 years and was the owner of King Diesel, which provided diesel engine service primarily for the marine industry. King, who had been contracted to maintain the diesel-powered electrical generators at the central Maui landfill, noticed that the landfill was being swamped with tons of used restaurant grease from trucks that serviced local hotels and restaurants. He grew increasingly concerned about the potential environmental and health problems caused by the grease, and in 1995 he decided to do something about it. He eventually contacted Daryl Reece, a researcher at the University of Idaho who had worked with Dr. Charles Peterson as both an undergraduate and a graduate student on ways of making biodiesel fuel from discarded cooking oil. King had found the answer to the grease problem—and a business partner as well.

King and Reece formed Pacific Biodiesel Inc., and without outside financing they proceeded to build the first biodiesel plant on the Pacific

Rim and the first commercial facility in the United States to rely entirely on used cooking oil as its feedstock. "We couldn't get any normal financing because nobody had heard of biodiesel," King recalls. "And we didn't have a market for the product because no one knew what it was. It probably wasn't a very wise decision, but we didn't know any better at the time."[14] After the plant was completed at the Maui landfill site, the trucks that had previously pumped used cooking oils into the landfill found that it was more economical to deliver their loads to the biodiesel facility.[15]

Initially King's vision for the venture didn't extend beyond Hawaii. But that changed almost immediately. "When we held the grand opening for the first plant, there were people who had come from all over the place, and we got press coverage from all over the world," he says. "I quickly realized that this wasn't just something that was going to apply to Maui; there were some much larger implications to this." When the company set up its Web site, it began to receive inquiries from people in nations around the globe. The following year Pacific Biodiesel built a similar plant in Nagano, Japan, for a fast-food restaurateur (see page 154).

Shortly after the completion of the project in Japan, King and Reece turned their attention to an even larger problem at the Maui landfill—grease-trap waste (brown grease) from sewer grease traps in restaurants and other grease sources. Reece designed a custom processor for grease-trap waste, and Pacific Biodiesel was able to create its own boiler fuel while diverting 140 tons of grease-trap sludge from the landfill every month. Building on its successful operation on Maui, Pacific Biodiesel built a 1,500-gallon-per-day biodiesel plant in Honolulu that processed 25,000 gallons of grease-trap waste per day. The Honolulu facility now processes used cooking oil as well.

"It took a little bit of time to convince people that this fuel wasn't going to ruin their engine or cause problems, so it was a very slow, methodical, customer-by-customer movement of the product into the market for the first three years," King says. "We put a lot of time into education about what the product was. Then it started getting easier, but it has only been since 2004 or so that biodiesel has become a common word and people finally seem to know what it is."[16]

THE SOYBEAN FACTOR

Unlike the European biodiesel industry, which was largely a creation of national-government-sponsored research and energy policies, much of which was focused on rapeseed, the biodiesel industry in the United States has had its longest and strongest support from Midwestern soybean farmers. The United States is the largest producer of soybeans in the world, with an average annual production of about 3 billion bushels. The nation is also the largest consumer and exporter of soybeans on the planet. Because the demand for soy meal had generally risen faster than the demand for soy oil for many years, the soybean industry was faced with the dilemma of what to do with the surplus oil. Much of this surplus has been directed toward the biodiesel market, especially in recent years, which largely explains the enthusiasm that soybean growers and processors have had for the fuel. One bushel of soybeans is needed to produce about 1.5 gallons of biodiesel.

The American Soybean Association

It should come as no surprise that one of the organizations that took an early interest in biodiesel was the American Soybean Association (ASA) and its associated marketing, research, and communication arm, the United Soybean Board (USB). Originally founded in 1920, the ASA has worked for many years to promote the interests of soybean farmers. In 1991, the association implemented a national soybean checkoff of 0.5 percent on the per-bushel market price of soybeans when the crop is first sold to help fund market promotion, research, and educational programs. Two years later, the ASA and the USB began to implement checkoff-supported programs, and the ASA became heavily involved in a wide range of legislative, research, and development activities for what they call "soydiesel."

"We saw a wonderful opportunity to use a homegrown crop to meet our country's energy needs, and to do so in a way that helps farmers while improving air quality and making the country safer," says Neil Caskey, the former special assistant to the CEO of ASA. "Over the years, soybean farmers have invested about $50 million through their checkoff in building a biodiesel market. This is something that they believe in, and they are putting their money where their mouth is. This is one of the reasons why this initiative continues to grow and get a lot of support. Also, when people see that farmers are willing to use biodiesel on their own farms in their own trucks and equipment, they understand that they are truly committed."[17]

The National Biodiesel Board

In 1992, the USB decided to promote an American biodiesel project and founded the National SoyDiesel Development Board, which was succeeded in 1994 by the National Biodiesel Board (NBB). The NBB, headquartered in Jefferson City, Missouri, has developed into the national trade association representing the biodiesel industry and is the coordinating body for research and development in the United States. The NBB also works with a broad range of industry, government, and academic entities on a wide range of biodiesel-related activities. Members of the organization include state, national, and international feedstock-growing and processing organizations; biodiesel suppliers, marketers, and distributors; and technology providers—all big players in the industry. The NBB also supports quality assurance initiatives, especially its voluntary BQ-9000 program, which is designed to ensure that participating producers consistently meet ASTM standards. The NBB also acts as a clearinghouse for information.[18] The organization's Web site (www.biodiesel.org) offers a vast amount of constantly updated information about biodiesel and the industry.

Joe Jobe, the chief executive officer of the NBB, has high praise for the farmers who have worked tirelessly to support the biodiesel industry. "The support of the soybean farmers has been absolutely critical," he declares. "If the soybean farmers had not decided to invest—through their state and national soybean checkoff programs—in biodiesel research and development programs, we wouldn't have a National Biodiesel Board. We wouldn't have a legally registered fuel designated as an alternative fuel, the only alternative fuel in the country to have complied with the health-effects testing requirements of the 1990 amendments to the Clean Air Act. We also wouldn't have one of the best-tested alternative fuels in the country, with more than 50 million successful road miles and countless off-road and marine hours in virtually every diesel engine type and application. Soybean farmers have driven the development of the U.S. biodiesel industry, both through their investments of their money and through their grassroots support of biodiesel."[19]

"The National Biodiesel Board is an exceptional organization that has supported our industry in so many ways," says Gary Haer of West Central Soy. "The resources and technical information that they provide to the industry and the general public are remarkable, and they have really assisted in the growth of this market. The soybean industry and the NBB have helped make biodiesel the success story that it is today."[20]

OTHER ORGANIZATIONS

A number of other organizations have supported the development of biodiesel over the years. The U.S. Department of Energy and some of its affiliates, such as the Alternative Fuels Utilization Program and the National Renewable Energy Laboratory (NREL) in Golden, Colorado, have had biodiesel initiatives. NREL, in particular, has conducted research into many aspects of biodiesel, the most notable of which is the lab's extensive research into algae as a feedstock in Roswell, New Mexico. Less familiar organizations, such as the Fat and Protein Research Foundation (FPRF) and the National Renderers Association (NRA), also have benefited the biodiesel industry with their research efforts to develop new outlets for waste oils and fats from plants and animals.

HOMEBREW

In addition to the activities of soybean farmers and commercial producers, for many years there has been a parallel grassroots biodiesel movement in the United States propelled by a group of dedicated biodiesel enthusiasts who generally refer to themselves as the "B100 Community," due to their preference for running their vehicles on B100 whenever possible. Like their counterparts in many other countries, these individuals grew tired of waiting for the government to establish a comprehensive national biodiesel initiative and went ahead and started to make and use biodiesel on their own. Quite a few small-scale producers scattered across the country have been making biodiesel from used cooking oil in kitchen blenders, old 55-gallon drums, and an assortment of other funky, low-tech pieces of equipment for many years. Quality levels of this homebrewed fuel vary widely. And it should be noted that if homebrewers use this biodiesel to fuel their vehicles for on-road use, they are legally required to pay all highway taxes. Some do. Some don't.

The widely distributed homebrew feedstock sources, such as individual fast-food restaurants (especially those located in rural areas), do not lend themselves well to supplying large amounts of used oil to industrial-size biodiesel production facilities. But these scattered feedstock sources are a good match with equally scattered backyard "homebrewers," who often produce just enough fuel for their own use or, perhaps, for a few close friends. Although used cooking oil is a limited resource that is increasingly hard to find in large quantities, it does offer biodiesel producers, especially smaller producers, an inexpensive (sometimes free) alternative to more expensive virgin vegetable oils. And producing fuel from local resources for local use makes a lot of sense from an energy-efficiency and environmental standpoint.

Fat of the Land

Many knowledgeable people in the homebrew sector credit Dr. Thomas Reed of Colorado (mentioned earlier) for developing and popularizing a simple, easy-to-follow recipe for making biodiesel in small quantities from used cooking oil. As his recipe spread by word of mouth, especially in the back-to-the-land and counterculture communities, more and more people began to experiment with making their own fuel from old grease. In 1993, independent San Francisco filmmaker Nicole Cousino and her team of documentary filmmaker friends (Sarah Lewison, Julie Konop, Florence Dore, and Gina Todus) discovered Reed and his recipe. "I first heard about running vehicles on vegetable oil a few years earlier from an inventor in upper New York State named Louis Wichinsky," Nicole Cousino recalls. "He was the one who put me in touch with Tom Reed at the Colorado School of Mines. I had never taken chemistry, so Tom and a friend of his, Agua Das, spent perhaps forty or fifty phone calls patiently explaining to me what we needed to do in order to make the fuel."[21]

In July 1994, armed with their newfound knowledge, the five women filmmakers took a pioneering grease-powered trip across the country from New York City to San Francisco in a GMC diesel van, stopping at an endless succession of greasy spoons and other restaurants, making biodiesel along the way, and interviewing a number of early biodiesel pioneers, including Dr. Reed (who graciously allowed them to make biodiesel in his laboratory). They also produced a humorous documentary about their adventure titled *Fat of the Land*. Although this classic 1995 cult film was unabashedly humorous, it nevertheless raised serious questions about the stranglehold that petroleum has on the U.S. economy, American energy usage, and sustainability. "Even today, the whole idea is still amazing; there is a certain radical element to being able to run your car on waste vegetable oil," Cousino says. "We also wanted to make the point that this was accessible science; that if we could do it, anybody could do it. Tom Reed often referred to this as 'bucket science' or 'kitchen science,' and it really is."[22] *Fat of the Land* is a must-see for biodiesel fans who want to learn about the early history of the homebrew movement in the United States (see "Organizations and Online Resources," page 321).

The Veggie Van

It would be difficult to talk about the grassroots biodiesel movement in the United States without mentioning the Veggie Van and its creators, Joshua Tickell and Kaia Roman. In 1995, while she was finishing the production work on *Fat of the Land*, Sarah Lewison received a phone call from Kaia Roman, who asked many questions about the cross-country trip. "We talked to Kaia a lot," Lewison recalls. "We tried to be helpful and gave her our recipes and told her about Tom Reed and Agua Das."[23] After gathering as much information as they could, the pair organized a similar cross-country venture of their own. In 1997 and 1998, they managed to create a highly effective media buzz about their adventure as they toured the United States in a 1986 Winnebago that had been repainted with bright, impressionistic sunflower designs. In tow behind the van was a crude biodiesel processor on wheels, dubbed "The Green Grease Machine," capable of making biodiesel from used cooking oils. Along the route of this 10,000-mile-plus odyssey, Joshua and Kaia stopped regularly at fast-food restaurants to refuel, and they took advantage of the publicity generated to promote biodiesel at every possible opportunity. (They also visited Dr. Reed in Colorado.)

After the trip ended, Joshua Tickell wrote his popular how-to-make-it-yourself book, *From the Fryer to the Fuel Tank: The Complete Guide to Using Vegetable Oil as an Alternative Fuel*. A second tour with the Veggie Van added more than 10,000 additional miles to the vehicle's odometer. Joshua Tickell continues to travel at home and abroad, lecturing extensively on the benefits of biodiesel. In 2006, he published a new book, *Biodiesel America*, about his favorite subject. He also has finished working on a feature documentary film about biodiesel, *Fields of Fuel*, which premiered to critical acclaim at the 2008 Sundance Film Festival in Utah.[24]

Biodiesel Online

Since the mid-1990s, small-scale biodiesel production has continued to expand across the country. But because homebrewers tend to be geographically scattered, communication and exchange of information was initially a problem and progress was slow. However, with the spread of the Internet and the proliferation of online resources—especially various biodiesel community forums—the physical barriers to the sharing of information have disappeared. Some of those early forums have evolved into highly popular online resources such as the Biodiesel Discussion Forum (biodiesel.infopop.cc) and BiodieselNow.com (www.biodieselnow.com). They typically contain hundreds (or even thousands) of postings on a wide range of regional, national, and international biodiesel topics. Today, online biodiesel discussion has become a major part of the homebrewing community. "It's so exciting now that it's possible to have this kind of information so readily available," Nicole Cousino says. "It makes such a difference. When we first started, there was no easy way to disseminate and share biodiesel information. Now we can get information on how to make the fuel, as well as questions and answers and concerns and warnings. And it's all shared for free; I think that's really important."[25]

10

The Main Players

The U.S. biodiesel industry has come a long way from its humble beginnings in kitchen blenders and garbage cans. As part of that process, the number of biodiesel plants in the United States grew from 1 in 1993 to 25 in 2004. Since then, the number has catapulted to approximately 170 as of January 2008, with more scheduled to open in the near future. Biodiesel production grew from a few hundred gallons in the 1980s to 500,000 gallons in 1999. Since then, production has multiplied exponentially, to approximately 450 million gallons (1.35 million metric tons) in 2007. In late 2007, however, the National Biodiesel Board estimated total U.S. biodiesel *capacity* to be a startling 1.85 *billion* gallons, reflecting the huge stampede in recent plant construction. Although this represents remarkable progress in the development of the U.S. industry, it also raises serious issues about overcapacity in a market that has not yet generated enough demand to absorb all of the biodiesel that *could* be produced. Nevertheless, since approximately 35 *billion* gallons of diesel fuel are consumed for on-road use annually in the United States, it's clear that the biodiesel industry still has a long way to go before it begins to have a significant impact on the nation's fuel sector.

THE PROCESS

As the U.S. biodiesel movement has matured, more sophisticated and efficient process technologies have been developed that have transformed

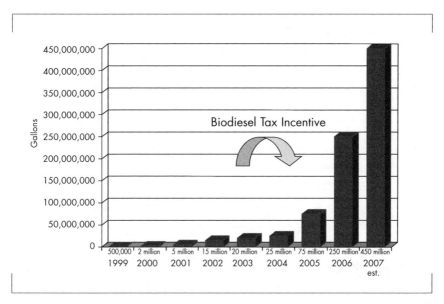

Figure 7. U.S. Biodiesel Demand National Biodiesel Board

production from a backyard science project into a full-fledged industrial-scale sector of the energy economy. In the early years, all biodiesel was made using the batch process. Increasingly, biodiesel is being produced by the continuous process, which was first developed specifically for biodiesel by Oelmuhle Leer Connemann in Germany in the early 1990s. Since that time, both processes have been fine-tuned somewhat by a number of technical developments, but the basic strategies remain the same. Although there are exceptions, biodiesel plants that produce more than 10 million gallons of biodiesel per year generally use a continuous-process design, while plants that produce less than 10 million gallons typically use the batch process.

"The batch process tends to lend itself better to smaller operations because it can be run as a single-shift operation very easily," explains Jon Van Gerpen, a professor and head of the Department of Biology and Agricultural Engineering at the University of Idaho, Moscow campus, who has been involved in biodiesel research since the early 1990s, "whereas with continuous flow you pretty much are committed to round-the-clock operation, and most small plants are not prepared to do that."[1]

Regardless of the process, there are a number of different strategies for treating the feedstock triglycerides, according to Van Gerpen. "Some people start the process with the triglycerides and make biodiesel directly from them, while others first convert the triglycerides to free fatty acids and then convert them to biodiesel, although there are not too many people in this country who are using that second strategy," he says. But Van Gerpen acknowledges that the second approach is useful when the feedstock is already high in free fatty acids, as may be the case for animal fat or used cooking oil.

Although there is a good deal of attention being focused on new advances in biodiesel process technology, Van Gerpen believes that it tends to be somewhat misplaced. "There's a lot of new technology being promoted, but the real issue for biodiesel is that the process doesn't cost that much compared to the total cost of production," he maintains. "Process is just not where the largest cost is, and coming up with a better process is not going to significantly affect the price of the product. My feeling, generally, is that process changes are not going to have much impact on commercialization."

So what will have an impact? Feedstocks and government policy, according to Van Gerpen. "The thing that has the most dramatic impact on the market is what the government decides to do in terms of subsidies," he says. "And secondarily, the feedstock issue—whether there are any new feedstocks on the horizon such as mustard seed or something else that has the potential to increase the amount of oil that is available." Under current market conditions, most biodiesel feedstocks—especially virgin oils—are simply too expensive for biodiesel to be price-competitive with petrodiesel, according to Van Gerpen. Although price is not everything, it certainly is the main issue that most consumers tend to look at first.

In 2004, the average-size commercial biodiesel refinery in the United States had an annual production capacity of about 2 million gallons. As of the end of 2007, that average capacity had grown to more than 11 million gallons, reflecting the trend to larger and larger plants. Although construction costs vary somewhat, a ballpark figure for a new biodiesel plant is around $1 to $1.50 per gallon of annual production capacity.

However, some extremely large plants have been built recently for less than $1 per gallon due to their economies of scale.

Traditionally, biodiesel production costs generally have run somewhere between $1.50 and $2.75 per gallon. The cost of feedstock now makes up approximately 85 percent of those costs in the largest and most efficient plants, since their other production costs per gallon have been reduced to approximately 20 to 35 cents per gallon. In the past, at about 20 cents a pound, soybean oil was the lowest-cost virgin oil in the United States. It takes about 7.35 pounds of oil to make 1 gallon of biodiesel, and using 20 cents a pound as the multiplier, this brings the feedstock cost to roughly $1.50 per gallon.

However, in response to growing global demand and tight supplies, soybean oil has risen to approximately 60 cents per pound recently, increasing the feedstock cost to more than $4.41 per gallon. Add to that the other costs of production, plus a profit margin, and the price at the pump for B100 easily exceeds $4.95 per gallon. This clearly demonstrates just how difficult it is for biodiesel to compete with petrodiesel given the recent high prices for vegetable oils. The only common feedstock that has been low enough in price to allow biodiesel to be relatively competitive is "yellow grease" (recycled cooking oils), at about 12 cents per pound. Unfortunately, the price of yellow grease nearly doubled in 2007, making even this feedstock less attractive than in prior years. Rendered animal fats are another traditionally low-cost feedstock, but they also have been subject to recent price increases. There is a reason for this trend.

More than a third of the existing biodiesel plants in the United States are now able to switch feedstocks from virgin oils to recycled oils or vice versa without having to make changes to the hardware system in their production facilities. This is a trend that undoubtedly will accelerate as the industry continues to look for more effective ways to level out the ups and downs in feedstock prices. Ironically, the growing number of multi-feedstock plants has, at least in part, been responsible for the large price increases in the traditionally cheaper feedstocks such as yellow grease and animal fats. As the price of soybean oil increased in recent years, many of these multifeedstock plants switched to the cheaper feedstocks, but since

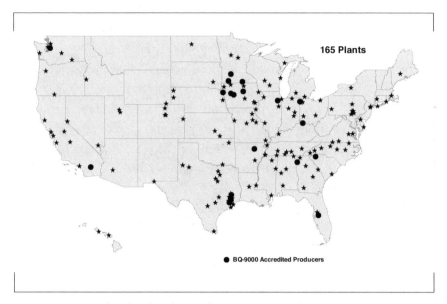

Figure 8. Commercial Biodiesel Production Plants National Biodiesel Board

yellow grease and animal fats are relatively limited in supply, it didn't take much additional demand to send their prices soaring. This demonstrates the limits of these conventional feedstocks and the need to develop alternatives, especially algae and inedible oils.

LARGE U.S. PRODUCERS

The proliferation of commercial biodiesel refineries across the nation in the past few years has been truly remarkable. Spurred by federal (and in some cases, state) tax incentives, new biodiesel plants have been opening on nearly a weekly basis. As of late 2007, almost every state in the nation had at least one plant, while some, like Texas and Iowa, boasted 23 and 13, respectively. Next in line are California, Georgia, and Tennessee, each with 8 plants; followed by North Carolina with 7; Missouri, Ohio, Pennsylvania, and Washington, each with 6; Indiana and Minnesota, each with 5; Alabama, Colorado, Mississippi, and Virginia, each with 4; Illinois, Kansas, Michigan, Nebraska, and Wisconsin, each with 3; and

Arkansas, Florida, Hawaii, Kentucky, New Jersey, Nevada, New York, Oklahoma, Oregon, South Carolina, and Utah, each with 2.

In terms of production capacity, however, the top three states are Iowa, Texas, and Illinois, and the top three refineries are Imperium Grays Harbor in Hoquiam, Washington (100 million gallons, multifeedstock); Green Earth Fuels of Houston, Galena Park, Texas (86 million gallons, multifeedstock); and ADM, Velva, North Dakota (85 million gallons, canola).

Some of the early U.S. production facilities have already been described in Chapter 8. The remaining list of more recent plants is far too long for individual profiles. Consequently, what follows is a sampling of a few companies that demonstrate varying approaches to ownership, production, feedstock, or marketing strategies.

Griffin Industries

During the initial flurry of biodiesel plant construction in 1996, a family-owned rendering company in the state of Kentucky became interested in the potential that biodiesel offered. Founded in 1942, Griffin Industries of Cold Spring, Kentucky (about eight miles south of Cincinnati, Ohio), is the nation's second-largest rendering company. Like other companies in the business, Griffin takes what most people don't want—inedible waste from food plants and used cooking oils—and turns these ingredients into animal feed, organic fertilizer, industrial oils and fats, and, in recent years, biodiesel.

After carefully studying the market for a number of years, Griffin Industries decided to take the plunge into biodiesel in 1998 with the construction of a 1.6-million-gallon-capacity plant. The Austrian firm BioDiesel International Anlagenbau GmbH (BDI) from Graz provided the multifeedstock batch-processing technology for the facility. "We've been involved in biodiesel longer than anyone else in the rendering industry," says Dennis Griffin, chairman of Griffin Industries and one of five sons of company founder John L. Griffin who are still active in the business. "We spent about two years in research going around the world looking for a process that would work with high-free-fatty-acid feedstocks, which are primarily produced in our industry. It took a while to find that technology, but we finally discovered it in Austria. They had been at it longer than we had, so we took advantage of that experience and hoped to eliminate some of the trial and error that they had started out with."[2]

The plant, which opened in February 1998, is a state-of-the-art facility that is viewed as a prototype for larger plants to come in the future. "Even though it's a small batch plant, it's highly automated," Griffin says. "That's what made it cost so much, but we wanted to put the full operating system to the test in a small plant for shaking out all of the issues that we thought might be there with a much larger facility." The plant has performed as designed, and the support from BDI has been good, according to Griffin. "As I recall, on day four after we started it up, we were making ASTM-quality [i.e., meeting the standard set by the American Society for Testing and Materials] product, which is almost unheard of on starting up a brand-new process," he says. The plant is so automated, in fact, that only two people need to be present while it is in operation. And the fact that even two people are there is mainly for safety reasons, rather than for workload reasons, according to Griffin. The company is an Accredited Producer, under the National Biodiesel Board's BQ-9000 voluntary program, to ensure that Griffin's biodiesel output is consistently produced in accordance with ASTM specifications.

Griffin produces its Bio G-3000 Premium Biodiesel brand at its Butler, Kentucky, plant for customers in about 15 states, including underground mining companies and trucking firms, and even for trials in the river transport and railroad sectors. But Griffin's biggest market is school buses. "Our main customer is the school bus industry," Griffin says. "They are becoming the primary users because of the diesel exhaust issue with the kids." And Griffin sees considerable room for additional growth in this segment of the market. Much of Griffin's biodiesel sales are of a B20 blend. "We have quite a few B20 customers, and that dramatically increases in the nonattainment areas during the summer months," he continues. "But we're trying to get away from that market to try to keep our production facilities running twelve months a year, not just four or five months during the summer for the B20 business." In order to accomplish this, Griffin is actively promoting a wider range of biodiesel blends for a variety of different purposes.

One of the biggest challenges for Griffin has been the company's ongoing effort to get government regulators as well as other feedstock organizations to understand that "feedstock neutrality" is vital for the future success of the biodiesel industry. Put simply, this means that regulations (or regulators) should not erect unfair barriers against, or provide unequal incentives for, one particular biodiesel feedstock over another. Animal-fat-based biodiesel has not received a fair share of support for many years, according to Griffin. "It's been a constant battle," Griffin acknowledges. "For a long time we were the only ones who had an interest in this or had a plan or any involvement, so it was kind of a solo battle for the first three or four years." Today, more and more people in the industry are coming to understand the importance of this issue and are actively working to level the playing field. (For more on this issue, see Chapter 11.)

Despite these problems, Griffin Industries has plans to expand its biodiesel activities significantly in the coming years.

World Energy Alternatives LLC
World Energy Alternatives LLC holds a unique position in the U.S. biodiesel industry as both an early pioneer and a leading supplier. World Energy's story begins in Massachusetts in 1994, when businessman James Ricci and a group of other investors purchased an idle oleochemical plant located in Quincy from the Procter & Gamble Company, renaming their venture Twin Rivers Technology. Most biodiesel literature credits Twin Rivers with being one of the first nondedicated commercial producers of biodiesel in the United States, but this is simply not true. From the start, the new owners *intended* to gradually shift the plant to the production of biodiesel, but after about two years the company had made very little progress with its biodiesel initiative—and absolutely no biodiesel of its own. "In fact, they had actually gone backward," recalls Gene Gebolys, who was hired to manage the biodiesel operation in 1996. "They had become a pretty good fatty-acid manufacturer, which is what P&G had

done there previously, but they were not having much luck in the biodiesel arena. We bought biodiesel from Italy, Germany, and a number of U.S. oleochemical companies. The thinking was that, regardless of who made it, the focus had to be on selling the product."[3] In its earliest days, however, Twin Rivers had mistakenly positioned their product as an alternative to other alternate fuels—especially compressed natural gas (CNG)—rather than as an alternate to petroleum diesel. As a result, environmental groups in both Boston and New York, who supported the use of CNG in transit fleets, were campaigning actively against Twin Rivers' introduction of biodiesel.

Gebolys worked hard to rescue the troubled biodiesel division and, in late 1997, succeeded in lining up a number of promising contracts with the Boston Transit Authority and the New Jersey Transit Authority. But the environmental activists remained adamant in their opposition, and by early 1998 Gebolys came to the conclusion that something drastic had to be done. "The biggest obstacle was that nobody had any real idea about what it was that we were selling, or any understanding of the value of what we were offering," he says. "And without the help of our natural allies in the environmental movement, we were handicapped right at the starting gate." Gebolys reluctantly recommended selling or closing the faltering biodiesel initiative in order to save the larger and more successful oleochemical operation. He then tried to find a buyer for the biodiesel division. When that plan failed to attract any serious interest, Gebolys decided to set up a new company of his own that was focused on energy rather than oleochemicals. "Despite all the problems, I thought it had promise, but not at Twin Rivers," he says. Gebolys spent some time looking for financial backing for his plan and eventually attracted the support of Gulf Oil. With several biodiesel contracts spun off from Twin Rivers in hand, Gebolys launched World Energy in 1998.

At first World Energy bought all of its biodiesel from Procter & Gamble in Cincinnati, Ohio, and Gebolys spent the next few years building a customer base and a market. In 2000, World Energy entered into an agreement with OceanAir Environmental of California to buy the entire output of the former NOPEC plant in Lakeland, Florida (which

OceanAir was about to purchase from NOPEC). Subsequently, World Energy contracted with AGP in Sergeant's Bluff, Iowa, and several other producers for a portion of their annual biodiesel production. Then, in the summer of 2002, OceanAir broke their contract with World Energy, according to Gebolys. Legal action soon followed, and the case was eventually resolved with a settlement in favor of World Energy, which took over operation of the 18-million-gallon-capacity, batch-process Lakeland plant in the fall of 2003. "It does seem to be the Bermuda Triangle of the biodiesel industry," Gebolys says of the troubled former OceanAir (formerly NOPEC) plant. "But early in 2004 we made a significant investment in capital improvements to make the Lakeland plant a top-notch, modern facility," he says.[4]

After establishing itself as a leading U.S. biodiesel company, World Energy turned its attention to a number of new markets. World Energy was the first U.S. biodiesel company to become involved in significant trade to Europe and went on to be active in South America and Southeast Asia. The company was honored in 2004 as "Outstanding National Stakeholder of the Year" by the U.S. Department of Energy, and received the Platts Energy Award as "Global Downstream Business of the Year" in 2005. And in 2006, World Energy earned BQ-9000 accreditation. With U.S. offices in Ohio, Georgia, Florida, and Massachusetts, and international operations in Argentina, Belgium, the Netherlands, and Malaysia, World Energy is now a leading global marketer, distributor, and producer of biodiesel.

The fact that the company is involved in so many different aspects of the biodiesel industry in so many regions of the globe gives World Energy a number of advantages in the marketplace, according to Gebolys. "We are the only nonregional player in the United States," he says. "We are able to pull our product from many plants, so we have stable supply relative to the rest of the industry, which tends to go up and down. Also, we cost-average between these multiple plants, so we tend to have very stable pricing relative to other companies, who tend to be subject to swings in feedstock costs and so on. We're just more stable in supply and price and also more reliable. If someone has a problem with their plant we just get our product from a different plant. By almost any measure, our participation has been a stabilizing force in the growth of the industry. Internationally, we operate across geographic and political borders to reduce the risks inherent in sudden economic and regulatory changes."

But there is another important reason for the company's general stability: feedstock flexibility. World Energy has the ability to supply biodiesel from both virgin and recycled feedstocks, which allows it to average out some of the worst price fluctuations in the commodity markets—especially the recent price spike in the cost of soybean oil. "The fact that prices have been all over the place recently has been somewhat problematic, but we're in the energy business, and price swings are not unusual," Gebolys notes. "In fact, there has been better price stability in biodiesel than in petroleum. So, in terms of feedstocks, we don't have a particular ax to grind one way or the other; we just want what's best for our customers. We don't say that virgin and recycled are the same, because they're not, but we also don't say that one is inherently bad while the other is inherently good, either."[5]

Imperial Western Products

Headquartered in Coachella, California, Imperial Western Products has been in business since 1966 as a diversified feed commodities business. These commodities include feeds, oils, soaps, methyl esters, and glycerin. Imperial Western has been developing its batch-process biodiesel production facility in Coachella since 1999. "We designed and built the system ourselves from the ground up," says Bob Clark, sales manager for Imperial Western's biodiesel division. "We got into biodiesel because we were already large processors of used restaurant cooking oil."[6]

The facility, with an annual capacity of 12 million gallons, went into commercial production in 2002. Other than the self-designed aspect, the biodiesel plant has fairly standard elements found in most batch facilities: a receiving area where the feedstock is unloaded; reactors where all of the ingredients of the recipe come together and interact; a separation tank, where the biodiesel and glycerin settle out and go their separate ways; a wash tank, where residual impurities are removed from the biodiesel; a centrifuge, where the water is removed; and, finally, a flash evaporator, which eliminates the alcohol. The glycerin by-product of the reaction is processed and then consumed either locally or sold for export. When the plant (which makes up about one-third of the total oleochemical facility at Coachella) is running at full capacity, it is staffed by about 10 employees, according to Clark.

Imperial Western makes two different types of biodiesel under the Biotane brand name: Imperial Biotane, from recycled restaurant oil, and Supreme Biotane, from virgin vegetable oil. In early 2004, however, due mainly to the high cost of virgin soybean oil, the company was producing biodiesel only from used oil. "At the moment, our feedstock is mainly recycled cooking oil," Clark says, "but we can use any other virgin seed oils with no problem. That's the beauty of the batch system; you can switch very easily from one feedstock to another." Imperial Western's marketing area is "everything west of the Rockies," and the company's primary customers are governmental fleets, both state and federal, including the military. The company is an Accredited Producer, under the National Biodiesel Board's BQ-9000 voluntary program.

Although there are no immediate plans for further expansion of the biodiesel division, Clark says it might be a possibility at some point in the future "if the state of California becomes more supportive of biodiesel." The state's largest concern, according to Clark, is nitrous oxide (NO_x) emissions, which have been associated with the combustion of biodiesel in diesel engines. "But with recycled cooking oil as a feedstock, you really don't have any significant NO_x increase, particularly at a B20 blend level," he maintains. Another problem with certain state agencies and some members of the general public, according to Clark, is a lingering aversion to anything that even mentions diesel. "We did ourselves a great disservice by ever calling biodiesel 'bio*diesel*'—we should have called it something else," he says. "The word *diesel* just has an unfairly bad association in the minds of some people. The perception with the general public is negative, and it shouldn't be." Another problem with some state officials, according to Clark, is that they oppose anything that prolongs the use of fossil fuels. "And biodiesel, in its most widely used form, depends on a mutual support structure with the petroleum industry," he says.[7]

Blue Sun Biodiesel

Blue Sun Biodiesel of Denver, Colorado, is an unusual, vertically integrated agriculture energy company that was founded in 2001. Blue Sun's objective is simply to provide the best diesel fuel on the market. In order to accomplish this goal, the company has embarked on an interesting series of market and technology-based initiatives that begin with the customer and then work backward to the feedstock—the exact opposite strategy as that employed by the larger sector of the U.S. biodiesel industry.

"What we really need to do is focus on the market, and that is what Blue Sun has done," says Jeff Probst, the company's president and CEO. "What we're doing is generating demand in the marketplace based on our customers' needs and then pulling the product through the distribution network from the feedstock suppliers—a 'market pull' approach. That, basically, tells us how to build a viable business and a viable industry. And if you don't start in the marketplace, you're going to build something that isn't going to fit—you end up with the proverbial square peg in a round hole."[8] Probst, who previously worked for many years in the new product and technology division at Duracell, says that he and his colleagues always started with manufacturers to find out what the companies needed for their devices and then worked backward to develop their batteries. "That was a winning strategy at Duracell, it's going to be a winner here at Blue Sun, and it's going to be a winner for biodiesel too," he declares.

As it turned out, what the market was looking for was quality and performance. Over a period of several years, Blue Sun conducted market research into biodiesel quality from numerous U.S. producers and came to the striking conclusion that about 60 percent of the fuel tested did not meet even minimum ASTM standards. In response, in November 2006, Blue Sun unveiled its new, improved biodiesel fuel, Blue Sun Fusion™ B20, the result of four years of market research and product development. Blue Sun compiled hundreds of independent laboratory tests and more than 400 million miles of performance data to develop their blended fuel. To create Fusion B20, Blue Sun started with their B100, mixed in petrodiesel fuel, and enhanced it with a proprietary additive. The company maintains that Fusion B20 tests well above the ASTM standard to guarantee superior performance.[9]

To maintain these high standards, Blue Sun is involved with every aspect of their product, from the oilseed to the company's authorized distributors. Proprietary technologies for oilseed extraction, biodiesel production, and filtration ensure that the company's customers will receive quality fuel. All distributors are required to follow strict protocols to maintain fuel quality during the critical final steps in the distribution chain. These strong commitments to quality brought about an endorsement from Isuzu, the worldwide diesel engine manufacturer, which singled out Blue Sun's Fusion B20 as "the best diesel fuel in the world" in 2007.

Authorized distributors have access to Blue Sun blending terminals, which incorporate proprietary blending and filtration with high-speed automated loading. Essentially, the company has set up a franchise system of blenders, dealers, and retailers who move the fuel from the producer to the customer. Many fleets now use Blue Sun fuel, including Safeway, Waste Connections, and Corporate Express. Even at high altitudes and cold climates, the company provides fuel for many mountain customers, such as the Aspen Ski Company. In addition, a number of government agencies use Blue Sun fuel, including the City of Colorado Springs, the U.S. Department of Defense, and the City and County of Denver.

Blue Sun is committed to supporting American farmers, preferring to use oilseed grown in the United States. Currently working with regional farmers on the high plains of Colorado, Kansas, Nebraska, New Mexico, and Oklahoma, Blue Sun is reducing the cost of biodiesel fuel through the development and production of low-cost oilseed crops for dryland agriculture. Development efforts are focused on high-oil-content crops, such as camelina and canola. To further this research, Blue Sun has received grants from the U.S. Department of Energy, U.S. Department of Agriculture, and the State of Colorado's Advancing Colorado's Renewable Energy (ACRE) initiative.

One of the main strategies Blue Sun is following is to encourage the production of industrial oilseed crops that will not impact the food supply. The idea is to grow biodiesel crops on unused cropland during the winter-wheat crop rotation, providing farmers with additional income while not taking cropland out of food production. "Without taking away any winter-wheat production, we can develop a $1 billion industry easily right here in Colorado, Nebraska, and part of western Kansas," Probst predicts. "And it could go up to $2 to $3 billion, and there would be positive impacts across the board from planting these new crops. We hope to set this up so 75 percent of that money would be going into our rural communities that are struggling right now, instead of sending it overseas to the Middle East."

Probst admits that the Colorado market has been a challenge, but he is enthusiastic about his company's success so far. "We're getting a very good response from our substantial marketing and educational efforts," he reports. "I think people are smart and want to do what's right, as well as what's economical, and if you provide them with the opportunity they will participate."[10]

In January 2007, Blue Sun announced a new joint venture, ARES Blue Sun Development, LLC, with the ARES Corporation, Burlingame, California, to collaborate on biodiesel process technology and to build biodiesel production facilities nationwide. The first plant, located in Clovis, New Mexico, is scheduled to be completed in late 2008. Blue Sun also expects to complete a merger in the first quarter of 2008.

Renewable Energy Group

Renewable Energy Group, Inc. (REG) was founded in 2002 with the formation of a partnership between West Central Soy of Ralston, Iowa, and engineering and construction company Todd & Sargent, Inc. of Ames, Iowa, along with an alliance with process equipment builder Crown Iron Works of Minneapolis, Minnesota. West Central's previous success in biodiesel production at its Ralston plant, along with Todd & Sargent's many years experience in the development of commercial-scale biodiesel facilities, provided the strong foundation for this relatively unique venture that has capitalized on replicating a successful model again and again. The two companies had worked together prior to the formation of REG; Todd & Sargent had, in fact, built West Central's Ralston plant.

REG used West Central's proven design for an energy-efficient, continuous-flow process that recovers excess methanol and produces no process wastewater, and combined it with an integrated package of services that offer turnkey plant construction, operating-staff training, and startup assistance. With this carefully integrated approach, REG can have a 30- or 60-million-gallon biodiesel refinery ready to begin operations in as little as nine months after construction and other permits are approved. Once the plant is operating, REG also can provide a core group of experienced operators until the new staff is fully trained and ready to take over the plant. REG also provides testing laboratory setup and equipment to ensure that the biodiesel produced at its plants meets national quality standards. In addition, West Central can provide the new facility with feedstock and marketing assistance as well as plant managers, if needed.[11]

Since REG began operations in 2003, it has been involved in numerous biodiesel projects. The first REG-built and -managed project was the SoyMor refinery in Albert Lea, Minnesota. This 30-million-gallon facility opened in 2005, just in time to fill Minnesota's new 2 percent biodiesel mandate. (This mandate created a 16-million-gallon market for biodiesel in the state almost overnight.) SoyMor subsequently achieved BQ-9000 accreditation from the National Biodiesel Board.

REG's next build/manage project was Western Iowa Energy, LLC in Wall Lake, Iowa, which opened its 30-million-gallon plant in 2006. In the same year, REG announced ambitious plans to expand its biodiesel production, both owned and managed, to 640 million gallons per year at a total of 15 plants by 2009.

This announcement was followed quickly by three more 30-million-gallon Iowa facilities in 2007, including Central Iowa Energy, LLC in Newton; Iowa Renewable Energy, LLC in Washington; Western Dubuque Biodiesel, LLC in Farley; plus a 60-million-gallon plant owned by East Fork Biodiesel, LLC in Algona. REG also has two 60-million-gallon plants of its own under construction in New Orleans, Louisiana, and Emporia, Kansas, that are scheduled to begin operations in 2008, as well as an additional 60-million-gallon facility in Cairo, Illinois, that is in the preconstruction phase. There are three other 30-million-gallon build/manage plants in preconstruction in Rock Port, Missouri, Freeport, Illinois, and Rock Rapids, Iowa, and REG appears to be well on its way to achieving its ambitious goal.[12]

REG now accounts for about 27 percent of total U.S. biodiesel production, and in September 2007, the company was listed on *Inc.* magazine's Top 500 Companies, the first biodiesel company to make the list of fastest-growing private companies in America. REG came in at number 200 overall with annual revenue of $178 million.

Imperium Renewables

Imperium Renewables, Inc. was founded as Seattle Biodiesel, LLC in 2004 by John Plaza, a former commercial airline pilot. It was while he was on an overnight flight across the Pacific that Plaza realized that he could drive his car for 43 years on the same amount of fuel his plane was consuming on that one trip. It was a life-altering experience. He eventually quit his job, mortgaged his house, sold his motorcycle and boat, and embarked on a search for a better way to make biodiesel.[13] After a lot of research and development work, Plaza was convinced that he had come up with a better way, and he founded Seattle Biodiesel to commercialize his new design. "We created our own technology because I had limited investment resources when I started the business," Plaza says. "We looked at existing technologies in the dairy, petroleum, and other industries which deal with the movement and mixing of fluids and tried to combine those strategies into a cost-effective, high-quality production system."[14]

Seattle, Washington, had (and still has) the highest per-capita consumption of biodiesel in the nation, and Plaza believed that this represented a significant opportunity for his new biodiesel venture. Plaza collaborated with Saybr Contractors, the leading petroleum facility contractor in the Northwest, to construct Seattle Biodiesel. In early 2005, the 5-million-gallon facility opened for business, the first commercial-scale company to produce ASTM-quality biodiesel in the Pacific Northwest. The company eventually hopes to produce biodiesel that will be less expensive than petroleum diesel, and is working hard to achieve that goal. Seattle Biodiesel has been involved in several pilot projects

Figure 9. Imperium Grays Harbor Imperium Renewables

with Northwest farmers to develop oilseed crushing and refining capacity in Eastern Washington. Imperium Renewables, Inc. now operates the Seattle refinery as a wholly owned subsidiary, and the plant has a new focus on testing next-generation feedstocks.[15]

Along the way, Plaza assembled a team of financial, production, and marketing experts who quickly realized that the same traits that had made Seattle Biodiesel an immediate success—quality, competitive price, and product availability—were what customers everywhere were looking for. Anticipating a new B2 requirement in the state of Washington, Imperium embarked on a bold new venture. In October 2006, after attracting a remarkable amount of investment capital, work began on Imperium's new 100-million-gallon refinery, the largest in the nation, at the Port of Grays Harbor, in southwest Washington. The huge new $78 million facility that was completed in August 2007 also has the capacity to store up to 17 million gallons of biodiesel and feedstock at any one time in a group of large new tanks. In addition, the facility is able to produce biodiesel from numerous different feedstocks—even simultaneously.

"We wanted to create an independent biodiesel production facility that would give us the most flexibility and options," Plaza says. "Our technology has improved over the original Seattle Biodiesel system and now provides a full conversion of the vegetable oil, full recovery of any excess methanol, and the final product really shines above other biodiesel in this country produced with traditional biodiesel technology. It's an impressive facility, and I think it shows that this industry can really be a meaningful contributor to our national energy policy of reducing our dependence on imported oil." Imperium selected the Port of Grays Harbor because of its strategic connection to rail, highway, and water transportation.[16] The company hopes to expand the Grays Harbor plant by an additional 300 million gallons in the next few years.

Imperium's vision now extends beyond Washington State, and includes tentative plans for additional 100-million-gallon plants in Hawaii, Pennsylvania, and Argentina. Plaza admits that when he first started Seattle Biodiesel he had no idea how successful the venture would be, or how quickly it would evolve and grow into Imperium Renewables. "I have to be honest and say that I didn't have a clue," he says. "The timing was just right and there was unquestionably an element of pure luck involved." But Plaza is also quick to praise his coworkers for the part they have played in the company's spectacular rise. "This has been accomplished by a really tremendous group of hard-working people who are very focused on making Imperium the success that it has become; I am forever indebted to them."[17]

SMALL PRODUCERS

Although the market is increasingly dominated by larger and larger producers, a discussion of the biodiesel industry would not be complete without at least mentioning the role played by the small producer. Making biofuels from local resources for local consumption makes a lot of sense from an environmental perspective, although making money at it can be a real challenge. These days, a small producer probably would fall somewhere in the range of between 100,000 and 500,000 gallons per year. Consequently, it's fairly safe to say that with a production capacity of between 15,000 and 30,000 gallons, John Hurley's Dog River Alternative Fuels Company in Berlin, Vermont, is one of the very smallest commercial biodiesel producers in the nation. Hurley, a logger and sawmill operator who manages 1,200 acres of Green Certified Forest land in Berlin, was concerned about global climate change and the impact his own diesel-powered equipment was having on water and air quality. After reading Joshua Tickell's book, *From the Fryer to the Fuel Tank*, in 1998, Hurley began to think about ways he could integrate biodiesel into his forestry and lumber operation on Chase Mountain.

In October 2000, the Vermont Sustainable Jobs Fund sponsored a conference, "Building an Ecological Economy." As part of that initiative, a challenge grant of $1,500 seed money was offered to someone who would put together a proposal to use the waste from any business to create a useful product. Hurley immediately thought about the possibility of making biodiesel from used cooking oil. "I thought, 'What the heck?' So I sent in an application," Hurley recalls. "Frankly, I didn't think I would get the grant." But he did. And that was the catalyst for Dog River Alternative Fuels. After discussing his plan with a number of other interested conference participants, Hurley and the group formed Dog River Alternative Fuels Company LLC the following April. "We became the first commercial biodiesel production company in the state of Vermont," Hurley says. The company's admittedly tiny pilot plant, housed in a small wooden shed behind Hurley's lumber-drying kilns, uses a standard batch process and on a good day can turn out perhaps 160 gallons, according to Hurley. "But we generally didn't make more than 120 gallons," he admits.[18]

Despite the modest output, Dog River managed to land an 18-month contract to supply biodiesel for a shuttle bus trial at the University of Vermont in Burlington. Dog River even has received a Governor's Award for Environmental Excellence in Pollution Prevention for its work with biodiesel. But for a small producer like Dog River, joining the National Biodiesel Board or paying for ASTM testing is extremely difficult. This is the same dilemma faced by many other small producers across the nation. Recently, Dog River has been selling small quantities of biodiesel only for off-road (mostly farming and logging) vehicles and home heating. Still,

Hurley is generally optimistic about the future of the biodiesel industry in Vermont and hopes eventually to tap Vermont oil feedstock crops grown on fallow land that is not used for food production. "It may require some creative efforts to do that," he acknowledges. But he's excited about trying.[19]

COOPERATIVES

There is a final variation on the small producer theme: the biodiesel co-op. By all accounts, there are quite a few co-ops of various types in existence (or springing up) around the country, but not too many of them actually have produced much biodiesel. However, they do seem to spend a lot of time and energy discussing, debating, and planning to make biodiesel. And because there are many different ways of achieving that end, there are many, many variations on the co-op theme. Nevertheless, biodiesel co-ops generally seem to fall into one of three main categories:

- very small-scale producer
- bulk buyer
- large volunteer producer group

The first category, very small-scale co-op producer, usually involves perhaps two experienced individuals who take care of all fuelmaking chores for themselves and a small group of other subscribers. This model, which offers relative safety, quality control, and ease of management, has been fairly successful—as long as it stays small.

The second category, the bulk-buy co-op, eliminates production chores completely (as well as possible production accidents) and normally offers less-expensive fuel to subscribers. Some bulk-buy co-ops use a large central storage tank where loads are delivered and where subscribers pick up their fuel. Other co-ops use distributed smaller storage tanks or barrels in multiple locations. Regardless of the storage strategy, a bulk-buy co-op still requires quite a lot of coordination and oversight to work smoothly.

Finally, there is the kind of co-op where everybody gets trained to make fuel on a volunteer basis. Of all the models, this one potentially presents the greatest challenges, according to people who have been involved with them. This is especially true for groups of more than five fuelmakers. The logistics involved with trying to train, maintain, and coordinate a large group of people to make consistently high-quality biodiesel safely are daunting. The Berkeley Biodiesel Collective in California, a producers' co-op founded in September 2002, tried and tried to make this model work, but ultimately failed. After a thorough reorganization of its efforts, the collective did eventually find a successful model that involved a major training initiative and a decentralized approach involving very small groups combined with a bulk-buyer's club for the purchase of commercial biodiesel.[20]

Piedmont Biofuels

There is, however, one successful biodiesel cooperative that stands out from the rest. Piedmont Biofuels is a member-owned cooperative located in Pittsboro, North Carolina. It currently has about 226 members. The group has been leading the grassroots sustainability movement in North Carolina by using and encouraging the use of clean, renewable biofuels. Members are entitled to buy biodiesel from the co-op or to learn how to make their own using co-op equipment. The co-op has five retail outlets for biodiesel and is working actively at establishing more. Piedmont Biofuels also has a B100 fuel terminal (the first of its kind in the state) as well as a 1,600-gallon fuel delivery truck, which it uses to deliver biodiesel in the region. Most of the biodiesel the co-op sells is for "on-road" uses, meaning that the price of the fuel includes all state and federal taxes. A smaller amount of biodiesel is sold for "on-farm" vehicles and equipment, home heating, marine applications, generators, and other "off-road" uses, which are not subject to road taxes. When sold for "off-road" purposes, the fuel is dyed to indicate that road taxes have not been paid. Regardless of its end use, though, all of the biodiesel sold by the co-op meets the ASTM (D-6751) quality specification.

The co-op also sells biodiesel reactors (the equipment in which the biodiesel is made). Local, microscale biodiesel production and consumption

Figure 10. Piedmont Biofuels Co-op Members. Piedmont Biofuels Cooperative

makes a lot of sense in the co-op's view, and it is committed to spreading the knowledge and technology necessary to produce quality biodiesel in small quantities (under 250,000 gallons per year). The members of the co-op design and build small biodiesel reactors, consult, and train their customers on how to use them properly. The first reactor the co-op sold went to North Carolina State University, as part of the Solar Center's Alternative Fuel Garage. Since then, the co-op has designed and built a number of units for its own use, along with a mobile reactor for the U.S. Department of Energy through the State Energy Office and Central Carolina Community College. The co-op also delivered one reactor, which was mounted on a trailer, to the North Carolina Zoo.[21]

The co-op has a strong educational program. It created and runs the Biofuels Program at Central Carolina Community College. Co-op members have taught, demonstrated, and lectured around the country—from Solar Energy International in Colorado to NC State to the University of Florida. The co-op holds periodic workshops and conferences on everything from reactor design-build to straight vegetable-oil conversions to co-op formation. "This is by far the biggest thing that we do," says Lyle Estill, one of the co-op's co-founders. "More time, money, and effort is spent on education than anything else, but it's extremely important. The number of gallons of fuel that we make is really insignificant compared to the number of people that go away thinking about energy in a different way."[22]

And, of course, the co-op has been making biodiesel at its backyard facility in Moncure. People who want to learn how to make biodiesel out of waste vegetable oil can join the co-op and then join in with an experienced homebrewing crew. Homebrewing at the co-op is a voluntary activity that rises and falls with the enthusiasm of the membership. It's legal to make your own fuel and use it to drive on public highways. In order to stay legal, however, the co-op pays road taxes on all homemade fuel consumed on behalf of its members. Dealing with regulations has been one of the largest challenges the co-op has faced. "We're out there speaking, blogging, writing books, shooting our mouths off, and as a result, we're just way too public to try to do anything under the table," Estill says. "We've been visited by the IRS three times, North Carolina Revenue twice. When we plug into the regulatory framework, what happens is the town makes a lateral pass to the county, which then passes to the state, the state passes to the feds, and we end up in this regulatory no-man's-land. Consequently, the regulator sitting on the other side of the table doesn't know what to do. It drives us crazy."

Despite the many hurdles, Piedmont Biofuels has just made the transition from being a small, backyard cooperative biodiesel producer to being a small cooperative industrial producer with its new Piedmont Biofuels Industrial LLC venture. "After years of successfully resisting the urge to go into commercial biodiesel production, we have finally succumbed," Estill admits. Located in an abandoned chemical plant at the edge of Pittsboro, the multifeedstock batch-process facility is designed to produce about a million gallons of biodiesel annually. "We are attempting to collect our feedstocks from a 100-mile radius and we're going to distribute our fuel within that area," Estill explains. "In our view, it's infinitely preferable to have 100 separate 1-million-gallon plants scattered around the country in small towns rather than a single 100-million-gallon plant.

We're not interested in being the next big fuel monopoly; we're just trying to help our town to meet its own needs." This model, with local feedstocks and local customers, makes a lot of sense for small towns everywhere, and could become a part of the increasingly popular *relocalization* movement that is beginning to spread across the nation in response to peak oil and global warming concerns.

In early 2008, however, Piedmont's Industrial venture ran into the same feedstock cost problems that have hit most other commercial-scale biodiesel producers and has been struggling to remain profitable. Nevertheless, by virtually anybody's standards, Piedmont Biofuels has been a remarkable success. Trying to replicate that success, however, is not as easy as one might assume. The national landscape is littered with "the remnants of biodiesel coops," according to Estill. "The main thing that we have done that many others have failed to do is survive," he says. "No one got mad and took their reactor and went home. No one fled when the IRS came to town. Leif [Forer] and Rachel [Burton] and I have stuck together, and that's one of our greatest strengths. We're the garage band, and we haven't broken up yet. I joke that what you need for success with a biodiesel co-op is easy. Start with a magnetic, female diesel mechanic . . . and then I always qualify it by saying if you want to be like Piedmont Biofuels, be careful what you wish for. I think we're probably going to have a board meeting every single Sunday this month; it's a lot of hard work."[23]

For a list of co-ops in the United States, see "Organizations and Online Resources" (pages 324–325).

A MATURING INDUSTRY

At the other end of the spectrum, along with the proliferation of new large-scale biodiesel production facilities, one of the most significant recent developments is that the big agricultural and petroleum oil companies in the United States, which generally ignored the biodiesel movement for years, are now becoming actively engaged. In November 2003, Jerrel Branson, former president of Best BioFuels (which was constructing a new biodiesel plant in Texas at the time), attended a board meeting of the National Biodiesel Board. "I looked around the room, and guess who showed up for the first time? ADM, the 800-pound gorilla of the agricultural oil business," he reports. But Archer Daniels Midland, an agribusiness giant and leading soybean processor, wasn't the only heavyweight in attendance. "Bunge Oil was there, Cargill, Tyson, Shell, and Ashland were all there too. The big boys are finally showing up, and it will become a big-boy business. This will no longer be a mix-it-in-a-garbage-can-to-fill-the-VW-microbus movement. Nor will it even be the folks designing a million-gallon-a-year plant. It's going to be the folks constructing large refineries and building real volume. Frankly, that's great for the industry, and that's the most significant change: the recognition that the industry is here, it's real, and we better be ready to do business."[24]

The National Biodiesel Board's CEO, Joe Jobe, agrees. "I feel good about this development, and I want more of it," Jobe says. "We are working in every way to educate the petroleum industry at all levels because the only way biodiesel will be truly successful is to integrate into the existing national liquid petroleum infrastructure. Without that key piece of the puzzle, biodiesel is just theoretical. It's pretty easy to see that the same infrastructure and vehicles that deliver diesel fuel can deliver biodiesel. But more important, biodiesel blends with petroleum diesel, and a blend strategy makes all the sense in the world. So I think this development is extremely important."[25]

Not everyone shares this enthusiasm for the large corporate model, but there is no doubt about it: The big players are now involved. Shell Oil Company is already an associate member of the NBB, as are some very large regional petroleum marketers. Oil companies such as Gulf and BP have started marketing biodiesel. Chevron has had a minority investment in a biodiesel refinery in Galveston, Texas. And Archer Daniels Midland—already the largest biodiesel producer in Europe—recently opened a new 85-million-gallon biodiesel refinery in Velva, North Dakota, and also signed an agreement with Volkswagen to collaborate on the development of next-generation biodiesel fuels.

11

Biodiesel Politics

Although most Americans have been unaware of the biodiesel industry until fairly recently, those who have been involved for years understand that many different individuals and groups have played a part in the growth of the industry. Although in Europe the biodiesel movement developed essentially from the top down as a result of national government policy, biodiesel in the United States developed mostly from the bottom up, resulting in some uniquely "American" features in the U.S. biodiesel industry—and its politics. As mentioned previously, American soybean farmers have been the strongest and most persistent advocates of biodiesel for many years. Although there undoubtedly have been times when these farmers must have wondered if their efforts (and $50 million in financial support) would ever succeed, their persistence, accompanied by a lot of political clout in Washington, D.C., has finally been paying off. However, the initial dominance of soybean interests in the industry also has led to some rather bizarre political and practical dynamics that the industry continues to struggle with today, not the least of which is the simple fact that soybeans are not very high on the list of preferred oil-producing crops (see Chapter 3).

But if you look a little closer at the industry, the internal and external political picture becomes even more complex. As the industry has expanded rapidly in recent years, the trend toward larger and larger production facilities (and the increasing participation of larger corporate players) has inevitably caused some of biodiesel's most ardent supporters in the "homebrew" and small-scale producer sectors (self-described as the "B100 Community") to feel increasingly threatened, marginalized, or even shut out. This is unfortunate, but not surprising. As a result, in the United States today there are two distinct segments of the biodiesel industry developing simultaneously. One follows the large agribusiness model, while the other pursues a small-scale, local model. Both groups produce biodiesel, and they are not necessarily mutually exclusive. Each has a productive and potentially complementary role to play as the industry evolves, but the tensions between them present the industry with one of its more difficult, ongoing challenges.

Despite these internal difficulties, biodiesel has begun to attract a good deal of external political support in recent years from both sides of the aisle in the U.S. Congress. "That's one of our strongest selling points," says Neil Caskey, formerly with the American Soybean Association. "If you look at some of the policy successes that we've had on Capitol Hill, a lot of them are because biodiesel appeals to both political parties; it's not something that only Republicans or only Democrats are going to support. In light of the current energy situation we have, with increasing petroleum prices, if you have something that can address that issue, help farmers, and help the country all at the same time, it's an easy sell."[1] Increasingly bipartisan support makes the political case for the biodiesel industry easier to sell at the state level as well. It's hard for lawmakers, regardless of their political affiliations, to argue against helping farmers, creating new jobs, developing local economies, and strengthening energy security, to say nothing of numerous environmental advantages—all in one package.

MAJOR BIODIESEL MILESTONES

Although a series of important legislative and regulatory initiatives have helped propel biodiesel to its current status as the fastest-growing renewable fuel in the United States, a number of initiatives stand out as being major milestones: the ASTM standard, the Energy Policy Act,

Environmental Protection Agency (EPA) approval, several Presidential Executive Orders, and more recent energy and farm legislation.

The ASTM Standard

When the biodiesel industry was still in its early developmental stage, the quality of the fuel was, at best, uneven. Trying to convince engine manufacturers to approve biodiesel use in their engines without some sort of common national standard was virtually impossible. In order to address this issue, in June 1994 a task force was formed in a subcommittee of the American Society for Testing and Materials (ASTM) to develop an ASTM standard for biodiesel. In June 1999, the preliminary standard was approved, and in December 2001 it was published in its final version, D-6751, for pure biodiesel (B100) in blends with petrodiesel up to 20 percent by volume (B20). Higher levels of biodiesel are allowed, but officially only after discussion with individual engine manufacturers. (There are, however, plenty of individuals and an increasing number of fleets that use B100 all the time in their vehicles.)

The publication of the ASTM standard was a major milestone in the development of the biodiesel industry, according to Charles Hatcher, the former regulatory director for the National Biodiesel Board in Jefferson City, Missouri. "The importance for fleet managers was enormous; it assured them that the fuel had to meet a specification and that it was going to measure up to certain qualitative and quantitative measures," he says. The ASTM standard also helped to move biodiesel and the industry that produced it to a new level of legitimacy in the eyes of many people. In some respects, the new standard was "like a coming-out party for biodiesel," according to Hatcher.[2]

The Energy Policy Act

The Energy Policy Act of 1992 (EPAct) was passed by Congress to reduce U.S. dependence on imported petroleum by requiring certain fleets to acquire vehicles able to operate on nonpetroleum fuels. The act sparked interest in a wide range of alternately fueled and powered vehicles and resulted in federal subsidies for research into electric vehicles, ethanol, methanol, and propane, as well as other alternate-energy technologies. In 1998, the EPAct was amended to allow fleets to meet a portion of their annual alternate-fuel-vehicle (AFV) requirements through the purchase and use of biodiesel in existing diesel vehicles. Previously, biodiesel and biodiesel-fueled vehicles had been excluded from the program. With this change, fleet managers now could comply with the EPAct by using an alternate fuel rather than by buying alternate-fuel vehicles.

The effect of the amendment on fleet usage of biodiesel was dramatic. In March 1999, three major fleets were known to be using B20 for EPAct compliance. By December of the same year, that number had increased to 25 and included the Ohio Department of Transportation, the U.S. Postal Service, the General Services Administration, Alabama Power, and the U.S. Department of Agriculture. This represented a more than 700 percent increase in biodiesel users in just nine months.[3] On January 11, 2001, the final rule concerning the use of biodiesel to fulfill EPAct requirements was published in the Federal Register. This final rule amended Titles III and V of the Energy Policy Act of 1992, providing biodiesel-fuel-use credit to fleets that otherwise would have been required to purchase an alternate-fuel vehicle. Many fleets found this new rule to be a very attractive alternative.

If the ASTM standard was the coming-out party for biodiesel, then the amended Energy Policy Act of 1998 has been a strong driving factor since then, according to many observers, since fleets and fleet managers have been the ones who have purchased the bulk of the fuel for many years. The U.S. military alone accounts for at least 6 million gallons of biodiesel use and has been the single largest buyer in the United States.

EPA Approval

Although the ASTM standard and the Energy Policy Act have been major milestones for the biodiesel industry, they certainly are not the only ones. In March 1998, biodiesel became the only alternate fuel in the nation to complete the Environmental Protection Agency-required Tier I Health Effects testing under section 211(b) of the Clean Air Act successfully. The testing clearly demonstrated biodiesel's significant reductions in virtually all regulated emissions, except for a slight increase in NO_x (although these NO_x findings have been contradicted by recent tests conducted by the National Renewable Energy Laboratory that found NO_x emissions to be lower under certain circumstances). In May 2000, a Tier II subchronic inhalation testing study was completed as well. This testing demonstrated biodiesel's decreased threat to human health, especially when compared to the use of petrodiesel fuel. The total cost of the Tier I and II tests totaled more than $2.2 million and was funded by the National Biodiesel Board.

As important as these tests were, the EPA has taken an additional step that has given biodiesel an enormous boost. In January 2001, the EPA finalized a rule requiring that sulfur levels in all diesel fuel for on-road vehicles be reduced from 500 parts per million to only 15 parts per million (a dramatic 97 percent reduction) by 2006. Virtually all on-road diesel fuel sold in the United States since October 15, 2006 now complies with the new rule. Although this is good news for emissions, the downside of this reduction is that the refining process used to reduce the sulfur content in diesel fuel also reduces the fuel's natural lubricity agents, potentially causing excessive engine wear. Biodiesel is uniquely positioned to address this thorny issue. By adding to petrodiesel enough biodiesel to make up just 1 percent of the fuel, lubricity increases up to 65 percent.

On May 11, 2004, the EPA took the sulfur reduction initiative one step further when it announced new regulations aimed at cutting the amount of smog-causing chemicals and particulates that come from off-road, diesel-powered vehicles and machinery such as ferry boats, harbor and river tugs, farm tractors, railroad locomotives, and heavy earth-moving equipment at construction sites. These vehicles account for about one-quarter of all the smog-causing nitrogen oxide and nearly half of the particulates from mobile sources, according to the EPA. First proposed a year earlier, the new EPA regulation requires petroleum refiners to lower the amount of sulfur in diesel fuel for such engines to 500 parts per million by 2007 (now implemented) and to 15 parts per million by 2010. These changes will allow manufacturers to build cleaner-burning engines, since the fuel will contain much less sulfur, which damages catalytic converters and other emissions-control devices. The cleaner fuel, combined with the new engine standards, should reduce smog-causing nitrogen oxide and particulates from off-road vehicles and equipment by more than 90 percent, according to the EPA.[4] These new regulations have added even more momentum to the greater use of biodiesel already generated by the previously mandated reductions in sulfur levels in petrodiesel for on-road vehicles.

Executive Orders

Presidential executive orders can be useful tools for a wide variety of national agendas. Several executive orders have had a substantial positive impact on the biodiesel industry. On August 12, 1999, President Clinton signed Executive Order 13134, which called for the increased use of farm products, including agriculturally based biodiesel, instead of fossil fuels. Implementation of the order would effectively triple the use of bioenergy and bio-based products by the year 2010. President Clinton followed that up in April 2000 with Executive Order 13149, which called for a 20 percent cut in petroleum use by federal fleets. The continued strong increase in the use of biodiesel in federal fleets has unquestionably been encouraged by these two orders.

ENERGY LEGISLATION

National energy legislation also has played a crucial role in the growth of the biodiesel industry, and over the years, the National Biodiesel Board (NBB) and the American Soybean Association (ASA) have been prominently involved in helping to shape that legislation. The hotly debated, and long-delayed federal Energy Bill of 2002 had its roots in the now famous (or infamous) energy plan that resulted from the closed-door meetings between Vice President Dick Cheney and a group of energy industry officials shortly after the Bush administration took over the White House in 2001. Containing huge subsidies for the coal, oil, natural gas, and nuclear industries, the Energy Bill also contained a few tidbits for renewable energy. One of those tidbits was a tax credit for biodiesel.

Republican Senator Chuck Grassley from Iowa has been a staunch supporter of biodiesel. The centerpiece of his efforts in support of the industry was his work on legislation to create biodiesel tax incentives that would include an excise-tax credit for biodiesel when blended with petrodiesel. Grassley and Democratic Senator Blanche Lincoln of Arkansas played a key role in getting the legislation included in the Energy Bill that the U.S. Senate approved in 2002. However, the differences between the House and Senate versions of the Energy Bill could not be worked out in conference committee before the 107th Congress adjourned.

In the years that followed, the biodiesel tax provision was included in a wide variety of Energy Bills and other pieces of legislation, only to get bogged down in a series of legislative stalemates. Eventually, the biodiesel tax credit ended up in HR 4520, also known as the American Jobs Creation Act of 2004, which was passed by Congress and subsequently signed into law by President Bush in October 2004. The huge, 650-page document had something in it for just about everyone—including the biodiesel industry. The long struggle was over, and the industry's greatest legislative achievement had finally arrived. The ASA, the NBB, and virtually all of the other major players in the industry had worked hard to ensure passage of the tax credit. "This tax incentive generated strong bipartisan support because it truly is a win for all Americans," said former NBB chairman and ASA first vice president Bob Metz of South Dakota after the legislation was signed into law. "Our nation has a direct interest in taking steps to promote renewable fuels, like ethanol and biodiesel, which lessen our dependence on foreign oil. Biodiesel has many benefits that are important to all citizens."[5]

The tax credit amounts to one penny per percentage point of biodiesel blended with petrodiesel for "agri-biodiesel," such as that made from virgin vegetable oils, and a half-penny per percentage point for biodiesel made from other sources such as used cooking oil. The credit is taken at the blender level, with the intent of lowering the cost of biodiesel for consumers. As expected, with the implementation of the tax credit, biodiesel consumption and production began to soar. Although implementation of the credit was considered a major victory for the industry, the legislation included an expiration date for the credit of December 31, 2006.

Not surprisingly, extending the tax credit soon became the industry's number one legislative priority. In May 2005, a bill to extend the credit to 2010 was introduced in the Senate and subsequently was included in the Energy Bill of 2005. The bill ultimately passed with a compromise extension date of 2008. The legislation, signed into law by President Bush in August 2005, also contained a number of additional biodiesel provisions, including a 7.5-billion-gallon Renewable Fuels Standard, support for demonstration and testing projects, and a credit for alternate refueling stations among its many provisions.

Not resting on its laurels, the industry renewed its efforts to extend the tax credit yet again. HR 6, The Energy Independence and Security Act, which was passed by the House on December 6, 2007, contained another extension of the tax credit through 2010. However, the bill also included an expansion of the Renewable Fuels Standard to 36 billion gallons by 2022, as well as a renewable requirement for diesel fuel to be met by biodiesel and other biomass-based diesel fuels. The Energy Bill was finally passed by the Senate and signed by President Bush on December 19, 2007. Despite its many other shortcomings, the Energy Bill did contain a modest expansion of the Renewable Fuels Standard, increasing a specific renewable requirement for diesel fuel to be met by biodiesel and other renewable biomass-based diesel fuels. By increasing the minimum renewable requirement in the so-called "diesel pool" from 500 million gallons in 2009 to 1 billion gallons in 2012, the new legislation is expected to help create a more stable domestic market for biodiesel.[6]

Renewable Diesel

The biodiesel tax incentive contained in the federal JOBS Act of 2004 was viewed by the biodiesel industry as a major accomplishment. However, there was a relatively obscure provision added to the legislation at the last minute (subsequently incorporated into the Energy Policy Act of 2005) that has come back to haunt the biodiesel industry, and sparked a major political controversy within the larger energy industry. The provision referred to "thermal depolymerization" (TDP), which, simply stated, involves turning virtually any carbon-based waste material, including vegetable oil, animal fats, plant material, etc., into usable hydrocarbon fuels.

Initially, few people seemed to pay much attention to this relatively obscure strategy. But behind the scene a powerful group of oil and chemical companies soon began to heavily lobby the federal government to expand the definition of the TDP provision for their benefit. As a result, in April 2007, the Internal Revenue Service agreed to a new, expanded definition that included the conventional petroleum refining process, allowing fuel made from the TDP process to qualify for the same dollar-per-gallon incentive that was originally created for biodiesel production from agricultural resources. This meant that the oil companies could add small amounts of vegetable oils or animal fats to their existing refining process and qualify for the credit on a new product they called "*renewable diesel*," also more specifically referred to as "*co-processed renewable diesel*."

It was no coincidence that on April 16, 2007, shortly after the IRS announcement, ConocoPhillips and Tyson Foods, Inc. announced a plan to collaborate on the production of "renewable diesel" from pork, poultry, and beef animal fats, with production of up to 175 million gallons per year by the end of 2008. The biodiesel industry reacted quickly and firmly. "It's our belief that this credit was developed to help a specific emerging technology, and not to further subsidize existing petroleum refineries," said NBB's CEO Joe Jobe. "This is bad energy policy, bad agricultural policy, and bad fiscal policy. If Congress lets this stand, our government will be handing over U.S. taxpayer money to some of the richest companies in the world, and it will not provide many of the benefits that the biodiesel tax incentive has given back to America."[7]

In May 2007, bipartisan legislation was introduced in the House of Representatives, titled the "Responsible Renewable Energy Tax Credit Act," designed to prevent the large oil companies from exploiting the tax incentive. Under the legislation, only producers making renewable diesel solely from renewable sources as originally defined would be eligible for the credit. The bill was referred to the House Committee on Ways and Means. Stay tuned.

Subsidy Issues

In late 2000, the USDA's Commodity Credit Corporation (CCC) launched its U.S. Bioenergy Program. The program, which was intended to spur the construction of new biodiesel facilities, provided reimbursements to biodiesel and ethanol producers for converting a wide range of commodities into bioenergy. The program, which had the strong support of the National Biodiesel Board, also was intended to provide a transition until biodiesel tax-incentive legislation was approved by Congress. As originally written, the biodiesel part of the program offered the subsidy only for soybean oil, which producers of biodiesel from other feedstocks widely condemned as being unfair. In the 2002 Farm Bill, the program was reauthorized and the list of eligible feedstocks was broadened to include recycled fats, oils, and greases. But the revised subsidy structure still favored soy-based biodiesel, with payments approximately twice as high as those for biodiesel produced from used cooking oil and animal fats. This remained a sore point among some biodiesel producers who used recycled cooking oil or animal fats as feedstocks.

The soybean portion of the CCC program was tied to the market price for soybeans. The price was set four times each year. The quarterly subsidies were paid for new plant production on a production basis to those companies that had qualified as biodiesel producers. The greatest payments were received in the first year, and company participation in the program ended after four years. The dramatic increase in soybean-oil prices caused the subsidy payments to soar from about $1.50 per gallon in the fall of 2003 to $2.50 per gallon for many producers in 2004. The price swings in feedstocks were somewhat dampened by the correlating price swings in the CCC subsidy program, giving producers a bit of a cushion to the ups and downs in the commodity markets.

The program payments to producers were generally passed on to consumers and reduced the retail price of biodiesel by $1.00 per gallon (or more). This price cut was the single largest contributor to making market acceptance of biodiesel possible. It also has been largely responsible for the dramatic increase in U.S. biodiesel production (and plants) in recent years. "The CCC subsidies have really helped to build the market," says Bill Ayres, formerly with Ag Environmental Products.[8] The CCC program finally expired in August 2006, and the NBB and American Soybean Association as well as many others in the industry worked hard to get Congress to authorize a successor program in the 2007 Farm Bill.

In July 2007, the House of Representatives passed the Farm Bill which, among its many provisions, included an expansion of the CCC Bioenergy Program as well as a Biodiesel Education Program. Once again, the Bioenergy Program would support the production of biodiesel and other bio-based renewable fuels using domestic feedstocks by providing farm payments to help U.S. biodiesel producers offset rising feedstock costs.[9] In December, the Senate passed similar legislation, although there were a number of limitations on the CCC Bioenergy Program in the Senate version that would make it difficult for biodiesel producers to use it. The differences between the House and Senate versions of the Farm Bill remain to be worked out.

In recent years, the NBB's focus on the three key elements in its federal policy efforts—a biodiesel tax incentive, a Renewable Fuels Standard, and reauthorization of the CCC Bioenergy Program—have been largely successful. Nevertheless, the NBB continues to work diligently to ensure that these achievements remain on track in the years ahead.

Splash and Dash
In an issue related to subsidies and the Energy Bill of 2005, the controversy known as "splash and dash" recently has created a good deal of political drama on both sides of the Atlantic Ocean. As mentioned in Chapter 5, the tax credit for blended biodiesel has been used by some producers to ship tankerloads of biodiesel from places such as Indonesia or Malaysia, to a U.S. port, where 1 percent or less of petrodiesel fuel is added in order to qualify for the U.S. blending credit. The B99 shipment then is sent on to a European port, where it qualifies for yet another tax

credit from various EU member nations. In addition, much U.S.-produced biodiesel that also qualifies for the tax credit has also found its way into the European market. The European biodiesel industry, represented by the European Biodiesel Board, has strongly opposed this practice, claiming that it seriously threatens its survival. This claim is not unreasonable considering the volume of biodiesel involved. By the end of 2007, the flood of biodiesel into the EU from the United States had risen to 700,000 tons (210 million gallons, or about 46 percent of total U.S. production) since the beginning of the year. That compares to only 90,000 tons the previous year.[10] In December 2007, after repeated lobbying efforts by various EU officials and calls for U.S. Congressional action failed to change the policy, the EBB voted unanimously to institute legal proceedings to stop what they consider to be illegal dumping. The European Commission is investigating the charges and may adopt import duties to stop the practice.

The U.S. view of the controversy is somewhat different. Many in the industry, including the NBB, agree that the shipment of biodiesel from, say, Indonesia or Malaysia to a U.S. port to take advantage of the blending tax credit before sending the cargo on to a European destination was not the original intent of Congress (or the U.S. biodiesel industry), and they are actively working to close this loophole. However, they also believe that using the tax credit for biodiesel legitimately produced in the United States before shipping the fuel to foreign customers accomplishes Congressional (and industry) intent, which was to stimulate the growth of the U.S. biodiesel industry. "Having trade flows open is important to help national energy security," says the NBB's Joe Jobe, who favors the retention of the tax credit for both export and domestic consumption of U.S.-produced biodiesel.[11] World Energy Alternatives President Gene Gebolys agrees. "The original intent of the legislation was to foster a biodiesel industry in the United States; it has, and continues to do so. Ending the tax credit would kill the industry in the United States," he says. "I think that what is going on in the European market has a lot more to do with the changes in the German tax code than exports from the United States."[12]

Although there had been some calls in Congress to eliminate the tax credit for biodiesel not produced and consumed domestically, that sentiment disappeared in recent discussions in various committees, and the 2007 Farm Bill reaffirmed the tax credit for exports. European producers are not happy with this decision.

FEEDSTOCK ISSUES

The subject of feedstock neutrality is unquestionably one of the more important issues facing the U.S. biodiesel industry. But feedstock neutrality has not always been a high priority for the industry. The fact that the NBB was originally named the National SoyDiesel Development Board is an undeniable indication of the organization's—and the industry's—roots. Around 1999, responding to criticism about a perceived soy bias, the NBB changed its by-laws to state clearly that the organization is "feedstock-neutral."

"No one feedstock will grow the market or be able to supply a fully implemented biodiesel market, so it really comes down to the need for cooperation at some point," Charles Hatcher says. "We simply can't take all of the soybean oil off the grocery store shelves, nor can we use all of the recycled cooking oil, or all of the animal fats, because there are other products and markets that are based on those commodities. In addition, cooperation is absolutely required so we can go through upturns and downturns in various commodity markets and still be able to provide a product." Despite the official neutrality of the NBB, it was hard for some in the industry to swallow the lingering price differential in government programs between soy and other feedstocks. What's more, when the Energy Policy Act was amended to include biodiesel in 1998, there was a federal guideline that limited fleets to purchasing only soy-based biodiesel. Producers using other feedstocks complained bitterly about being shut out of the market, and the limitation was eventually relaxed to include biodiesel made from virtually any feedstock, as long as it met the ASTM standard. To its credit, the NBB was involved in the process to have the soy limitation removed from the federal purchasing guidelines.

World Energy's president Gene Gebolys is philosophical about the feedstock wars. "It's important for us to understand that we absolutely need political support, and that the soy interests are in the best position to deliver that support. And we can't expect them to do that without getting their fair share of the return," he says. "After all, they are the ones who have been forking out the cash for this industry from the beginning and exerting their political influence. If we are not politically savvy enough to understand that if we don't have public support for our product, then we don't have an industry, and if we aren't savvy enough to understand that farmers have more political clout than biodiesel producers, then we are pretty foolish. Willie Nelson doesn't play his guitar for the producers." Having said that, Gebolys also understands that the soy interests need the cooperation and support of other feedstock producers to help stabilize the market. "Ultimately, the success of soy biodiesel is going to be dependent on the ability to get other feedstocks to dampen price swings and to help create a predictable market," he says. "You can't have an industry if people don't know whether they can get the feedstock or even roughly what it is going to cost them. That's ultimately where we are headed—a plurality of feedstocks."[13]

STATE INITIATIVES

In addition to national initiatives, the industry in general, and the NBB in particular, has been very active with a wide range of state-related programs, especially in recent years, according to Manning Feraci, the NBB's vice president of federal affairs. "State-based legislative activity continues to grow both in terms of volume and complexity," he says. "The result is that state activity remains critically important to the biodiesel industry. For the fifth consecutive year, we have seen an increase in the level of interest in statehouses to promote biodiesel. In 2007, we saw a tripling in the number of biodiesel-related bills introduced (298) from the number tracked in 2003. These bills were introduced in 45 states, with 56 of them being sent to governors, more than twice the success rate as compared to 2003. State-based initiatives are not only positively impacting awareness but also helping expand the supply and demand for biodiesel."[14]

In addition to state legislative efforts, critical regulatory initiatives also are playing an increasingly important role in the development of the industry, says Scott Hughes, the NBB's director of governmental affairs. "We are seeing a significant focus by states on fuel quality standards and enforcement, air quality improvements, as well as addressing greenhouse gas emissions," he notes. "These efforts can have real impacts on our industry in terms of creating potential opportunities for growth, as well as creating potential barriers for the industry. Remaining engaged with stakeholders and regulatory bodies will continue to be essential as we move forward."[15]

Although this state-based activity began in the 1990s, laws approved in Minnesota, Illinois, and Washington State are good examples of this growing trend.

Minnesota
Although there have been a number of state biodiesel initiatives, Minnesota's groundbreaking "2 percent" legislation is unquestionably one of the most sweeping, and the first of its kind in the nation. On March 15, 2002, the state mandated that all diesel fuel in the state contain 2 percent biodiesel (B2). The implications of this seemingly insignificant percentage of biodiesel are enormous. Minnesota uses more than 831 million gallons of diesel fuel each year, and 2 percent of that equates to a new market demand for more than 16 million gallons of biodiesel. The mandate included the stipulation that there would be at least 8 million gallons of in-state production capacity to ensure that local farmers and businesses would benefit. The mandate would begin 30 days after the state's commissioner of agriculture certified that the capacity threshold had been met. The law finally went into effect on September 29, 2005. The new law spurred a flurry of biodiesel activity across the state. The Minnesota Department of Agriculture has reported that, since then, renewable-fuels

production has led to a 13 percent increase in demand for the state's soybean crop and a 31 percent increase of in-state soybean processing.[16]

The B2 mandate was not without its challenges, however. By December, with the arrival of cold winter temperatures, there was a rash of biodiesel-related problems, but mainly plugged fuel filters that were traced, at least in part, to off-spec biodiesel. A temporary waiver of the B2 mandate was initiated while fuel supplies were tested and off-spec fuel was eliminated from the supply chain. Meanwhile, the new in-state biodiesel producers tightened up their quality assurance procedures. "It wasn't all biodiesel, but there certainly were some issues with some out-of-spec biodiesel," said Ed Hegland, an Appleton, Minnesota, farmer and member of the NBB governing board. "There is much biodiesel in the state that is being used very successfully."[17] The startup glitches were resolved eventually and the B2 mandate went back into effect.

In August 2007, Minnesota Governor Tim Pawlenty took a bold step when he announced a new initiative for the state to raise its biodiesel blend level from B2 to B20 by 2015. "Minnesota has led the nation in unleashing a renewable energy revolution," Governor Pawlenty said. "Other states are starting to catch on and it's time for us to continue to blaze the trail to a cleaner, more secure energy future. Increasing the level of biodiesel in diesel fuel means that more of our energy will come from farm fields rather than oil fields, and that's a good thing."[18] The governor signed a bill containing the new biodiesel initiative in May.

Illinois

In June 2003, two important pieces of biodiesel legislation were signed into law by Illinois Governor Rod Blagojevich. One bill provided for a partial tax exemption on biodiesel blends, while a second bill offered grants of up to $15 million for the construction, modification, or retrofitting of renewable energy plants with a minimum production capacity of 30 million gallons of biodiesel a year. As a result, the Illinois Soybean Association, anticipating an increased demand for soybeans, estimated that the legislation would add five cents per bushel to the price of Illinois soybeans. With the state's annual production of soybeans at about 450 million bushels, the legislation was expected to boost the state's economy by more than $22.5 million.[19] In addition, the governor signed an executive order that required increased use of biodiesel and ethanol by state employees.

In May 2005, Governor Blagojevich signed new legislation that gave biodiesel yet another boost by expanding biodiesel blends eligible for state rebates and grants. The legislation changed the Alternate Fuels Act to make biodiesel blends of B20 or higher eligible for the state EPA's Alternate Fuels Rebate Program and the Department of Commerce and Economic Opportunity's Alternate Fuels Infrastructure Program. The legislation, however, did not contain a statewide biodiesel-blending mandate. And, in August 2006, the governor unveiled a dramatic new $1.2-billion energy plan for the state, with a goal of replacing 50 percent of the state's motor vehicle fuel supply with biofuels by 2017. Among its many provisions was $25 million to aid in the construction of five new biodiesel plants and $200 million for new ethanol plants.[20]

Washington State
The State of Washington imports all of its diesel fuel, which amounts to nearly 1 billion gallons annually. In March 2006, Washington became the second state in the nation to pass a B2 mandate, when Governor Christine Gregoire signed a new law that established some of the most aggressive renewable-fuel requirements in the nation. The bill, which had strong bipartisan support in the state legislature, called for diesel fuel to contain a minimum of 2 percent biodiesel by November 30, 2008, or sooner if there is sufficient in-state production capacity to meet the requirement. In addition, state agencies must use B20 to fuel their diesel-powered vehicles by June 1, 2009. The bill also contained an initial 2 percent ethanol mandate. In response to some of the early startup problems that surfaced in Minnesota's B2 mandate, the Washington State legislation included a number of more flexible options that would allow fuel suppliers and service stations some discretion on how to best meet the mandate.[21]

In addition to Minnesota, Illinois, and Washington, there has been significant biodiesel legislative activity of various types in New York, Pennsylvania, Maryland, Virginia, Georgia, Mississippi, Missouri, Indiana, Arkansas, Iowa, Kansas, Colorado, Arizona, California, Hawaii, and several other states.

SMALL-PRODUCER ISSUES

One of the most challenging political issues in the biodiesel industry is internal. As mentioned earlier, there has been a certain amount of ongoing tension between small producers and some of the larger producers, as well as with the National Biodiesel Board. Prior to 2004, annual membership dues starting at $5,000 for the NBB were viewed by many small producers as deliberately exclusionary, and the NBB was seen by some as simply a mouthpiece for large producers and soybean farmers. In July 2004, the NBB did, in fact, address the membership issue by deciding to create a new membership class, "small producer nonvoting associate member," for producers making less than 250,000 gallons per year, and also lowered the minimum annual dues from $5,000 to $2,500. This was still a fairly high threshold for most small producers, but the move was seen as an honest attempt to address this issue.

A related bone of contention has been the issue of access to the health-effects data of the EPA Tier I and Tier II tests that the NBB paid for. Access to the data greatly facilitates biodiesel registration with the EPA. The NBB holds the EPA registration, and access to that registration is normally gained through membership in the association and by paying yearly dues. In theory, there is nothing that would prohibit a nonmember from doing their own testing and obtaining their own EPA registration— except for the extremely high costs involved—which automatically eliminates all small (and even most large) producers. Some small producers continue to maintain that this is simply not fair. Some of these producers have been so frustrated with the situation that there have been some discussions of forming their own industry organization that would be more in tune with their needs. Many of these smaller producers attend an annual gathering known as the Sustainable Biodiesel Summit (sustainable-biodiesel.org), where the emphasis is on local, community-based business models rather than the large, multinational corporate model.

In terms of actual production, the small producers admittedly make up only a tiny fraction of the total biodiesel output for the nation. Nevertheless, the NBB's Joe Jobe acknowledges the valuable role that the B100 community plays in promoting biodiesel at the grassroots level. "These (small producers) bring a very important element of education to communities," he says. "They're the ones who will talk to the school boards, and other local groups."[22] However, many in the B100 community are uncomfortable with the widespread use of GMO soybean seeds and chemical fertilizers and herbicides by most Midwestern soybean farmers. They are also leery of the increasing domination of the market by huge, multinational agribusiness companies, and despite ongoing efforts to work together, the two groups maintain a fragile and uneasy alliance.

One producer that has distanced itself from the NBB for many years is Pacific Biodiesel in Hawaii. But that situation has changed, according to Bob King, the company president. "We've had a long history of not participating in the NBB because of how they did business in the past, but I'm really excited about the new NBB," King says. "Joe Jobe and [former NBB chairman] Bob Metz are great guys, and they are somehow able to work with this extremely diverse group of constituents and do some good things. That's just really great for the industry, and I'm really pleased with where the organization is at now. I have to give them a lot of credit for what they have done."[23]

Since King is located in Hawaii, he says that he has managed to stay out of the small producer/big producer squabbling on the mainland. But he still has some thoughts on the issue. "My opinion has always been that small production plants are going to be the industry-opening plants," he explains. "So, in an area that does not have any biodiesel, this is the size you start off with. Then, after things catch on, the industry should move on to big plants. It's going to be a different group of people who are going to be doing those larger plants; it's a different mind-set and a different production process, but I don't really have a problem with either group because they each have a role to play. As long as the company is doing a good job for its customers, the community, and the environment, then it doesn't matter if the company is big or small. I really hope that everybody gets over the infighting that's going on and realizes that we're all doing a good thing and there's plenty of room out there for everybody."[24]

But tensions between small and large producers are not unique to the United States, according to Dr. Rico Cruz of the Environmental Science and Technology Program of the Confederated Tribes of the Umatilla Indian Reservation in Washington State. Dr. Cruz, who is originally from the Philippines, is a pioneer in the production of biofuels and simple-process technology. Since the early 1990s, he has been involved in a wide range of biodiesel projects in the United States and in countries such as India, Uganda, Bulgaria, Russia, Azerbaijan, and the Philippines. "Some of the bigger biodiesel producers are definitely trying to eliminate the smaller producers," Cruz maintains. "In many of the countries that I have visited the biodiesel producers have not had the support of their national governments. But once a large producer monopolizes the production, they try to get all of the government support. I've seen it happen, it's still happening, and I think that's a problem."[25] This is a matter of personal concern for Dr. Cruz, since most of the biodiesel projects he has helped set up are small-scale, community-based operations. Dr. Cruz is credited with bringing biodiesel to the Philippines as well as to the Nez Perce Indian Band in the Pacific Northwest.

THE ETHANOL CONNECTION

The ethanol industry has had a significant impact on the development of the biodiesel industry in general, and on biodiesel politics in particular. The ethanol industry had a long head start on the biodiesel industry, with a well-organized and powerful lobby composed traditionally of Midwestern corn farmers, who have been able to attract and maintain substantial government support via their considerable political muscle in

Congress. Up until the mid-1990s, if you were talking about a liquid bio-fuel in the United States, you were basically talking about ethanol. Since then the industry has continued to prosper and grow. To put this in per-spective, in 2006 the U.S. ethanol industry produced 4.9 *billion* gallons, while the U.S. biodiesel industry made about 250 million gallons of biodiesel. This 20-to-1 ratio is a clear measure of the political success of the ethanol industry, as well as some indication of the future potential for the biodiesel industry.

Nevertheless, the biodiesel industry has been able to use many of the same political strategies developed by the ethanol industry to gain gov-ernment support, in part because some of the same people have been involved in both industries. And these people learned some lessons from their prior experience with ethanol. "Farmers who grow soybeans also grow corn, and a lot of them have been involved in the development of the ethanol industry," says the NBB's Joe Jobe. "One of the lessons learned from the ethanol industry was that some of the technical issues on blending, storage, water absorption, and so on weren't totally worked out. So ethanol got a bad name and developed a bad reputation that to some extent is still following it around 30 years later. So the farmers who started biodiesel in the U.S. took a very long approach and wanted to get all of the technical, legal, and regulatory issues settled before they worked on commercialization."[26]

All of this experience and attention to detail has created a reasonably successful political record for the biodiesel industry at the national level. It is expected that this momentum will continue in the future. And what works at the national level can sometimes be transferred to the state level. Even in Hawaii—about as far from the Midwest as you can get—biodiesel producers have been able to make use of the momentum gener-ated by the ethanol industry. "The ethanol industry has already done a lot of the groundwork for the biodiesel industry," Daryl Reece of Pacific Biodiesel says. "The state of Hawaii already had a tax credit for ethanol, so all we had to do was to piggyback biodiesel on the existing state tax credit for ethanol, and it was passed easily."[27] Similar strategies have worked in other states as well.

STRAIGHT VEG

Of all the political issues related to biodiesel, probably the most con-
tentious is the struggle between the industry and individuals who promote the use of straight vegetable oil (SVO) as a diesel fuel substitute.
This strategy, referred to by many enthusiasts as "straight veg," makes use
of waste vegetable oil (yellow grease) collected from restaurants to fuel
diesel engines *directly*, without first going through the transesterification
process. This bit of alchemy is accomplished by adapting the vehicle to
the fuel with a conversion kit (costing roughly $300 to $1,500 in the
United States). Basically, the typical conversion kit involves adding a
parallel fuel system that consists of a second fuel tank, a heater, an extra
fuel line to the engine, a filter, and a control that allows the driver to
switch back and forth between the two systems. The vehicle has to be
started with diesel fuel (or biodiesel) from the first tank, and then, after
the engine heats up and thoroughly warms the straight vegetable oil in
the second tank, the driver manually switches to the second tank. (Some
systems feature an automatic control for this.) At this point, the vehicle
is running on straight vegetable oil. When the vehicle needs to be shut
off, the process is reversed. The driver manually switches back to tank
number one (containing the diesel fuel) for a few minutes to ensure that
all the straight vegetable oil has been purged from the fuel line and
engine.

Although this strategy is relatively simple, the debate about it is anything but. Virtually all the big players in the biodiesel industry in the
United States and Europe warn about possible engine damage and com-
plain that straight-veg users are breaking the law by not paying fuel taxes
and giving the biodiesel industry a bad name when problems do occur.
There is some justification for this view, since most of the general public
is still not educated enough about various fuels to understand the finer
points of these different strategies. On the other side, the straight veg
camp (which also includes many backyard biodiesel producers in the
B100 community) tends to have a conspiratorial mind-set about the
biodiesel industry's intentions and complains that the industry is

spreading false information about straight veg. Unfortunately, the print and broadcast media cause additional confusion when articles or news reports sometimes confuse biodiesel and straight vegetable oil in their reporting. This happens frequently.

The straight-veg strategy appeals mainly to people who are mechanically inclined and who can remember to switch between fuel tanks at the appropriate moments. (This is definitely *not* a good strategy for the absent-minded.) If you enjoy spending your weekends under the hood of your vehicle—or under the cover of a dumpster scrounging for the smelly, greasy, used cooking oil—this may be the strategy for you. However, many people who have experimented with SVO, especially those who use their vehicles primarily for around-town driving, eventually give it up because the system never warms up enough to be of much use, and they find that it simply is too much trouble. In addition, there may be some potential long-term problems, especially if the straight veg is used carelessly. And since there is a shrinking supply of easily available used cooking oil, straight veg always is going to be a somewhat limited option. Nevertheless, the lure of essentially free fuel is compelling.

Ironically, all the players involved in this controversy have the same general goal in mind: a shift from a fossil fuel to a renewable biofuel. But they follow different paths to reach that common goal.

$$\boxed{12}$$

Recent Developments

In the summer of 2002, as she performed around the country with Lyle Lovett, folk/rock/blues singer Bonnie Raitt was doing more than entertaining her many fans. She was the first U.S. entertainer to adopt biodiesel for use on tour. On what was dubbed the Green Highway Tour, the nine-time Grammy Award-winning artist traveled part of the route in a biodiesel-powered bus, visiting 42 cities along the way and performing at major venues while raising awareness about alternate fuels.

"It's no accident that we're in danger of losing both our ecological and our economic well-being at the same time," says Raitt, who created the Green Highway concept along with colleagues Kathy Kane and Harvey Wasserman. "I feel too many government and corporate policies are inseparably shortsighted, and we've created Green Highway to demonstrate that working in harmony with nature can offer real solutions for preserving both our planet and our prosperity."[1]

The tour was powered by a combination of biodiesel, solar energy, wind power, and hybrid-vehicle technologies. At each stop, the tour set up exhibits on alternate energy and handed out information to concertgoers. More recently, in her 2005 Souls Alike Tour, Raitt continued to use B20 in two of her tour buses and two tractor-trailer trucks. She even hosted a special benefit event for biodiesel education at her sold-out performance in Knoxville, Tennessee, in December 2005. She continues to be a strong, high-profile advocate for biodiesel.[2]

Willie Nelson has been a longtime supporter of American farmers, and in 1985 he organized the first Farm Aid benefit concert. It should come as no great surprise that Nelson was an early biodiesel convert. He now powers all of his tour buses and cars with biodiesel. In December 2004, along with several other business partners, Nelson went on to found Willie Nelson Biodiesel, which promotes and sells its "Bio-Willie" brand biodiesel blend at truck stops around the nation.

Even actress Daryl Hannah is a strong advocate for biodiesel. Hannah, who starred in memorable films such as *Splash*, *Blade Runner*, and *Steel Magnolias*, has been a longtime environmentalist and biodiesel activist. She has educated many thousands of people about biodiesel through her appearances on television shows such as *The Tonight Show* and *The O'Reilly Factor*, as well as through numerous magazine interviews. Hannah was honored at the 2004 Biodiesel Conference & Expo in Palm Springs, California, for her volunteer advocacy of biodiesel as well as longtime use of the fuel in her own personal vehicle. "We have the technology to reduce greenhouse gases and grow much of our own fuel," Hannah said. "We have the technology to make sure no kid goes to school breathing dangerous toxins. We have the technology now, and that technology is biodiesel."[3]

In December 2006, Hannah and biodiesel advocates Kelly King and Annie Nelson (Willie Nelson's wife), along with other members of the biodiesel community, founded the Sustainable Biodiesel Alliance (SBA). The SBA is a nonprofit organization dedicated to promoting sustainable biodiesel practices, including the harvesting, production, and distribution of biodiesel fuels. Since its creation, the Alliance has grown to include family farmers and farm organizations such as Farm Aid, the Institute for Agriculture Trade and Policy (IATP), members of the 25 x '25 coalition, environmental organizations, renewable energy experts, and a wide array of other groups. The SBA is developing a list of sustainable biodiesel principles for biodiesel production and distribution from field to fuel pump.[4]

BIODIESEL USE EXPANDING

Although celebrities who use and endorse biodiesel attract a lot of media attention and are unquestionably helpful in raising public awareness, there are a lot of other people all across the United States who have gotten the message about the fuel's many advantages as well. Recent market research conducted for the National Biodiesel Board found that familiarity with biodiesel among consumers has increased from 27 percent in 2004 to 45 percent in 2007. Even former President Jimmy Carter has been an active proponent of biodiesel, working in his home state of Georgia to bring together industry representatives and legislative and government leaders to discuss making biodiesel production a reality there.

The pace of first-time biodiesel use has accelerated in the past few years to the point where news of yet another school district, city bus system, or municipality somewhere in the nation switching to B5 or B20—even B100—has become almost a weekly occurrence. It would be impossible to list them all here, but what follows is a brief summary of recent biodiesel news and events that will present a reasonably good cross-section of current U.S. biodiesel activity.

New Biodiesel Pumps Open
In the past, it was possible to purchase biodiesel—if you knew where to find it—from a small producer or farmers' co-op. But for all practical purposes, biodiesel was not readily available to the general public. As of 2001, that began to change. Sparks, Nevada (near Reno), appears to deserve the credit for having the first public biodiesel pump in the nation—but just barely. Officially opened on May 22, 2001, by representatives from Western Energetix and the Nevada Energy Office, the Western Energetix Cardlock location at 655 South Stanford Way in Sparks was the first public filling station to distribute biodiesel made from recycled cooking oil to the general public. The used cooking oil came from local casino resorts and restaurants and was made into fuel by Biodiesel Industries, Las Vegas, with the assistance of the Nevada Energy

Office and a grant from the U.S. Department of Energy. "The support we have received from the State of Nevada in getting this project from a dream to reality has been tremendous," said Russ Teall, president of Biodiesel Industries. "I hope this can show communities all across America that biodiesel can be made and used almost anywhere. When government and industry work together to solve problems, it's amazing what can be accomplished in a relatively short time."[5]

Less than 24 hours later, America's second public biodiesel pump was opened officially in San Francisco, California. The ribbon-cutting ceremony took place at the Third Street facility of Olympian Inc., with local, state, and company officials in attendance. World Energy Alternatives of Chelsea, Massachusetts, supplied the fuel. "As an oil company operating in today's economy, we believe offering biodiesel not only improves our corporate image but also increases our competitive advantage," explained Tom Burke, Olympian's Division Manager of Cardlock and Mobile Fueling. "When renewable fuel becomes the rule and not the exception, we will already be a recognized provider."[6] Many other forward-looking fuel companies have been having similar ideas.

Since these initial openings, public biodiesel pumps have been sprouting up all across the nation like mushrooms, from California to Connecticut and from Maine to Missouri—and a lot of states in between. As of late 2007, there were more than 1,570 retail biodiesel locations nationwide (compared with 300 in 2004). The largest number, however, are still clustered in the Midwest. The National Biodiesel Board offers a map showing U.S. public biodiesel pump locations by state at www.biodiesel.org/buyingbiodiesel/guide. Scroll down to the "Guide for Consumers" section and click on "View Guide" and then "Locate a Biodiesel Retailer in the U.S." for a National Map of Retail Fueling Sites and then individual states for details.

Biodiesel in Schools
There are many niche markets for biodiesel, but school buses, in particular, make a lot of sense. The fact that there are about 500,000 school buses in the United States—nearly six times as many buses as all the

nation's public transit buses combined—is reason enough. But the fact that the 25 million children—especially young children—who ride those buses tend to be more susceptible than adults to the toxic and potentially cancer-causing and asthma-aggravating emissions from petrodiesel has been an even more compelling reason for school boards and parents across the nation to insist on school buses being switched to biodiesel fuel. The federal government has been helpful in this process. Congress initially included $5 million in the Environmental Protection Agency's budget for Clean School Bus USA, a cost-shared grant program designed to help school districts in cleaning up their bus fleets. The fact that the EPA received more than 120 applications requesting almost $60 million is a clear indication of just how popular the program has become.[7] Since its inception, the program has been funded on an annual basis at between $5 and $7.5 million, a relatively small amount considering the magnitude of the work that still needs to be done with the nation's aging school bus fleet. In recognition of that fact, a complementary National Clean School Bus Grant program was established by Congress in 2005 and authorized at $55 million per year for 2006 and 2007.[8]

In 1997, the Medford, New Jersey, school district was the only one in the nation to run its fleet on biodiesel. But today thousands of school buses use the fuel. The Clark County, Nevada, school district now powers more than 1,200 of its buses with biodiesel, making it the largest school bus fleet in the nation (and possibly the world) to use biodiesel. Clark County soon may be edged out of the number one position by the Cook-Illinois Corporation based in Chicago, which, in September 2007, announced its intention to fuel its 1,400-bus fleet with biodiesel. This would bring the total number of school buses powered with biodiesel in Illinois to well over 3,000, spread across at least 36 separate school districts. In Indiana, a growing number of school districts have switched fuels, including Franklin Township and Wayne Township with 88 and 160 buses, respectively, now burning B20, while in the state of Kentucky at least 35 school systems are now running their buses on biodiesel. Hundreds of districts nationwide have made the switch to biodiesel, and the numbers are growing daily. The City of Denver, with the largest school bus fleet in Colorado, switched to biodiesel in 2005. Two other large Colorado school districts, Littleton Public Schools and Jefferson County Public Schools, have been using biodiesel successfully since 2004, and Littleton has reported decreased maintenance costs since making the change.

Although the details are still somewhat preliminary, there is increasing evidence that other school bus fleets also are saving money by using biodiesel, even though the fuel costs more than petrodiesel. How is this possible? The savings are in reduced maintenance costs and increased mileage per gallon. Some of the strongest evidence comes from the Saint Johns Public Schools in Michigan, where careful maintenance records have been kept from both before and after biodiesel was adopted in April 2002. Saint Johns was the first Michigan school district to switch its entire fleet of buses (totaling 31) to B20. The main cost savings have been due to extended intervals between oil changes, according to Wayne Hettler, garage foreman and head mechanic for Saint Johns. "I'm convinced," he says, "that we are able to extend the oil changes because the B20 burns cleaner and isn't dirtying the oil as quickly. We're using oil analysis to determine oil change times. We solely credit biodiesel for cleaning up the oil, thus saving the district the costs of oil, filters, labor, and the like. I challenge other fleets to 'read' their fleet records and make these cost-saving changes after switching to B20." Longer fuel-pump life due to biodiesel's higher lubricity and increased miles-per-gallon rating are also cited by Hettler as adding even more savings. "Pre-April 2002, our fleet's mileage averaged 8.1 miles per gallon. Now we average 8.8. That's a huge difference in MPG for buses," said Hettler. A combined savings of $3,500, even after the extra cost of the biodiesel was deducted, was realized by the district for the two-year period.[9] If savings can be achieved by this school bus fleet, it seems reasonable to assume that other fleets can do the same.[10]

On the East Coast, the Warwick, Rhode Island, school district not only uses biodiesel in its entire 70-bus fleet but also has been successfully heating three of its school buildings with biodiesel since 2001 (see Chapter 4). But the school district has gone even further by integrating biodiesel into its classroom curriculum. Their program is modeled after the high-school curriculum on alternate fuels developed by the Northeast Sustainable Energy Association (NESEA) called "Cars of Tomorrow and the American Community."[11]

More Mass Transit Use

Transit buses use a lot of fuel, consuming on average about 10,000 gallons every year. Of the more than 81,000 active transit buses in the United States, about 80 percent are still powered by diesel fuel; roughly 12.9 percent are running on compressed natural gas (CNG); 1.9 percent run on liquefied natural gas (LNG); 1.8 percent are hybrid electric, and a small number are either fully electric or fuel-cell-powered. Despite the increasing numbers of natural-gas-powered buses in recent years and the fact that orders for new diesel buses gradually have been declining, the size of the present diesel-powered fleet has declined only slightly.[12] In any case, diesel-powered transit buses offer a substantial market opportunity for biodiesel. Among the many city bus fleets in the United States currently using biodiesel are those in Cedar Rapids, Iowa; Cincinnati, Ohio; St. Louis, Missouri; Oklahoma City, Oklahoma; Olympia and Seattle, Washington; Raleigh, North Carolina; Springfield, Illinois; Minneapolis, Minnesota; and Oneonta and Albany, New York.

In 1993, Five Seasons Transportation & Parking (FST&P) in Cedar Rapids, Iowa, became one of the first bus fleets in the United States to use biodiesel, but in 1996 the program was dropped due to high fuel costs. Since then the cost of biodiesel has come down while the interest in renewable fuels has gone up. In March 2001, the program was revived and 60 buses began burning B20 again. Five Seasons was Iowa's first mass transit system to convert its entire diesel-powered fleet to biodiesel. Ag Environmental Products of Lenexa, Kansas, supplied the fuel. Using a combination of emissions-control technologies and biodiesel, Five Seasons' 25-year-old bus fleet was able to run cleaner than brand-new buses. "Part of the reason behind the decision to use soy-based biodiesel is that we're always concerned about emissions, and this helps us help the environment," said Roger Hageman, FST&P maintenance manager. "We are very conscientious about supporting Iowa's farmers, and using B20 will benefit them while decreasing our dependence on foreign oil."[13]

St. Louis, Missouri, has been the location for another long-running biodiesel transit bus program. For more than 10 years, the Bi-State Development Agency, a mass transit provider for the St. Louis area, conducted extensive testing of biodiesel for the U.S. Department of Energy and the National Renewable Energy Lab. The results of the tests were favorable and demonstrated a significant reduction in vehicle emissions without an impact on fuel economy or performance. Use of the B20 fuel posed no operational problems in the transit buses that participated in the tests. Bi-State (now called Metro) noticed that, in addition to reducing vehicle emissions and the release of particulate matter, B20 had such good lubricity that it increased injector life and decreased the need for vehicle maintenance, a finding that generally has been confirmed by other fleet tests. Passengers riding the B20-fueled buses appreciated the absence of the acrid smell and black exhaust smoke normally associated with diesel buses. Metro hopes to incorporate biodiesel into its entire fleet of diesel buses.[14]

Diesel-powered transit buses are going to be around for many years to come. In order to meet the new stringent EPA low-sulfur rules that went into effect in 2006, many transit authorities are switching to ultra-low-sulfur fuel and are installing diesel particulate filters. Switching to biodiesel is an extremely cost-effective method of meeting the new EPA rules while potentially reducing maintenance costs and possibly increasing mileage.

Commercial Trucking

As mentioned in Chapter 4, the commercial trucking industry in the United States has been slow to adopt biodiesel due to the highly competitive nature of the industry and the generally higher cost of biodiesel. However, this has begun to change in recent years. The high-profile promotion of biodiesel by singer-songwriter Willie Nelson unquestionably has been a factor in this shift of opinion in the trucking industry. In addition, with a rising demand for "green" shipping, more and more truckers are using biodiesel blends in an effort to meet their customers' preferences. Another factor is the Environmental Protection Agency's new "Go and Grow" program that educates the trucking industry about the benefits of biodiesel and matches product shippers with truckers that use the biofuel. The new program, announced in August 2007, has been attracting a lot of interest from trucking companies nationwide. Tom Verry, outreach director for the National Biodiesel Board, has been working with the EPA, truck brokers, and other businesses in the supply chain. He says that a growing number of transport companies are looking for truckers who use biodiesel for their environmentally conscious customers.[15]

Another recent development that demonstrates the growing popularity of biodiesel in the trucking industry was the rollout of new mapping software that allows truckers to locate truck-accessible biodiesel fueling sites along their routes with a laptop computer. Announced in August 2007, the software, known as ProMiles XF, is available on CD and features address-to-address truck routing for professional drivers and fleets. The fueling locations are clearly marked for pre-route planning or on-the-road searches. "I've been a ProMiles user for years and I rely on it for all of my routing," says Tony Hamilton, a company driver for Dixie Midwest Express in Alabama. "Having biodiesel locations available helps me to use my fuel of choice and incorporate it into my planning instead of going out of my way to search for it when on the road." The NBB, ProMiles, NREL, and the Oil Price Information Service are working together to produce a continuously updated list of truck-accessible biodiesel locations to be included in the mapping software.[16]

One national trucking company that has embraced biodiesel is Decker Truck Line, Inc. of Fort Dodge, Iowa. With nine terminals in five states, Decker is the first major trucking company in the country to compare a soy B20 biodiesel blend to regular diesel in a comprehensive over-the-road test covering two million miles. The company has been using the biodiesel blend in 10 of its trucks from Fort Dodge to either Chicago or Minneapolis. An additional 10 trucks running on petrodiesel are the control group for the study. In addition to Decker, partners in the Two Million Mile Haul include the Iowa Soybean Association, the NBB,

USDA, Iowa Central Community College, and Renewable Energy Group. Initiated in the fall of 2006, the test had completed 350,000 miles by March 2007 and the test partners released some encouraging preliminary results. "What we've observed so far is great performance in the particularly cold winter we just experienced, and reduced maintenance and engine wear benefits that equal or outweigh the slightly higher cost of the biodiesel blend," says Dale Decker, industry and government relations director for the company.[17]

Approximately 1.5 million miles were completed by the end of the first year of the test, and although a slight decrease in fuel efficiency in the B20 group of 2.2 percent was detected, it was not considered to be statistically significant. There also was a higher number of fuel-filter-plugging episodes in the B20 group, but when fuel blending procedures were changed, the number of plugged fuel filters dropped significantly.[18] Diesel engine manufacturer Caterpillar and many trucking companies, as well as independent truckers, government agencies, and original equipment manufacturers have been following the results of the tests with great interest. It is expected that the use of biodiesel by the commercial trucking industry in the United States will continue to grow in the years ahead.

Biodiesel on the Railroad

Railroads in the United States generally have been slow to adopt biodiesel. But, as mentioned in Chapter 4, a small but growing number of U.S. railroads—mostly short lines and industrial switching operations—are beginning to use biodiesel blends to fuel their diesel locomotives. The TriCounty Commuter Rail Authority in southern Florida, and the Minnesota Prairie Line Railroad in Minnesota were among the first. More recently, in October 2006, Gerdau Ameristeel Inc. in Knoxville, Tennessee, switched to using a B5 blend in its fleet of nearly 30 locomotives. In August 2007, Eastman Chemical Company switched five onsite locomotives at its Kingsport, Tennessee, facility to a 20 percent biodiesel blend. The locomotives are in use at the Eastman facility and around-the-clock operations move an average of 12,000 railcars within the plant each month. The company had previously moved to B10 a few months earlier and experienced no problems with the initial move.[19]

In January 2008, in a rather unusual move, the Tri-City & Olympia Railroad Co. announced that it intends to make its own biodiesel to fuel its locomotives. Headquartered in Richland, Washington, the 20-mile short line railroad has already purchased $175,000 worth of storage tanks and other equipment for its planned biodiesel production facility that it hopes to locate in Richland. The refinery, to be operated by its Green Diesel Inc. subsidiary, will also reportedly experiment with a wide range of feedstock oils in an effort to find the best feedstock at the lowest price for its needs. There should be a ready supply, since the railroad hauls 26 different kinds of oils used in ConAgra's potato processing plants in the Mid-Columbia region. "We're going to test different types of those oils, to see which are economically sound," says David Samples, the railroad's business development director.[20] If the TC&O's bold venture into biodiesel proceeds as planned, it will set this little railroad apart from virtually all of the competition.

Biodiesel and Uncle Sam

From the U.S. Marine Corps Base in Camp Lejeune, North Carolina, to Everett Naval Station in the Puget Sound area of Washington State, military installations across the nation are using biodiesel—and a lot of it— in their diesel-powered vehicles. The U.S. military, in fact, is the single largest user of biodiesel in the country, consuming more than 6 million gallons annually. The U.S. Army, Navy, Air Force, and Marines all use B20 in their nontactical vehicles. Of the four branches of the military, the Marine Corps uses B20 in the largest number of locations. "We use biodiesel to help us meet our federal alternative-fuel requirements and to reduce our petroleum fuel consumption to meet the Executive Order directing the government to do so, and on a third level it is just the right thing to do," said Tim Campbell, Headquarters Marine Corps GME program manager. "We've had no reported maintenance issues. I asked the bases to contact me with their experiences, negative or positive, with biodiesel. I received only positive feedback." Most of the military installations get their biodiesel through the Defense Energy Support Center (DESC), which coordinates the federal government's fuel purchases. DESC is the largest single purchaser of biodiesel in the country and for the 2003–'04 period had requirements for 5.2 million gallons of B20 for both military and civilian locations across the country.[21]

The U.S. Marine Corps Base at Camp Lejeune, North Carolina, has used biodiesel since 2002 in approximately three to four hundred pieces of equipment, including buses, Caterpillar tractors, and bulldozers, and consumes approximately 148,000 gallons of B20 a year. Scott Air Force Base in Illinois, located about 30 miles east of St. Louis, Missouri, serves as headquarters for 10 Air Mobility Command bases throughout the nation. Scott AFB has used B20 since April 2001 and consumes about 75,000 gallons every year. Biodiesel can be found on at least 40 Air Force bases nationwide. Located in the Puget Sound area, the Everett Naval Station in Everett, Washington, has used about 50,000 gallons of B20 annually since 2001. The change to biodiesel was virtually seamless, according to transportation director Gary Passmore. "Older equipment took a filter change, but newer equipment needed nothing," he said. "It went so smoothly that no one really noticed."[22]

The Navy probably deserves the credit for one of the most interesting biodiesel developments. On October 30, 2003, there was a ribbon-cutting ceremony at Navy Base Ventura County in Port Hueneme, California, that could have big implications for both the biodiesel industry and the military. The ceremony, attended by about a hundred people, marked the unveiling of the Navy's first mini-refinery for converting used restaurant cooking oil into biodiesel. The refinery, developed by Biodiesel Industries of Santa Barbara, California, in cooperation with the Naval Facilities Engineering Service Center (NFESC) at the base, is small enough to fit on the back of a pickup truck and can produce the fuel in 200-gallon batches. "This is the culmination of four years of working with the U.S. Navy," says Russell Teall, CEO of Biodiesel Industries. "Our research and development of the Modular Production Unit has been completed and implemented in our civilian plants in Las Vegas and Australia. Now, with the cooperation of NFESC, we hope to continue making improvements so that it can soon be deployed at military installations around the world."[23]

Since the official ceremony, the base has been collecting used cooking oil and transforming it into biodiesel fuel for use in its vehicles at the naval facility. The base intends to use about 20,000 gallons of biodiesel in its own vehicles annually and will also be producing about 20,000 gallons each for the nearby Channel Islands National Park and for Ventura County. The mini-refinery not only offers the naval facility a convenient way of disposing of a solid waste product but also provides a measure of energy security, according to Kurt Buehler, chemical engineer at NFESC. "If petroleum gets cut off, we can keep the base running on biodiesel," he notes. "So, in addition to reducing dependence on foreign oil, producing our own biodiesel could provide a tactical advantage in case of crisis."[24] This same idea is beginning to occur to planners and commanders throughout the U.S. military, and this creates a huge potential market for biodiesel and biodiesel technology providers. The concept of an "all-American" fuel in the national defense sector is a compelling argument.

In January 2005, the Navy issued a memo directing that all U.S. Navy and Marine nontactical diesel vehicles to be fueled with a B20 blend as of June 1, 2005 as part of a larger effort to increase the use of domestically produced, renewable fuels by the military.

Even the U.S. Coast Guard has gotten into the act with a biodiesel trial in one of its 41-foot boats. In 2004, the Coast Guard's Office of Naval Engineering decided to conduct an in-depth engineering study to see if biodiesel could be used successfully in some of its 1,400 boats under 65 feet long—many of which are potentially well suited to being powered with biodiesel. The 41-foot utility boat, nicknamed "soy boat," is powered by two 400-horsepower diesel engines with separate fuel tanks, which allowed for one engine to be fueled with B20 and the other with regular petrodiesel—and for a convenient side-by-side comparison of their performance.[25] The test, conducted over a period of two and a half years at the Coast Guard Academy at New London, Connecticut, was a success, according to Professor Andrew Foley from the academy's engineering department. "The study was conducted by three groups of undergraduates," he says. "The emissions from the biodiesel-fueled engine were much cleaner, and overall there was no difference in performance."[26] This was followed by a provision in the Coast Guard and Maritime Transportation Act of 2006, directing the Coast Guard to conduct a much larger feasibility study of biodiesel use in new and existing Coast Guard vehicles and vessels in its entire fleet.

Biodiesel in the Park

The highly publicized 1994 "Truck in the Park" project conducted in Yellowstone National Park (the first national park in the nation to test biodiesel) has led to additional biodiesel use in the park. The project now has been expanded to include tour buses, garbage trucks, and heavy equipment as well as boilers. Beginning in the spring of 2002, all of the park's 300 diesel-powered vehicles began running on B20. In October of the same year, the first public biodiesel pump in Montana was opened at the Econo-Mart in West Yellowstone. In addition, biodiesel and ethanol are now available to the general public at gas stations within the park itself. "We're stewards protecting this national treasure, and using biodiesel is one way we can best do that," said Jim Evanoff, management assistant at Yellowstone. "I have talked with hundreds of visitors about biodiesel use at the park, and the majority of our visitors are really interested in renewable fuels. This is an educational program as much as it is an environmental one."[27]

But Yellowstone is not the only national park to adopt biodiesel. Mammoth Cave National Park in south-central Kentucky was the first national park to have virtually all of its vehicles powered by alternate fuels. All of the park's transit and support vehicles run on either biodiesel or ethanol. The park's tractors, backhoes, graders, and even riding lawn mowers are running on B20. Two ferries on the Green River, which carry about 300 cars and light trucks each day during the summer, run on biodiesel as well. The park's tour buses, however, run on propane. The 10,000 to 12,000 gallons of biodiesel consumed by the park annually are supplied by Griffin Industries of Cold Spring, Kentucky.[28] The biodiesel projects at Yellowstone and Mammoth Cave have been great successes with park employees and visitors alike, and the National Park Service has since introduced biodiesel to more than 50 parks across the country—including Harpers Ferry National Historic Park, Everglades National Park, Glacier National Park, Yosemite National Park, Hawaii Volcanoes National Park, and many more—through the Green Energy Parks Program.

One of these parks, the Channel Islands National Park, located off the coast of Southern California, has been able to demonstrate the many benefits of using biodiesel in a marine environment. The park consists of five islands and the surrounding mile of ocean, totaling 249,489 acres. The islands' isolation has protected them from development, but presents park management with a real challenge in maintaining energy services. But the greatest challenge of all was marine transportation. The park's boat fleet used more than 70,000 gallons of diesel fuel annually. In August 2000, the park implemented its biodiesel program and now uses B100 in the vessels *Pacific Ranger* and *Sea Ranger II*, as well as in diesel equipment, including stationary power generators and forklifts, on the islands. The use of biodiesel and other renewable resources makes the islands petroleum-free. "We are an environmental organization, and we should be willing to be in the forefront in demonstrating things that have a positive environmental impact," said Kent Bullard, maintenance supervisor at the park. "It has been seamless. We haven't had any performance issues; the biodiesel is performing just as well as diesel."[29]

In December 2003, Yosemite became the first national park in the country to produce its own biodiesel onsite with the arrival of a small processing unit developed by Biodiesel Industries of Santa Barbara, California. Delaware North Parks and Resorts, the concessionaire that provides guest services in the park, now uses the processor to convert used restaurant cooking oil into biodiesel for Park Service vehicles.[30] Other parks with restaurant facilities are expected to follow this example in the future.

In 2005, the National Park Service used about 84,000 gallons of biodiesel at many of its parks, a dramatic increase over its initial early experiments with one tank full in one vehicle at Yellowstone National Park in 1994.[31]

Municipalities Go Green

For quite a few years, many cities and towns across the country—Keene, New Hampshire; Takoma Park, Maryland; Columbia, Missouri; Breckenridge, Colorado; and Missoula, Montana; to name just a few— have been switching some or all of their vehicles to run on biodiesel blends. Berkeley, California, was one of the first large cities in the nation to switch to biodiesel. In 2001, thanks to strong encouragement from the Berkeley Ecology Center, a local community and environmental organization, the city became the first to run its entire fleet of recycling trucks on B20. The success of the switch impressed city officials, and they decided to expand the initiative to virtually all of the city's diesel-powered vehicles, including fire trucks, school buses, and public works vehicles. In April 2007, the City of San Francisco announced that it was switching to B20 in its entire diesel fleet of more than 1,500 vehicles, and actually achieved its goal ahead of schedule by the end of the year. This dramatic shift makes San Francisco the largest city in the United States to use B20 fleet-wide. "Every city bears responsibility for taking local action to address our global climate crisis," said Mayor Gavin Newsom. "When it comes to the use of alternative fuels, renewable energy sources and greening our City fleet, San Francisco is demonstrating leadership and commitment on every front."[32] Seattle, Washington; Portland, Oregon; Grand Rapids, Michigan; and New York City all have switched to biodiesel blends in portions of their municipal fleets as well. These cities combined have cut their use of petrodiesel by at least 2.3 million gallons per year.[33]

Biodiesel Goes to College

Although there has been a good deal of biodiesel research at a number of colleges and universities in the United States, institutions of higher learning in general were slow to adopt the use of biodiesel on their campuses. In February 2004, however, Harvard University in Cambridge, Massachusetts, announced that it had begun to use B20 in all of its diesel vehicles and equipment, including shuttle buses, mail trucks, and solid waste and recycling trucks. Although a number of alternate fuels were studied, biodiesel was finally selected because it provided the greatest health and environmental benefits in the most cost-effective way, according to David Harris Jr., general manager of transportation services at Harvard. But there were other reasons for the switch as well. "Harvard is not a stand-alone campus," Harris said. "Our shuttle buses drive down the streets of Cambridge, past houses and other schools. We feel a responsibility to be a good neighbor and be as environmentally friendly as possible. Biodiesel helps us accomplish that using the vehicles we already have."[34] World Energy of Massachusetts supplied the biodiesel.

In April 2004, Purdue University in Indiana announced the introduction of B2 in as many as 80 of its vehicles, including nine university buses and dozens of diesel trucks. "Our decision to use biodiesel represents a balance between supporting the ag community and also keeping our diesel fuel costs reasonable," said Mike Funk, Purdue's director of transportation, about the relatively modest amount of biodiesel involved in the switch.[35] Yet Purdue's decision is definitely a step in the right direction, one that other institutions could follow easily without causing any major financial problems. Other institutions of higher learning using biodiesel include the University of Colorado; the University of New Hampshire; North Carolina State University; the University of South Carolina; Clemson University; Indiana University; the University of Michigan; Northwest Missouri State University; and, of course, the University of Idaho, where U.S. biodiesel research began. And the list continues to grow.

Biodiesel on the Slopes

The ski industry is one of the sectors of the economy that is most vulnerable to global warming. As early as 2000, The National Ski Areas Association, in partnership with the Natural Resources Defense Council, began to encourage ski resorts to use renewable sources of energy in their operations. In January 2001, Aspen Skiing Company released its Sustainability Report, the first formal sustainability report in the industry, which took a candid look at the company's environmental impact and explained what it was doing about it. Every year the company uses about

260,000 gallons of diesel fuel for activities like trail grooming by snow-cats. The emissions from these tracked vehicles create a good deal of air pollution, and Aspen decided immediately to begin the switch to B20. After initial tests proved to be a success, Aspen switched its entire fleet of snowcats to biodiesel in the winter of 2003. Eventually Aspen hopes to make biodiesel from used cooking oil from its restaurants.

In February 2004, New Hampshire's Cranmore Mountain Resort announced that it was switching to biodiesel to power all of its snow-grooming machines. The Cranmore Mountain biodiesel project was a collaboration between Cranmore, the New Hampshire Department of Environmental Services, and the Granite State Clean Cities Coalition. The 5,000 gallons of fuel used monthly in the initiative were provided by World Energy of Massachusetts. Cranmore was the first eastern winter resort to adopt the use of biodiesel. "Cranmore is passionate about taking measures to help the environment," said Jim Mersereau, mountain manager for Cranmore Mountain Resort. "We are proud to be the first resort in the East to use this alternative fuel."[36]

In the 2004–'05 ski season, the Sugarbush Resort in Vermont began using B20 in snow-grooming and snow-plowing equipment at one of its mountains. The initial tests were successful, and the use of biodiesel was subsequently expanded to Sugarbush's entire operation. The Breckenridge Ski Resort in Colorado also has been a leader in adopting the use of biodiesel, and even has hosted a workshop for fleet managers who were interested in learning more about the fuel. The number of ski resorts using biodiesel continues to grow steadily across the nation, with the Mount Sunapee Resort in New Hampshire being one of the latest to switch to B20 for its snow-grooming and snow-removal equipment, as well as B5 to heat its base lodges and buildings.[37] And in Maine, the Sugarloaf Ski Resort has added biodiesel to its operations as well. So much for concerns about the viability of biodiesel in severe winter-weather conditions.

Biodiesel Events
The speed with which biodiesel use has been growing in the United

States has been exceeded only by the speed achieved by a biodiesel-fueled dragster, which set the world renewable-fuel speed record using B100. On September 14, 2002, driver Mark Smith pushed "Wild Thang" to 211 miles per hour on the 660-foot racetrack at the Ozark International Raceway in Rogersville, Missouri. The dragster used biodiesel produced by West Central Soy of Ralston, Iowa, to achieve the Guinness World Record. "The biodiesel definitely impressed me," said Russel Gehrke of Seymour, Missouri, who helped prepare the car for its run. "It ran just as fast as conventional, but much cleaner. The crowd really liked it, too. It smells a lot better than diesel. When you're burning a gallon of fuel per second like we are, you want something that is environmentally friendly. I would definitely use biodiesel again." The National Biodiesel Board, of course, was excited about the new record. "Achieving a speed of 211 miles per hour with biodiesel just underscores that this is a high-performance fuel," said Joe Jobe, NBB executive director.[38]

In September 2003, a 2001 Volkswagen Jetta TDI running on B100 captured an impressive array of performance awards at the 2003 Michelin Challenge Bibendum held at Infineon Raceway in Sonoma, California. Michelin bills the Challenge Bibendum as the largest environmental vehicle event in the world. It is considered a performance event rather than a competitive event and is intended to display advancements in vehicle technologies. Consequently, entrants are rated only with A, B, C, and D letters. The Jetta, entered by American Biofuels, captured more top ratings than any other production class vehicle, earning an A rating in six categories, including energy efficiency, carbon dioxide, and range. It achieved more than 60 miles per gallon while clocking some of the fastest lap times in the fuel efficiency event. "Overall, the progress towards sustainable mobility by all of the participating technologies and energy sources is very impressive," said Patrick Oliva, director of Challenge Bibendum. "Each year, the variety of technologies and creative innovations displayed offer proof that sustainable mobility is within our grasp."[39]

In September 2007, a diesel-powered motorcycle named "Die Moto" set a new world speed record of 130.62 miles per hour (210.2 kilometers per hour) at the 2007 Bonneville Salt Flats speed trials in Utah. The bike was

Figure 11. "Die Moto" Chip Chipman

built from parts from a BMW R1150 RT; a custom-built frame, gearbox, and fairings were created to accommodate a two-liter turbodiesel engine from a BMW 3 Series car and fueled with B100. The unique custom motorcycle was designed and built by a team of environmentally conscious vehicle enthusiasts at "The Crucible," an industrial arts school in Oakland, California, and ridden by the school's founder and executive director Michael Sturtz. "It's great to know we have the fastest diesel motorcycle in the world—but it's about so much more than that," Sturtz says. "The challenge was to demonstrate the capability of biodiesel and call attention to the need for automotive technology to integrate environmental responsibility with performance. We've proved that style, speed, and environmental efficiency can come together in one vehicle with more than 210 km/hr already achieved with only 22 percent of the emissions of a standard diesel engine." The motorcycle is capable of running on biodiesel, straight vegetable oil, or regular petrodiesel fuel. The team that built Die Moto hopes to achieve even higher speeds once a few technical issues with the bike's computerized engine management system are sorted out.[40]

Growing Pains

Despite all of these exciting developments, the U.S. biodiesel industry is facing some real challenges. Like their counterparts around the globe, U.S. biodiesel producers are struggling with record-high prices for virtually all of their feedstocks, but especially for edible oils such as soy, canola, palm, and sunflower. The higher price for soybean oil, in particular, is mainly due to the fact that U.S. farmers have been planting more corn to meet the surge in demand for corn-based ethanol at the expense of soybeans. Since feedstocks now represent approximately 85 percent (or more) of the cost of production for biodiesel, it's hard to imagine how the industry can overcome this problem any time soon without a significant feedstock breakthrough. Tinkering with improvements to production process technology simply cannot overcome this fundamental feedstock problem.

One of the main differences between the U.S. biodiesel industry and its EU counterparts is that a growing number of European governments appear to be increasingly unwilling to continue special tax breaks and other supports for an industry, which, until fairly recently, was quite successful (admittedly, due mainly to those same tax breaks). Meanwhile, in the United States, Congress recently approved continued support for the blending tax credit and other programs, providing a safety net for the industry. But even this continued support is not enough to make the production of biodiesel price-competitive with petrodiesel, and many U.S. producers, especially smaller producers, have been losing between .05 and .20 cents on every gallon they make.[41] This does not auger well for the industry, at least in the short term.

"Near term, I think we're faced with an incredibly difficult situation," says John Plaza, the founder and president of Imperium Renewables, Inc. in Seattle, Washington. "This is mostly due to the present cost of feedstock. Soybean oil is at a 35-year high, palm oil is the highest it's ever been, and it's the same situation with canola oil. The basic ingredient of our product has gone up faster than the price of crude oil and at a higher percentage. This model is no longer sustainable for many companies, and throughout the Midwest biodiesel facilities are shutting down, shut down, or going bankrupt." As a result, the short-term potential for the industry

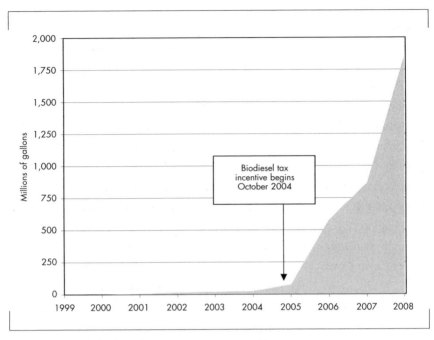

Figure 12. U.S. Biodiesel Production Capacity, 1999–2008 National Biodiesel Board

has been considerably reduced, according to Plaza.[42] Even Imperium has not been immune from the recent dramatic downturn in the industry. In December 2007, its CEO, Martin Tobias, who had been successful in attracting much of the venture capital for the company, stepped down, and Imperium announced that it was postponing its planned initial public stock offering due to "current market conditions." In early January 2008, the company also announced that it was cutting its workforce "to address short-term challenges and ensure the company's long-term growth."[43]

Many others agree that the industry faces significant challenges. "It's a tough time for biodiesel right now," says Jeff Stroburg, CEO and chairman of Renewable Energy Group based in Ames, Iowa. Stroburg says that the high cost of feedstock has hit the industry hard in 2007. "Which means that from a profit and loss perspective, we would be losing money on every gallon."[44]

But this is only part of the problem. The dramatic growth of the U.S. biodiesel industry in the past few years—fueled in large part by enormous amounts of speculative money that has been pumped into the industry recently—has resulted in a huge overcapacity for production that has far outpaced the ability of the market to absorb it. As of the end of 2007, the industry overall was operating at only about 20 percent of its total capacity, resulting in tight (or negative) profit margins, plant closures, and a dramatic reduction in new plant construction. "We obviously have an overcapacity situation right now, and we have been seeing the number of plants under construction shrink month by month, which is good," says Tom Bryan, editor of *Biodiesel* magazine, the industry's main trade publication. "Some of those plants have come online and some have just dropped off the list, but the main reason the number is shrinking is that they are not breaking ground for more new plants, at least not at the rate they were in recent years."[45]

"We have extremely high capacity to produce biodiesel, but it just is not sustainable right now at these levels of what the feedstocks cost," says Ed Hegland, chairman of the National Biodiesel Board. "And so it really is dire straits at this point."[46] Eventually, supply and demand will probably come back into balance, but in the meantime it's going to be a difficult period for many producers, especially if feedstock costs remain high.

PART FOUR

Biodiesel
in the
Future

13

Looking Ahead

In June 2008, the cost of gasoline reached a record-high price of $4.07 per gallon on average across the United States, and U.S. light crude hit $138.54 a barrel, the highest level in 24 years of recorded oil prices. Crude-oil prices jumped a stunning 58 percent in 2007, the largest annual gain in a decade. Just four years earlier light crude had been selling for $29.21 a barrel. These dramatic price increases were due to a wide range of domestic and international issues, but especially to a number of militant attacks on Western oil installations in Nigeria and other nations, continued declining supply from numerous older oil fields around the world combined with rising demand, along with continuing instability in the Middle East. Oil traders don't like instability of any kind, and the ongoing chaos in the Middle East and elsewhere has been giving them fits.

In the United States, the steady rise in gas prices at the pump has sparked increased interest in hybrid-electric as well as diesel-powered vehicles. As more and more people dumped their SUVs and other gas-guzzling cars, trucks, and motor homes in favor of more fuel-efficient vehicles, sales of diesel-powered cars and light trucks also began to accelerate around the country. And many people who bought those diesel vehicles were planning to use biodiesel as fuel, according to some dealers.

What does all this portend for the biodiesel industry? Although it's difficult to predict exactly where petroleum prices will be in the future, it's a fairly safe assumption that if demand remains strong, especially in countries such as India and China (which most industry analysts had underestimated), and reserves and production capacity remain limited (which seems likely), there will be a general upward trend in prices in the years to come. More and more oil-industry analysts recently have come to the view that oil prices probably have reached a new plateau and may not come down much at all (except, temporarily, during a recession), and that further price increases are inevitable. Some say that global oil production has peaked already. Whether this turns out to be true or not, the tipping point for petroleum may not be as far off as many had predicted.

But bad news in the petroleum sector has generally been viewed as good news in the biodiesel sector. The conventional wisdom about the price of biodiesel was that as the cost of petrodiesel increased in the future, the cost of biodiesel would rise at a slower pace, eventually making biodiesel less expensive. To the surprise of many, however, this has not happened. In fact, just the opposite has occurred, making biodiesel even more expensive relative to petrodiesel. This dramatic price increase is partly due to rising demand for biofuel feedstocks, but especially to rising international demand for edible oils such as soy, canola, and palm in the food industry. This is a crucial issue, since the food industry will probably always outbid the energy industry. A combination of global-warming-related bad harvests and strong demand has sent edible oil prices skyrocketing to record levels in 2008 around the world. The fact that increasingly expensive petroleum-based fuels and chemicals are used extensively in the production of most biodiesel feedstock crops has unquestionably also been a factor as well.

When edible oils became too expensive, many new multifeedstock biodiesel refineries switched to waste cooking oil or animal fats. Unfortunately, as mentioned earlier, these feedstocks are relatively limited in supply, and the dramatic upsurge in demand for them quickly raised their prices to record levels as well. This, combined with a huge overcapacity in global biodiesel production facilities, has led to a severe crisis throughout most of the global biodiesel industry in 2007 and 2008. Although the industry remains relatively robust in some countries such as Brazil and China, in others such as Australia, Malaysia, Indonesia, the United States, and the EU, most biodiesel companies are struggling. With the price of crude palm oil soaring to more than $1,000 a ton in early 2008 (up from $410 in 2006), it's hard to see how palm-based biodiesel, for example, can compete when the price of petrodiesel is at $780 a ton in Singapore.[1]

In an attempt to cut their losses, many biodiesel refineries (including some that have been mentioned in previous chapters) are operating at greatly reduced output or have closed altogether. Whether this is a temporary situation that will resolve itself once supply and demand come back into balance, or whether this is a longer-term condition is still not entirely clear. It would seem, however, that the industry is already beginning to bump up against the upper limits of where it can go with conventional edible oil feedstocks. "The prospects for the biodiesel industry are very slim, unless feedstock prices come down and we see mandates and subsidies in place," says Singapore-based biofuels analyst Chris de Lavigne.[2]

It does appear, however, that some biodiesel companies are better positioned to survive in a challenging market than others. In a survey conducted in June 2007 with hundreds of U.S. biodiesel professionals, it was found that two biodiesel plant sizes are better able to ride out hard times. The first group was composed of small regional operations in the 3-million-gallon range that rely on local feedstocks and produce biodiesel for local customers, eliminating most of the transport costs associated with larger operations. The second group included plants in the 30- to 80-million-gallon range that have the size and clout to negotiate long-term feedstock and fuel sale agreements, as well as deep pockets to help them weather temporary financial setbacks. In addition, the study found that multiple feedstock capabilities are now considered to be the minimum requirement for virtually all new facilities.[3] Many observers predict a significant consolidation in the industry, leaving only the largest and most well-managed companies standing after the dust settles. In any case, the industry faces numerous challenges in the years ahead.

Nevertheless, many other observers remain optimistic. According to a report published by Fuji Keizai USA in October 2007, "the global biodiesel market is estimated to reach 37 billion gallons by 2016, growing at an average annual rate of 42 percent." The report predicts that Europe will continue to dominate the market for the next decade, followed closely by the United States. In a separate report from BCC Research, biofuels industry fundamentals are viewed as good in the long term, and the world biorefinery market is projected to grow to annual sales of $155 billion in 2012, up from an estimated $84.7 billion in 2007.[4]

KEY ISSUES

Aside from current high feedstock costs, a number of key issues need to be addressed if the industry is going to achieve its full potential. Raffaello Garofalo, the secretary-general of the European Biodiesel Board, has some thoughts on these issues. First is quality. "Quality is really crucial because consumers purchase biodiesel not only because it is good for the environment, but because it performs well in their engines," he says. "Quality is also the way to gain acceptance by the auto industry as well as by our customers."[5]

The second concept is what Garofalo refers to as "progressive development," which relates to a steady, careful, long-term strategy for the industry, something that definitely has not been happening recently. "Biodiesel should not be seen as simply an adventure by people who produce biodiesel in their basement," he says. "It requires huge investments, as well as a long-term vision with big companies creating a real industry. The petroleum industry requires us to provide them with large quantities of biodiesel at the times and places where they need it. And that will take steady, progressive development instead of short-term considerations such as building a biodiesel plant in order to get subsidies and then ignoring quality; this is the worst enemy of biodiesel."

Finally Garofalo describes what he calls the "complementary approach" for the industry. "This means that biodiesel cannot be developed in opposition to other industries," he says. "Biodiesel needs to be combined with the contributions of all the other stakeholders, including the petroleum industry; we have to work together with all of them. The same applies to original equipment manufacturers, and of course we have to work with public authorities in order to get the taxation support that we need. I am convinced that the success of the industry requires the contribution and cooperation of everyone, from public authorities, car manufacturers, the petroleum industry, the vegetable-oil industry, the consumer, and even the other biofuels industries. Sometimes in Europe you hear people saying that bioethanol is better than biodiesel or the opposite. This is not helpful for us or for them. If we cooperate with all of these stakeholders, it will give biodiesel a better chance to develop, not only in Europe but worldwide."[6]

Many people in the industry also admit that there is a need for greater cooperation within the industry itself. The 2004 Biodiesel Conference & Expo held in Palm Springs, California, provided a venue to begin to address that issue, and many others as well. The Expo was a unique opportunity for about 500 participants to attend presentations, take part in panel discussions, and have one-on-one conversations with representatives of virtually every segment of the industry. Some of the key issues covered were: the dramatic growth of the industry, new products, research and development activities, the need for continued marketing and promotion, and the evolution of the NBB along with the industry. The mood was upbeat. Almost everyone who attended agreed that the gathering was an exciting and valuable event.

But for some, the Expo bordered on an emotional experience, particularly for those individuals who have been involved with the industry from the early years. Bill Ayres, formerly with Midwest Biofuels and Ag Environmental Products, attended the gathering along with Kenlon Johannes, CEO of the Kansas Soybean Commission & Association. (Johannes, a tireless supporter of biodiesel for many years, was the first executive director of the National SoyDiesel Development Board.) "Kenlon Johannes and I were sitting in the back of the room listening to Daryl Hannah and all these top people from the USDA and DOE talking about biodiesel," Ayres relates, "and it was kind of like watching your kid graduate from college; the industry really has matured and has so much going for it now. It's really exciting to see the growth in the industry, and I think it's going to do quite well in the future."[7]

Four years later, the 2008 Biodiesel Conference & Expo, held in

Orlando, Florida, had a somewhat different focus, and a different mood. The main theme, "Navigating a Changing Landscape," focused on ways to chart a course through a constantly evolving market. The gathering also provided an opportunity to take stock of progress to date, and to look at the many roadblocks that the industry is facing. "Despite reaching major milestones, variables in other industries have a huge impact on biodiesel's future," the promotional materials for the event said, hinting at the turmoil in the international vegetable-oil markets that have caused so many problems for the biodiesel industry. Candid observations about "current economic realities" were a feature of the keynote address in the first general session. Along with the usual technical subjects, discussions and workshops focused on the "next generation" of biodiesel feedstocks, and other topics such as algae and jatropha, aimed at moving the industry beyond its heavy reliance on traditional edible oils. Other topics included sustainable biodiesel production, food versus fuel, even "Biodiesel: Turning Failure into Success." Actress Daryl Hannah made an encore appearance and told the attendees: "We need to educate ourselves in sustainable feedstocks, production and distribution models. We need to take the lead and rebuild the faith and promise of biodiesel." Clearly, this was a more subdued event that focused on the industry's current problems, although there was still a strong undercurrent of optimism for the industry's long-term prospects.

QUALITY

The fact that quality was at the top of Raffaello Garofalo's list of key issues for the European biodiesel industry is no accident. Virtually everyone in the industry around the world acknowledges the central role that quality needs to play in the future of biodiesel. But it's also a fact that quality—or lack thereof—occasionally has been an issue for many producers, both large and small. Over the years, there have been repeated reports of off-spec biodiesel being made by producers in many different countries. In some instances, there may have been considerable justification for the claims, while in other cases the rumors appear to have been totally false. In the United States, a national fuel-quality testing project, co-funded by the NBB and the National Renewable Energy Laboratory, found that one-third of the biodiesel samples taken between November 2005 and July 2006 were out of spec as a result of incomplete processing. This was the same issue that caused some filter-clogging problems in Minnesota during the early implementation of the B2 mandate. Although fuel quality is always important, cold weather can intensify problems caused by off-spec fuel. Regardless of the particular circumstances surrounding these incidents, bad biodiesel is bad for the industry. "The standard myth from the industry's point of view is that big producers make better fuel than small producers," says Tom Leue, a longtime small-scale biodiesel producer and activist in Massachusetts. "But the reality is that they don't always do that. Nevertheless, small producers do need to 'get the religion' of quality control; we simply have to make sure that we produce quality biodiesel. There is no point in producing bad fuel, it doesn't help anybody."[8]

"Quality is definitely job one," says the National Biodiesel Board's CEO Joe Jobe. In response to ongoing quality concerns, the NBB has instituted a quality-assurance program in the U.S. called BQ-9000, Quality Management System Requirements for the Biodiesel Industry. Developed and implemented by the National Biodiesel Accreditation Commission, an autonomous nine-member group appointed by the NBB, the voluntary program is designed to ensure more consistent quality for biodiesel from producer to end user. There are two levels of certification: accredited producer and certified marketer. "You can apply for both if you are a producer as well as a marketer," Jobe explains. "If you are a certified marketer and buy from an accredited producer, then you can rely on some of the assurances of the accredited producer. But if you are a certified marketer and you buy from a nonaccredited producer, then the burden of testing and storage and so on is your responsibility. It's a pretty sophisticated program that is based on typical ISO 9000-type programs that are prevalent in almost any manufacturing industry. We already have a specification that assures quality biodiesel production. But BQ-9000 is intended to ensure quality from the plant gate to the fuel tank, because once biodiesel leaves the plant there are all kinds of things that can happen to it before it gets to the end user that can cause problems."[9] Germany's Association for Quality Management of Biodiesel (AGQM) and similar organizations in other countries around the world reflect the widespread recognition of the importance of quality control to the future success of the industry.

SMALL REGIONAL MARKETS

Because biodiesel is so easy to make on a small scale, there are numerous possible models that could be followed for local production by small organizations. This is especially true if their main focus is on producing fuel from local resources for local consumption, rather than trying to satisfy Wall Street investors. "I think that it would be relatively easy for almost every county in the nation to take a lot of their used cooking oil and even some of their virgin oils and use them in the production of biodiesel for their own fleets of vehicles," says Daryl Reece, vice president of engineering at Pacific Biodiesel. "Sure, it's going to cost a few pennies a gallon more to do that, but it's a good alternative to petroleum-based fuels. You don't have the transportation costs because everything's right there. And you can produce it with one or two people, and then it's utilized locally. Some of these municipalities could even use biodiesel to generate their own power. There are very good tax incentives for green energy power generation right now. There's a lot of things that could be looked at, and they're all very positive."[10]

Following a similar line of thinking, planners at the University of Missouri in Columbia, Missouri, have suggested the creation of what they call "community-size" biodiesel hubs. This strategy involves the production and use of biodiesel within regional agricultural/industrial centers or hubs. The production level for this model is in the 500,000-gallon-per-year range. This strategy, however, would require a good deal of planning and cooperation among all the participants in order to succeed.[11] In a variation on this theme, some independent producers are beginning to focus on niche or small regional markets that emphasize ecologically sound practices, such as avoiding the use of genetically modified crops and pesticides to grow bioenergy crops or using waste vegetable oil as a feedstock. By focusing on biodiesel as an organic and/or locally produced product that also helps stimulate the creation of local jobs, while keeping energy dollars circulating in the local economy, these producers are able to offer their environmentally savvy customers an attractive alternative to the large agribusiness model.

Another variation on this model recently has been promoted in Oregon, where in June 2007 the state established its new renewable-fuels standard. The legislation was somewhat unique because it also contained provisions for encouraging local feedstock production from a variety of crops, including canola, camelina, flax, and other plant matter. Local production of 5 million gallons of biodiesel will trigger the state's new B2 mandate. If local production reaches 15 million gallons, a B5 mandate will kick in. Smaller biodiesel producers in the state are promoting a "community-scale" industry based on local production from local feedstocks for local consumption. They say that this is a more sustainable model that especially benefits farmers and rural communities by keeping their energy dollars circulating locally. The state's petroleum distributors, on the other hand, maintain that this local model won't lead to enough production to satisfy the state's carbon-emission-reduction goals. It has been estimated that Oregon could produce between 10 and 20 million gallons of biodiesel a year with a local, community-based model, probably enough to satisfy the requirement for the B5 mandate. In any case, like smaller producers everywhere, Oregon's small biodiesel companies tend to be at a competitive disadvantage. They are faced with higher costs of production per gallon than their larger competitors, as well as the prohibitively high cost of fuel testing for small batches. In addition, the smaller producers recently have been having the same difficulties as the rest of the industry finding affordable feedstocks. Nevertheless, some of the small biodiesel companies in Oregon have managed to establish direct relationships with local feedstock suppliers to avoid the extreme price swings in the global commodities markets.[12]

These same ideas are increasingly popular at the community level in other parts of the nation as well. "I think it is inevitable that we start producing locally based fuel and that we utilize what have been badly managed resources," says Tom Leue. "I also think it is inevitable that biodiesel will take its rightful place on a solid footing alongside our dwindling petroleum resources. I don't think that biodiesel is the salvation of the world, simply because there isn't enough. But biodiesel is for those individuals and communities that can come to grips with their energy-use problems and who strive toward energy independence. They may not be able to get there completely, but at least they can move toward it. I don't have any illusions that we are going to solve all our problems with biodiesel, but I still think we've got to make the effort anyway. And I think that we can do it better than has been done to date by keeping it in the hands of local people rather than big industry."[13]

Piedmont Biofuels in Pittsboro, North Carolina (see Chapter 10) has been a co-sponsor (along with a number of other organizations) of the Sustainable Biodiesel Summit (SBS), usually held just before (and at the same location as) the National Biodiesel Board's annual conference. The SBS has a strong focus on smaller, community-scale business models, aimed at promoting greater environmental stewardship, shared economic development, and a stable, secure, self-reliant energy future for local communities.[14] This model is being studied in other parts of the country by various peak-oil-response and relocalization groups, to say nothing of locally focused initiatives like the nanohana projects in Japan or the community-based initiatives in India and South Africa (see Chapter 7).

In an energy-constrained future that could cause severe dislocations in the global economy, small-scale local fuel and energy production of all types ultimately may prove be the best strategy for community energy security. Many forward-looking communities around the world are beginning to initiate these types of projects, and biodiesel is often a part of the plan. (For additional information, see my 2007 book *The Citizen-Powered Energy Handbook* in the bibliography.)

GROWING A NEW ENERGY ECONOMY

A few optimistic observers believe that the United States has sufficient resources from fallow farmland, waste cooking oil, and all other sources to produce enough fuel to meet as much as 25 percent (13.8 billion gallons) of the nation's diesel needs. Most estimates, however, are more conservative and put the figure somewhere between 14 and 20 percent. Even more conservative observers say the figure is closer to 6 percent (3.3 billion gallons), given current agricultural practices. This latter estimate was more or less confirmed in March 2006 by the National Renewable Energy Laboratory, which estimated that current biodiesel feedstocks total about 2 billion gallons, and that an additional 1.8 billion gallons might be available by 2016.

In a 2002 German study on international biodiesel production potential, it was estimated that the upper limit of biodiesel replacement of petrodiesel worldwide was around 10 percent.[15] In some countries that figure might be lower, while in others it might be higher, depending on available feedstock and other factors, according to Werner Körbitz, chairman of the Austrian Biofuels Institute. "For example, Malaysia, which is producing a lot of palm oil, could reach a higher market share than, say, Egypt, a country that does not produce much oilseed," he says. "In Europe and North America, I think reaching 10 percent would be a challenge. Brazil and Argentina, on the other hand, might have an easier time since they grow a lot of soybeans. But then it all depends on whether you make more money selling the oil for food purposes or if you have surplus to sell to make biodiesel."[16]

A more recent report from 2006 suggests that with the addition of second-generation biodiesel feedstocks the EU might be able to replace between 12 and 18 percent of its current petrodiesel use.[17] Other recent studies estimate that with the assistance of major government support programs, combined with the development of second-generation feedstocks, it might be possible to replace 25 to 30 percent of global petrodiesel consumption. Obviously, there is no real consensus, and it's important to remember that crop failures due to global warming could render virtually all of these projections moot. Regardless of which figures are correct, it's clear that biodiesel is almost certainly never going to be a total replacement for all of the current petrodiesel market.

New Feedstocks

Trying to deal with that reality is a challenge. In the United States, simply requiring more fuel-efficient vehicles would almost certainly result in substantial reductions in petroleum use, but this is mainly a political, rather than a technical, problem. The gradual increase in fuel economy standards for cars and light trucks contained in the 2007 Energy Bill was a small step in the right general direction, but comes nowhere near what is really needed to make a significant difference. In addition, at some point the general public is going to have to realize that better vehicle mileage is simply not enough. It's clear that U.S. oil dependence has become so pervasive that it cannot be resolved without a combined strategy of various alternate energy sources in addition to serious conservation. And those necessary conservation strategies are going to alter many aspects of the global economy and people's daily lives fundamentally in the years ahead. That means that instead of just driving vehicles that get better mileage, people will have to figure out how to drive less. A lot less.

It's also increasingly clear that if biodiesel is going to play a significant role in this scenario, the U.S. biodiesel industry needs to lower the cost at the pump (relative to petrodiesel) as much as possible. And since about 85 percent of the cost of production is represented by feedstock, the cost of feedstock is the key factor that needs the most attention. This will not be easy, considering the recent high price of feedstocks and the growing competition between food and fuel. "For the consumers of the 55 billion gallons of diesel sold in this country every year the main issues are price, price, and price," says Jerrel Branson, formerly of Best BioFuels in Texas. "If the price was right, biodiesel would be at a pump near you tomorrow."[18] Jon Van Gerpen at the University of Idaho agrees. "Quite simply, biodiesel has been too expensive," he says. "If it had been a low-cost fuel, of course it would have happened immediately; people would have

bought all that the industry could make. But it has always been expensive, and so it's been an uphill battle."[19] Even with all the tax credits and
other incentives, biodiesel in the United States and many other countries
is still too expensive.

Since cost is such a key factor, people both inside and outside of the
biodiesel industry are questioning the wisdom of continuing to focus so
much attention on a feedcrop—the soybean—that is not very high on
the list of oil-producing seeds. "Given the very high price and low oil
yield of soybeans, it's just not economic to even think about making
biodiesel out of soybeans," Jerrel Branson maintains. "This is not the raw
material that will make a competitive product."[20] Dennis Griffin,
chairman of Griffin Industries in Kentucky, agrees. "Using an edible oil
to produce fuel has never made a lot of sense to me," he notes. "That's
why we wanted a multifeedstock type of production plant, so that as new,
more cost-effective feedstocks are found we can adapt to them—and they
certainly are not going to be in the edible oil class."[21]

Although most soybean farmers might take exception to these statements, nevertheless a lot of them recognize that, at the very least, the
industry needs to further diversify its selection of feedstocks considering
the current high prices for vegetable oils on the international market. And
these farmers are not necessarily opposed to this initiative, according to
NBB's Joe Jobe. "Soybean farmers have had a very broad vision of an
inclusive industry," he says. "They are often criticized for just the opposite,
but they have built an industry that is ultimately based on multifeedstock
markets." What's more, Jobe believes that as the biodiesel industry continues to expand, canola, mustard, and other alternate feedstock crops will
almost certainly play an increasingly significant role in the United States.
"I think we would have an agricultural response to that growth, so that
more oilseed varieties would be added to both cultivated and fallow acres,"
he predicts. "There is already some of this sort of response going on in
Colorado, Idaho, Hawaii, and other places in order to create higher-
yielding oil commodities. And this is part of the anticipated agricultural
response that would have a pretty substantial impact on our feedstock
capabilities." Nevertheless, Jobe acknowledges that even more needs to be

done about feedstock development, and he notes that the NBB has an ongoing program that is analyzing all available feedstock opportunities.[22]

Additional evidence that an agricultural response is already under way can be found in the Pacific Northwest, where rapeseed for biodiesel has been suggested as a good rotation crop for wheat farmers in Oregon and Washington. In Colorado and other High Plains states, canola and camelina have been tested as rotation crops by farmers in cooperation with Blue Sun Biodiesel. But the company's feedstock research is not limited to field crops, according to Jeff Probst, president and CEO of Blue Sun. "There are other technologies that need to be investigated and worked on, like algae," he says. "We can grow algae on a very cost-effective basis, but there is still a lot of development that needs to go on. People talk about these feedstocks now, but I think the private sector can actually make them happen. That's the really exciting part of this business; we don't have to rely on soybean oil. We want to grow our own industrial oilseed that doesn't have an impact on the food supply, and I think there is a lot more that can be done with that."[23]

The increasing role that the private sector is beginning to play in biodiesel research and development is a significant shift for the industry, which has relied on a good deal of government-supported research for many years. Admittedly, some of this new research is still partly funded by government grants, but industry's growing interest in biodiesel and second-generation feedstocks is viewed by most observers as a positive development. "One of the exciting things that's happening right now is that industry is seeing that there is a real benefit to using biodiesel," says Professor Leon Schumacher, a longtime biodiesel researcher at the University of Missouri in Columbia. "So they are getting involved in some of the research now, which is a good thing. They can really focus some effort on problems that perhaps haven't been solved previously and give them the extra push they need."[24]

In other countries, jatropha, which is capable of growing on marginal lands, has been receiving most of the attention as a key alternative to conventional oilseed feedstock crops. As we have seen on our global survey of the industry around the world in the previous chapters, millions of dollars are being spent on jatropha by governments and private companies in numerous countries. Some recent test plantings even have been made in the deserts in Egypt. Although most researchers stress that the economic viability of these experiments is still not entirely clear, they are generally optimistic about future prospects for jatropha. "We are really bullish about the future of jatropha in India, as the climate is perfect and there is plenty of land," says Sarju Singh, managing director of the UK-based D1 Oils India office. Others are a bit more cautious. "In theory, it looks good because it can be planted on marginal soil and at the same time it can satisfy your needs. But there may be some over-optimism," says Thomas Mielke, a Germany-based edible oils analyst.[25] Nevertheless, the research continues unabated, and in 2007, more than 40,000 liters (10,566 gallons) of jatropha B100 was used to fuel diesel Mercedes test vehicles successfully in India as part of an ongoing DaimlerChrysler demonstration project.

Regardless of what feedstock is used, trying to figure out ways of mitigating the gut-wrenching price swings in feedstock commodities that tend to cause so much instability in the biodiesel market remains a major challenge. "In both Europe and the United States, we have had these wild swings between profitability and loss," says Gene Gebolys, president of World Energy in Massachusetts. "This means you have a very hard time growing the capacity of the industry properly, because companies fail during long periods of loss, and then you have wild boom periods where everybody wants to build. This has already happened in Europe and the United States, and it's not a good way to grow capacity or grow a business. Right now, the way it is, it's really unstable."[26] Developing a broader array of feedstocks, especially inedible feedstocks grown on marginal land, is one way of smoothing out at least some of the larger bumps in the commodity markets.

Algae
Of all the research into new biodiesel feedstocks, none is generating more interest—and excitement—around the world than algae. This is because fast-growing algae, unlike conventional field crops, has the potential to

produce exponentially larger quantities of feedstock oil for biodiesel and other fuels without competing for limited amounts of cropland. An acre of good farmland can produce approximately 48 gallons of soy oil per year. An acre of marginal land set up for algae production could possibly produce 10,000 to 15,000 gallons (some say 30,000 or more) per year. In addition, an algae farm could be located almost anywhere. It can use seawater as well as pollutants from sewage treatment plants, power plants, or farms as nutrients. No wonder the international race is on to commercialize algal oil production.

As mentioned in Chapter 8, Aquaflow Bionomic Corporation, located in Marlborough, New Zealand, has grown "wild" strains of algae at a sewage treatment plant and processed them into biodiesel to fuel a demonstration vehicle in that nation's capitol of Wellington. In the United States, a similar project is taking place in Norfolk, Virginia, where scientists from Old Dominion University are growing algae in rooftop tanks at a sewage treatment plant run by the Hampton Roads Sanitation District next door to the university. This pilot project is capable of producing around 200 gallons of biodiesel a day. "Granted, that's not a lot," says Patrick Hatcher, the project leader and an ODU professor of chemistry and biochemistry. "But if you consider that this could be done at sewage plants across the country, then you're talking about a big, big thing."[27]

Yet there are challenges. Finding the right strains of algae to grow is one. Fine-tuning the oil extraction process is another. And the cost of production still has a long way to fall before it will be economically viable. The current cost of production is running around $20 per gallon according to a U.S. Department of Defense estimate. "If you can get algae oils down below $2 a gallon, then you'll be where you need to be," said Jennifer Holmgren, director of the renewable-fuels unit of UOP, an energy subsidiary of Honeywell International. "And there's a lot of people who think you can." Most observers believe that it will be many years before large-scale algal oil production really takes off; however, demonstration plants are probably only a year or two away.[28]

In October 2007, Chevron and the U.S. Department of Energy's

National Renewable Energy Laboratory entered into a collaborative research and development agreement for the production of liquid fuels from algae. "It's not backyard inventors at this point at all," says George Douglas, a spokesman for NREL. "It's folks with experience to move it forward."[29] Also in October 2007, AlgaeLink NV, a European leader in alternate energy production based in the Netherlands, announced the development of a new, patented, photobioreactor system for algae. Unlike other systems, AlgaeLink's relies on special UV-protected transparent sheets that can be assembled onsite into round, watertight tubes. This design greatly reduces transport costs for the reactors.[30]

Another interesting recent algae initiative is taking place in Wabuska, Nevada, where a company is making biodiesel from camelina and, if all goes well, from algae. Claude Sapp, principal for Infinifuel Biodiesel, has been working to turn the oldest geothermal plant in Nevada into a biodiesel processing facility. Using a former ethanol plant that has been retrofitted to make biodiesel, the refinery is powered by several geothermal electric generators. There is also plenty of hot geothermal water available, and eventually Sapp hopes to grow his own algae feedstock in geothermally heated ponds adjacent to the facility. Government researchers have been skeptical about growing algae in Nevada's desert climate because of the cool nights, but with geothermally heated water, Infinifuel can maintain a constant temperature in its algae ponds. "We can grow more algae and harvest it more often than we can dry crops," Sapp says. If the company's test ponds produce as much algae as hoped, there are plans to expand the algae operation to additional acreage nearby. Although the project is still in its early development phase, it sounds like a unique and potentially successful concept.[31]

Yet another pre-commercial algae project is in the works between U.S.-based Valcent Products, Inc. and Canadian Global Green Solutions, Inc. The two companies plan to build a pilot facility for the production of algae feedstock for biodiesel behind their research laboratory in Anthony, New Mexico. The companies claim to have made a breakthrough with their Vertigo system, which uses tall, clear plastic bags hung in rows in greenhouses to grow the algae oil. On December 12, 2007, the companies announced the results of a successful 90-day test, which indicated that a larger, one-acre facility could produce about 33,000 gallons of algal oil per acre annually. The companies plan to fine-tune the process and to expand the test facility.[32] There are many, many other companies around the world working on similar pilot projects. The real challenge is to bring all of this up to full-sized commercial scale and to make it economically viable.

"Algae is for the biodiesel industry what cellulosic ethanol is for the ethanol industry," says Joe Jobe. "We need to be investing more now in algae production. The U.S. Department of Energy is spending $2 billion this year on research related to cellulosic ethanol. Even if a fraction of those $2 billion were invested in algae it could produce some dramatic results. This needs to happen now if algae is going to come into commercialization even ten years from now."[33]

Agribusiness Concerns

Despite all the enthusiasm in the industry for various new feedstock crops, some observers—especially some in the environmental community, who otherwise are in favor of biodiesel—are troubled by the fact that corporate agribusiness is so heavily involved in supporting increased biodiesel use, and they are concerned about the genetically modified crops that often are used. Dr. Rico Cruz, of the Environmental Science and Technology Program of the Confederated Tribes of the Umatilla Indian Reservation in Washington State, has worked hard to establish small-scale biodiesel initiatives around the world, and he shares some of those misgivings. "I am concerned about the development of a monoculture," says Dr. Cruz. "What if there is a disease that wipes out the whole crop? My concern is about relying on one or two genetically engineered crops while having some other crops that have been here for thousands of years disappear because farmers have signed a contract with one of these big companies."[34]

Dr. Cruz is not the only one troubled by the GMO issue. A substantial proportion of the vegetable oil produced in the United States, especially from soybeans, comes from genetically modified crops. As far back as

2003, 81 percent of the U.S. soy crop was genetically modified, and by 2006, that figure had grown to 89 percent. It also has been estimated that 40 to 60 percent of the canola grown in the United States and Canada contains transgenic genes. And this trend probably will continue as large chemical and seed companies experiment with ways to "enhance" crops for the biofuels market. It should be noted, however, that the rapeseed grown in Europe does not rely on GMO crops, so it certainly is possible to produce significant quantities of biodiesel without them.

The possible use of GMO crops for biodiesel production in Oregon's Willamette Valley has had specialty vegetable seed producers in the region deeply concerned due to the potential contamination of their crops. "That (GMO seed) would be the nail in the coffin for the vegetable seed producers," says Jim Myers, an Oregon State University vegetable breeder.[35] The main problem is that most vegetable seed buyers, especially those outside the United States, will not tolerate GMO contamination of any sort. These same general concerns about GMO contamination of organic crops are widespread in other parts of the country as well, where organic farming is a fast-growing segment of the agricultural sector. "I understand the concerns with using GMOs in the biofuel supply," says Matt Atwood, an organic chemist and project manager for Biodiesel Systems of Madison, Wisconsin. "But fundamentally, as a scientist, you have to weigh the benefits against the detriments. Do I have a problem with GMO-only fuel crops? I feel the benefits far outweigh the negatives, and nobody really knows the full negatives yet. What we do know is that if we don't get this industry off the ground, and quick, there are likely to be much greater problems in the world than dealing with the issue of 'genetic drift.'"[36] The debate continues.

But some nagging concerns extend beyond the realm of genetic engineering. As biodiesel production and consumption continue to grow rapidly, additional questions are being raised about the implications of using food resources—all these soybeans and other edible-oil crops—for fuel rather than for food. Similar questions could be raised about the use of land and agricultural resources for the cultivation of inedible oil crops, unless they are grown on fallow or marginal lands using ecologically sound agricultural practices. As we've seen, in some parts of the world, biodiesel strategies are based primarily on the use of inedible crops grown on set-aside or marginal lands, eliminating or at least mitigating some of the concerns about using resources needed for food production, agribusiness involvement, and potential problems with land degradation.

Food versus Fuel

The food-versus-fuel debate has been ongoing for many years, but recently, as food prices have risen dramatically around the world, it has taken on a new urgency. The debate relates to an extremely complex series of issues that can be hard to untangle. One of the key factors, however, is the price of vegetable oils. If you understand what has been going on in the international vegetable-oil sector, then you will have a better grasp of how the growth of demand for biofuels has affected the vegetable-oil market—and vice versa.

Since 2000–'01, the world vegetable-oil sector has experienced the fastest rate of growth of any major agricultural commodity due to dramatically higher production and consumption. For 2006–'07, global edible oil production grew to about 37 billion gallons (124 million tons). Of this total, palm oil represented 31 percent, soy oil 29 percent, canola/rapeseed 15 percent, and sunflower seed oil 9 percent. The remaining 16 percent consisted of cottonseed, peanut, coconut, olive, and palm kernel oils. There have been three main trends that have driven the market: the expansion of the biofuels sector (ethanol and biodiesel), growing concerns in North American about trans-fatty acids (trans fats), and continued Asian economic and population growth.[37]

The concerns about trans fats have motivated many food manufacturers to shift away from partially hydrogenated soybean oil in favor of palm, canola, corn, and sunflower oils, driving up demand—and prices—for the latter group dramatically since 2003. This had virtually nothing to do with biofuels. However, more recently, the drop in demand for soybean oil for food in the United States has been more than made up for by rising demand in the biodiesel sector. For the 2006–'07 crop year, U.S. biodiesel production was estimated to account for 2.6 billion pounds of soybean oil, or 13 percent of total domestic soybean-oil use. If these trends continue, biodiesel easily could consume more than 25 percent of total U.S. soybean-oil production in the next few years. But there is another factor that complicates the picture even further—corn. The dramatic rise in demand for corn by the ethanol industry in the United States has had a direct effect on global food prices and an indirect impact on the vegetable-oil market. This is because between 2001 and 2007 about 17 million acres in the United States were shifted to corn production, mostly at the expense of soybeans and wheat. In 2007 alone, U.S. soybean acreage dropped 19 percent. It should come as no surprise, then, that the price of soybean oil has increased. Although it may be hard to get a grip on all of these statistics, another way to look at this is that the amount of grain needed to fill up the fuel tank on an SUV with ethanol would feed a person for a year, according to the World Bank.

There have been similar problems and pressures in other nations. In the EU, if all the biodiesel production capacity were actually used to make biodiesel, it would consume all of the rapeseed oil produced domestically—and more. In addition, the continued growth of demand for edible oils by Asian markets for food, most notably in India and China, has combined to create a perfect storm in the world's vegetable-oil market that has raised the prices of these oils to record highs, to the point where the biodiesel industry can no longer afford them and remain profitable.[38] Another way to put this in perspective is to look at how much vegetable oil is consumed for food and how much is consumed for biofuels. In 2003, about 1 percent of total world production of vegetable oils and fats was used for biofuels. In 2005, the biofuels percentage had risen to 3 percent. And in 2007, biofuels accounted for 7 percent and made up more than half of the increase in worldwide demand for vegetable oils, according to Oil World in Hamburg, Germany.[39] Nevertheless, 93 percent of edible oils was still used for food in 2007.

But increased prices for vegetable oils are not the only factor in higher food prices. Growing consumer demand for meat in countries like India and China also has resulted in higher demand for cereal grains to feed those animals, boosting prices for those feed grains. "India and China have definitely contributed to the recent price increases in vegetable oils due to their growing demand for food consumption," says Tun-Hsiang (Edward) Yu, international oilseed analyst from the Food and Agricultural Policy Research Institute at Iowa State University. "In China it's primarily soybean oil, while it's mostly palm oil in India. For different oils in different countries there are different scenarios, so it's a complicated picture."[40]

But there is yet another important factor that has driven recent price increases in oilseeds and grains: a sharp drop in supply due to adverse weather conditions in the United States, Canada, Australia, and the European Union. This recent decrease in supply, which amounted to 60 million tons, was four times as large as the recent increase in demand for biofuels, according to Loek Boonekamp, a division head in the Agro-food Trade and Markets Division at the Paris-based Organization for Economic Cooperation and Development. Consequently, the recent run-up in prices would have happened anyway, even without the rise in biofuel production, according to Boonekamp. "Closing your eyes and blaming the current high prices on biofuels is just too simplistic," he says.[41]

From all of this, it should be clear that the biodiesel industry is not solely responsible for the recent dramatic price increases for food or edible oils. Nevertheless, along with corn-based ethanol, biodiesel is unquestionably a contributing factor. "It's fair to say that biofuels are responsible for a portion of the increase, but it would be too strong to say that they are the dominant factor," Tun-Hsiang Yu, says.[42] And, given current trends, it is unlikely that this situation will change dramatically any time soon without the introduction of viable second-generation feedstocks that do not rely on edible oils or grains. Biofuels, supported by billions of dollars in government subsidies in the European Union, the United States, and other countries, could drive food prices 20 to 40 percent higher between now and 2020, according to the Washington-based International Food Policy Research Institute. Although not everyone agrees with these statistics, a growing number of nongovernmental organizations around the world have been raising the alarm as food prices continue to escalate.

Environmental Concerns

Although the use of corn for the production of ethanol (mainly in the United States) has generated a lot of criticism from a wide range of environmental activists, the use of palm oil as a biodiesel feedstock has also raised serious questions in recent years. The main problem centers on the conversion of tropical rain forests to oil-palm plantations in a number of countries, but especially in Indonesia. A number of environmental groups claim that tens of thousands of hectares of peatland and lowland forests are being cleared to plant oil-palm plantations to feed the burgeoning palm-oil market for food and biodiesel. Clearing forests and draining Indonesia's thick, carbon-storing peatland releases more than a billion tons of greenhouse gasses annually, according to Greenpeace Southeast Asia.[43] The habitat destruction involved with this activity also has endangered wildlife, most notably the orangutan.

But not everyone agrees that palm oil is the main culprit. "There's a misconception that oil palms are the reason for the destruction of the forests in Indonesia," says Pat Baskett, who runs PT Scofin, a palm-oil company that has been in business in Indonesia for 100 years. "Basically, the destruction of the forest is from timber operations." Even many environmentalists who work in the region agree that logging—legal or illegal—is a greater threat to the forests than most oil-palm plantations.[44] Regardless of who is right, groups such as the Roundtable on Sustainable Palm Oil (RSPO) have been organized to try to deal with these complex issues. As mentioned previously, the RSPO agreed to a set of principles and criteria governing oil-palm plantation development in November 2005. The group is now working on how to implement those principles and criteria.[45]

These concerns have been on the radar screen of the U.S. biodiesel industry for some time as well. "I think the issue of food versus fuel is very important," says John Plaza, Imperium Biofuels' founder and president. "But I think that deforestation is equally, if not more important. However the layers of complexity involved are not as simple as the short sound bites that are usually used in the media." Plaza says that his company was an early attendee at an RSPO gathering in Malaysia at about the same time that the palm-oil issue began to heat up in Europe. "We realized that this was an issue that we were going to have to understand and deal with, so we quickly joined the RSPO," he says. But Plaza maintains that the actual amount of palm oil used for biofuels relative to its use for food is very small. "To pin this palm-oil problem on biodiesel is irresponsible on the part of some in the environmental community because it's mainly an issue of food, not biofuels. The reality is that deforestation is caused by poverty. Even if the biodiesel industry decided tomorrow that it would not use palm oil ever again, would deforestation stop? Absolutely not."[46]

Joe Jobe agrees that dealing with the complex issues surrounding the palm-oil problem is important. "We passed a resolution at our November 2007 NBB board meeting to begin getting more involved in the development of sustainability issues," he says. "Also, one of the most encouraging things in the 2007 Energy Bill is that any biofuel that is going to be eligible is going to have to demonstrate that it meets a 50 percent life-cycle carbon reduction. Any feedstocks that are the result of deforested areas obviously would not meet these criteria. We are developing some tools to begin to deal with this issue, but we've got a long way to go. Nevertheless, we have now taken an official position that we are supportive of sustainable practices and that we are for the further development of those practices."[47] It's obvious that the biodiesel industry has been listening to its critics recently and is beginning to take some positive steps in the right general direction to address some of the concerns.

However, the industry, along with groups such as the RSPO, need to act quickly and decisively, because national governments—the EU in particular—are becoming directly involved in the growing environmental controversies surrounding biofuels. In January 2008, reflecting the recent shift in European public opinion, the EU proposed a new law that would prohibit the importation of biofuels from certain kinds of lands, specifically converted tropical forestland, wetlands, or grasslands. In addition, the draft law would require that biofuels used in the EU meet "a minimum level of greenhouse gas savings." Meeting those greenhouse gas requirements could be a real challenge for many present feedstocks depending on how the savings are interpreted. In any case, the new law primarily would affect palm oil imported from some Southeast Asian nations such as Indonesia and Malaysia, but it also might impact some imports from South America as well. These growing concerns are shared by many observers in other nations. "Different biofuels vary enormously in how eco-friendly they are," says William Laurance, a staff scientist at the Smithsonian Tropical Research Institute in Washington, D.C. "We need to be smart and promote the right biofuels."[48]

THE NEXT LEVEL

Although many observers believe that the U.S. biodiesel industry has been about 10 years behind the European biodiesel industry, there is no question that the industry in North America is catching up rapidly and represents the fastest-growing segment in the global arena. "The long view that has been taken by the farmers who supported the industry for so many years in this country has proven to be successful," says Joe Jobe, "because around 1999 we finally began to make the transition from the research and development phase to the commercialization phase. Since then, we've seen very rapid growth every year. The momentum and enthusiasm for biodiesel is growing, and the next challenge will be to get us from where we are now to the next level."[49] That next level, according to Jobe, includes broader public acceptance as well as the integration of biodiesel into future energy strategies at the governmental level and with the fuel and engine manufacturers. Although there have been significant advances in these areas, there is still plenty of room for additional progress. "Our overall broad vision for the industry involves varied markets and continued growth. The B20 market is currently our single largest market," Jobe continues, "but we also see a long-term growth in low-blend markets, because B2 or B5 is something that we can do as a nation to displace up to 5 percent of our diesel fuel. With the help of the tax credit, we can achieve those goals and displace 5 percent of petrodiesel fuel needs by growing it right here in the U.S. I'm very proud of our accomplishments, and we need to keep working hard on achieving even more."[50] The NBB now has set a target for biodiesel to meet 5 percent of the nation's diesel fuel needs by 2015, representing about 1.85 billion gallons.

The rapid growth in the biodiesel industry around the world—which in recent years has been breathtaking—has left little time to make sure that the growth was orderly and well managed. Some observers have been concerned that this could lead to problems. "We have to be careful that we don't go too fast," cautions Professor Leon Schumacher. "Any industry that is growing this quickly needs to try to control the process and make sure that the things that need to be in place are there."[51] Schumacher's concerns, which have been shared by many other longtime industry observers, have proven to be well founded. As we have seen in previous chapters, the construction of too much production capacity by overzealous investors has outstripped demand, and many of those investors around the world are now paying the price. Joe Jobe acknowledges that there will be some short-term challenges. "There's going to be some shakeouts and consolidations just like there are in any emerging industry," he says. "It will be painful as we go through that, but we'll come out on the other side and we'll continue pushing forward toward our goals. The long-term prospects for the industry are good. Ultimately, I believe that we will have an even more dramatic impact on our nation's energy supply because as we move further down the road with algae we will have greatly expanded production potential."[52]

Although many in the industry welcome any support they can get, the heavy reliance of the biodiesel industry on government subsidies, tax breaks, and research support is a matter of some concern. One thing that has become abundantly clear is that, although governments can give tax breaks and other incentives to the industry, they also can take them away. The recent reduction of support for biodiesel by some EU member nations, but especially Germany, has highlighted these concerns. "Not only in the U.S., but also in Europe, we are dependent on public support, and maybe too much on that support," says Raffaello Garofalo. "That will be a big challenge for the industry in the next ten to twenty years."[53] Jeff Probst agrees, saying, "I think the greatest challenge is to get out of the public sector and into the private sector. Everybody looks at the private sector as the bad guys who are concerned only about profits, but that's not true. The private sector is just as interested in cleaning up the air and getting ourselves weaned from imported foreign oil as any government official. What we really need is better public and private cooperation."[54]

THE BEST ALTERNATIVE

Virtually everyone in the industry around the world readily concedes that there isn't enough feedstock to allow biodiesel to completely replace petrodiesel. Some actually see this as a good thing, since replacing one total energy dependency—petroleum—with another—biodiesel—would hardly be an improvement. The primary strategy for the industry is to identify and fill key niche markets where biodiesel's use will do the most good, while reducing reliance on at least some petrodiesel. In addition, most in the industry also would admit that biodiesel isn't perfect. As we have seen, there are growing concerns about the environmental and social impacts of large-scale agribusiness activities associated with biodiesel production in a number of countries. In addition, the cold-weather storage, handling, and operational issues for biodiesel are probably the fuel's main disadvantages in regions with severe winter weather. But on balance, since there are very few viable alternatives in the liquid transport fuel sector (other than ethanol, which has its own set of problems), biodiesel does offer at least a partial solution to an increasingly endangered global economy facing a severe energy crisis in the years ahead.

"While biodiesel isn't the total solution, I think it probably is the best alternative fuel we have at the moment," says World Energy's Gene Gebolys. "If there is a broad recognition that making positive steps in the right direction with renewable fuels is something that is important, then biodiesel is going to succeed." Along the way, Gebolys predicts that there will be some rough spots, with periods of overbuilding or underbuilding of production capacity and instances of well-intentioned government policy that will have unintended consequences, and that may sometimes turn out to be counterproductive. "It's not always going to be pretty, but on balance, I think we are going to see generally steady growth of the industry," he says. "The biggest obstacle is inertia. We have to constantly try to move people away from what they did yesterday. I wouldn't say that there are too many obstacles in this industry that are unique or much different from those of any other new industry."[55]

Bob King of Pacific Biodiesel is optimistic about the industry's prospects, too, and says that many people are finally connecting the dots about petroleum dependency and foreign and domestic government policy. "I have one customer who has a bumper sticker on his vehicle that says 'Biodiesel: No War Required,'" he reports. But King is enthusiastic about biodiesel for many other reasons as well. "I understand economics, and I look at all the different alternative energy concepts that are out there—hydrogen fuel cells, ethanol, solar, wind—and I am just totally excited about biodiesel," he declares. "All those other fuels will be part of that future too, in their own niche, because we need all of them. But biodiesel just has so much going for it. It's easily produced and is used so efficiently in the diesel engine with technology that we already have. We hope to be long-term players in the future, but whatever happens I'm totally convinced that biodiesel is going to be a major portion of our energy future, so we're going to do what we can to make that happen. We're not hopelessly locked into a petroleum future if we don't want to be."[56] That's a good thing, since petroleum's future is looking increasingly unstable and unpredictable.

Gary Haer of West Central Cooperative in Iowa is also enthusiastic about the industry's future prospects. "I think that the biodiesel industry is poised for significant growth," he declares. "The product has exceptional lubrication and performance properties, is easy to use, and can be delivered through the existing infrastructure for petroleum fuels. And it's a reliable, renewable alternative fuel that comes from domestically produced resources. It just makes a lot of sense. We're able to keep our energy dollars in our own country instead of exporting them overseas to buy crude oil from foreign sources."[57]

In some countries, biodiesel offers additional advantages beyond good performance in diesel engines and energy security. "In our country, it's all about employment and empowerment for the low-income groups," says Darryl Melrose, owner of Biodiesel SA in South Africa. "Each country has its own initiatives and pros and cons regarding this fuel. In our case, we don't have a major air pollution problem, but creating jobs is an important issue here, and jatropha will play a big role. I don't see biodiesel being sold much cheaper here than regular diesel, and it probably won't have a major impact on our usage, given the amount of land we have available and our consumption. But it certainly will help with the bigger picture regarding employment."[58] Other countries, such as India, with large numbers of relatively impoverished rural populations see similar potential advantages with biodiesel.

A BRIDGE TO THE FUTURE

Many observers view biodiesel as a "bridge technology" that will help get the world from where it is today—almost totally dependent on liquid fossil fuels for transportation—to where it needs to be in the future—relying on a broad range of renewable energy strategies, conservation measures, and zero-emissions vehicles. Biodiesel is an immediate solution that fits current infrastructures, including those of highways, distribution networks, and vehicle fleets. It runs well in the millions of diesel vehicles already operating on roads, rails, and waterways around the world and can be stored in existing tanks at local filling stations and fuel dealerships. Biodiesel is unquestionably a good first step toward greater energy diversification and independence.

Gene Gebolys says that he used to agree with the bridge technology view, but now he's not so sure. "I don't know if it's a bridge technology or not," he says. "I just know that it's so much better than the alternative of doing nothing. What the future holds is anybody's guess. I think that the days of predicting that we are going to run out of oil in thirty years are over. We don't know exactly when we are going to run out of oil; we just know that the amount of oil is finite and that we are eventually going to run out. We don't know how much human activity is leading to global warming, but we do know that we are affecting that process with a reasonable amount of certainty. We also know, in general, that our reliance on foreign oil leads to instability in the Middle East. We know, in general, that these aren't good things, and therefore we are going to have to do whatever we can to begin to address them. I think biodiesel is a relatively painless way to do that, and I think it's going to be around for a long time. Whether it's a bridge to something else, I don't know. It may well be, but that's for the next generation to figure out. This generation should be focused on getting biodiesel into the mainstream."[59]

Daryl Reece of Pacific Biodiesel agrees about the long-term prospects for biodiesel. "I don't see the diesel engine being replaced anytime soon," he says. "I think we're going to have it around for many, many years in order to produce the torque and horsepower needed to run the many different types of equipment that rely on it. Gasoline-powered engines just can't match that kind of performance. That means there are good markets and excellent opportunities for biodiesel."[60] Nevertheless, it's important for everyone to understand that although the opportunities for biodiesel are substantial, they are definitely limited by available feedstocks. And no one should be under the illusion that biodiesel will solve all of our many looming energy problems.

The biodiesel industry is still young and in many ways like a recent college graduate. After years of study and hard work, the industry is finally embarked on a great adventure. And like any young person, the industry faces many challenges ahead. It needs to develop better feedstock choices from inedible oils such as jatropha and algae to mitigate the food-versus-fuel issue and to help level out the worst fluctuations in the commodities market. The industry needs to learn to work more cooperatively with the many different constituencies—both external and internal—that can help it succeed. Large producers, small producers, and homebrew advocates need to be allowed, and even helped, to find and develop their respective market niches. And all producers, regardless of their size, need to focus even more attention on producing a top-quality product—all of the time. The industry also needs to find the right balance between receiving government support and avoiding becoming totally dependent on it. It needs to take a steady, long-term approach of developing a stable, dependable, and ultimately profitable segment of the energy sector. The industry is now a global reality, and it faces global challenges. But those challenges can be met, and ultimately overcome, with greater international cooperation. Finally, the biodiesel industry needs to find—and settle into—its proper place in the new renewable energy economy of the twenty-first century that is being born even as the old fossil fuel economy comes to an end. Rudolf Diesel would be pleased.

ORGANIZATIONS AND ONLINE RESOURCES

AUSTRALIA

Biofuels Association of Australia
P.O. Box 2896
Taren Point NSW 2229
Phone: (02) 4861 5365
Web site: www.biodiesel.org.au
An industry group that provides useful information on biodiesel, ethanol, and more.

AUSTRIA

Austrian Biofuels Institute
Graben 14/2, Pf. 97
A-1014 Vienna
Austria
Phone: 43–1-53456–0
E-mail: Werner.Koerbitz@biodiesel.at
Web site: www.biodiesel.at
An international center of expertise on liquid biofuels (in German and English).

CANADA

Canadian Renewable Fuels Association
31 Adelaide Street East
P.O. Box 398
Toronto, Ontario

M5C 2J8
Phone: (416) 304–1324
Web site: www.greenfuels.org
An industry-association Web site that offers information on ethanol and biodiesel with links and more.

EUROPE

European Biodiesel Board
Ave. de Tervuren 363
1150 Brussels, Belgium
Phone: 32 (0)2–763–2477
E-mail: info@ebb-eu.org
Web site: www.ebb-eu.org
A nonprofit organization with the aim of promoting the use of biodiesel in the European Union.

GERMANY

Union for the Promotion of Oil and Protein Plants (Union zur Förderung von Oel- und Proteinpflanzen, or UFOP)
E-mail: ufop@wpr-communication.de
Web site: www.ufop.de
This educational and promotional Web site contains a wide range of resources with both a German and international focus (in German and English).

JAPAN

Nanohana Project Network
1273–5 Kamitoyoura, Azuchicho, Gamou-gun, Shiga Pref.
Shiga Prefecture Environment and Consumer Cooperative

Phone: 81–748–46–4551
E-mail: nanohana@nanohana.gr.jp
Web site: www.erca.go.jp/jfge/english/projects/P13.html
An organization dedicated to the local production and consumption of biodiesel from used cooking oil made from locally grown agricultural feedstocks.

UNITED KINGDOM

Allied Biodiesel Industries (UK)
Web site: www.biofuels.fsnet.co.uk/biobiz.htm
An organization representing the British biodiesel industry; the Web site offers information and links to suppliers, filling station locations, and more.

UNITED STATES

National Biodiesel Board
P.O. Box 104898
Jefferson City, MO 65110
Phone: 800–841–5849
Web site: www.biodiesel.org
The Web site is an excellent and extensive source of current industry information on biodiesel, news, technical information, links, and much more.

National Renewable Energy Laboratory (NREL)
1617 Cole Boulevard
Golden, CO 80401
Phone: 303–275–3000
E-mail: www.nrel.gov/webmaster.html
Web site: www.nrel.gov
The leading center for renewable energy research in the United States.

COOPERATIVES (U.S.)

The Berkeley Biodiesel Collective
Berkeley, California
E-mail: berkeleybiodiesel@yahoo.com
Web site: www.berkeleybiodiesel.org

The Biofuels Research Cooperative (Straight Vegetable Oil)
Sebastopol, California
E-mail: veggieoilcoop@yahoo.com
Web site: www.vegoilcoop.org

Boulder Biodiesel Cooperative
Boulder, Colorado
Phone: 303–449–3277
Web site: www.boulderbiodiesel.com

Brevard BioDiesel
Brevard County, Florida
Web site: www.brevardbiodiesel.com

Burlington Biodiesel Coop
Burlington, North Carolina
Web site: www.burlingtonbiodiesel.org

Connecticut Waste Vegetable Oil List
(formerly Connecticut Biodiesel Co-op)
Southington, Connecticut
Web site: ctbiodzl.freeshell.org

Corvallis Biodiesel Cooperative
Corvallis, Oregon
Web site: cbc.diversityxdesign.com/svo/

GoBiodiesel Cooperative
Portland, Oregon
E-mail: outreach@gobiodiesel.org
Web site: gobiodiesel.org/index.php?title=Main_Page

Grease Works! Biodiesel Cooperative
Corvallis, Oregon
Email: info@greaseworks.org
Web site: www.greaseworks.org

Olympia Biodiesel
Olympia, Washington
Web site: home.comcast.net/~olympia_biodiesel

Piedmont Biofuels
Pittsboro, North Carolina
Web site: www.biofuels.coop

Roaring Fork Biodiesel Coop
Roaring Fork, Colorado
Web site: www.rfbiodiesel.com

SoCo Biodiesel Co-op
Santa Rosa, California
Web site: www.socobio.org

Tacoma Biodiesel
Tacoma, Washington
Web site: www.tacomabiodiesel.org/index.php/Main_Page

OTHER BIODIESEL RESOURCES

Biodiesel America
Web site: www.biodieselamerica.org
A comprehensive site about everything biodiesel, including where to buy it, forums, and much more.

Collaborative Biodiesel Tutorial
Web site: www.biodieselcommunity.org/onlineresources/
A comprehensive listing of small-scale biodiesel and SVO resources.

Distribution Drive
Web site: www.distributiondrive.com/links.html
An exhaustive list of biodiesel and industry links.

Fat of the Land
Web site: www.lardcar.com
Web site for the classic 1995 film Fat of the Land, *describing the humorous adventures of five women who drove across the United States in a biodiesel-powered van.*

Journey to Forever
E-mail: keith@journeytoforever.org or midori@journeytoforever.org
Web site: journeytoforever.org
A highly educational Web site containing a wealth of information about biodiesel and many other sustainable-living subjects, including links to additional sources (in English, Spanish, Japanese, and Chinese).

Veggie Avenger
Web site: www.veggieavenger.com
Comprehensive information about biofuels and related technology, with news, glossary, numerous links, and much more about biodiesel and straight vegetable oil.

ONLINE DISCUSSION GROUPS

Biodiesel Discussion Forum
Web site: biodiesel.infopop.cc
A discussion forum and message board sponsored by the BioBeetle and the Veggie Van.

Biodiesel*Now*.com
Web site: www.biodieselnow.com
A comprehensive site that offers answers to virtually any question about biodiesel in the United States and around the world (foreign-language skills helpful for some message threads). Considered by many to be the biodiesel online discussion site.

Local B100 Forum
Web site: tech.groups.yahoo.com/group/local-b100-biz/
An advanced discussion list for very small-scale commercial biodiesel producers and distributors in the United States.

BIBLIOGRAPHY

Alovert, Maria "Mark." *Biodiesel Homebrew Guide: Everything You Need to Know to Make Quality Alternative Diesel Fuel Out of Waste Restaurant Fryer Oil, Edition 10.5*. San Francisco: self-published, 2005. Available online at www.localb100.com/book.html.

Campbell, C. J. *The Coming Oil Crisis*. Essex, England: Petroconsultants S.A., 2004.

Deffeyes, Kenneth S. *Beyond Oil: The View From Hubbert's Peak*. New York: Hill and Wang, 2005.

———. *Hubbert's Peak: The Impending World Oil Shortage*. Princeton, NJ: Princeton University Press, 2001.

Heinberg, Richard. *The Party's Over: Oil, War and the Fate of Industrial Societies*. Gabriola, BC: New Society Publishers, 2003.

———. *Powerdown: Oil, War Options and Actions for a Post-Carbon World*. Gabriola, BC: New Society Publishers, 2004.

Kemp, William H. *Biodiesel Basics and Beyond: A Comprehensive Guide to Production and Use for the Home and Farm*. Tamworth, ON: Aztext Press, 2006.

Mittelbach, Martin, and Claudia Remschmidt. *Biodiesel: The Comprehensive Handbook*. Graz, Austria: self-published, 2004. Contact: mittelbach_biodiesel@gmx.at.

Pahl, Greg. *The Citizen-Powered Energy Handbook: Community Solutions to a Global Crisis*. White River Junction, VT: Chelsea Green, 2007.

Tickell, Joshua. *From the Fryer to the Fuel Tank: The Complete Guide to Using Vegetable Oil as an Alternative Fuel*. 3rd ed. New Orleans: Joshua Tickell Media Productions, 2003.

Tyson, K. Shaine. *2004 Biodiesel Handling and Use Guidelines*. U.S. Department of Energy, 2004.

GLOSSARY

alkyl ester. A generic term for any alcohol-produced vegetable-oil esters or biodiesel.

aromatic. A chemical such as benzene, toluene, or xylene that normally is present in exhaust emissions from diesel engines running on petroleum diesel fuel. Aromatic compounds have strong, characteristic odors.

available production. The biodiesel-production capacity of refining facilities that are not specifically designed to produce biodiesel.

batch process. A method of making biodiesel that relies on a specific, limited amount of inputs for a single batch.

biodiesel. A clean-burning fuel made from natural, renewable sources such as new or used vegetable oil or animal fats.

biofuel. A fuel made from biomass resources, such as ethanol, methanol, or biodiesel.

bioheat. A name sometimes applied to biodiesel when it is used for heating purposes.

biomass. Plant material, including wood, vegetation, grains, or agricultural waste, used as a fuel or energy source.

bio-naphtha. A term used in some eastern European nations for biodiesel.

brown grease. The least expensive category of waste grease in the United States, usually produced from restaurant grease traps or rendering plant sludge.

Btu. British thermal unit(s), a quantitative measure of heat equivalent to the amount of heat required to raise 1 pound of water by 1 degree Fahrenheit.

carbon dioxide. (CO_2). A product of combustion and a so-called green-house gas that traps the earth's heat and contributes to global warming.

carbon monoxide. (CO). A colorless, odorless, lethal gas that is the product of the incomplete combustion of fuels.

catalyst. A substance that, without itself undergoing any permanent chemical change, facilitates or enables a reaction between other substances.

cetane number. A measure of the ignition qualities of diesel fuel.

cloud point. The point at which biodiesel fuel appears cloudy because of the formation of wax crystals due to cold temperatures.

coking. The formation of harmful carbon deposits on internal components of diesel engines.

compression-ignition engine. An engine in which the fuel is ignited by high temperature caused by extreme pressure in the cylinder, rather than by a spark from a spark plug. Diesel engines are compression-ignition engines.

continuous deglycerolization. One of a number of continuous-flow processes for making biodiesel.

continuous-flow process. A general term for any of a number of biodiesel production processes that involves the continuous addition of ingredients to produce biodiesel on a continual, round-the-clock basis, as opposed to the batch process.

co-processed renewable diesel. Renewable diesel that is produced when an oil company adds small amounts of vegetable oils or animal fats to the traditional petroleum refining process when producing diesel fuel.

dedicated production. The biodiesel-production capacity of refining facilities that are designed specifically to produce biodiesel.

diester. The French term for biodiesel, from the contraction of the words diesel and ester.

direct-injection engine. A diesel engine in which the fuel is injected directly into the cylinder. Most new diesel engines have direct injection (DI).

emissions. All substances discharged into the air during combustion.

energy crops. Crops grown specifically for their energy value.

energy-efficiency ratio. A numerical figure that represents the energy stored in a fuel compared to the total energy required to produce, manufacture, transport, and distribute it.

ethanol. A colorless, flammable liquid that can be produced chemically from ethylene or biologically from the fermentation of various sugars from carbohydrates found in agricultural crops and residues from crops or wood. Also known as ethyl alcohol, alcohol, or grain spirits.

ethyl ester. Biodiesel that is made with the use of ethanol.

fatty acid alkyl ester. Another term for biodiesel made from any alcohol.

fatty acid methyl ester (FAME). Another term for biodiesel made with methanol.

feedstock. Any material converted to another form of fuel or an energy product.

flash point. The temperature at which a substance will ignite (for biodiesel, above 260°F or 126°C).

fossil fuel. An organic, energy-rich substance formed from the long-buried remains of prehistoric organic life. These fuels are considered nonrenewable, and their use contributes to air pollution and global warming.

gaseous emissions. Substances discharged into the air during combustion, typically including carbon dioxide, carbon monoxide, water vapor, and hydrocarbons.

gasohol. A fuel blend of ethanol made from fermented corn and gasoline.

gel point. The point at which a liquid fuel gels (changes to the consistency of petroleum jelly) due to extremely low temperature.

glycerin. A thick, sticky substance that is part of the chemical structure of vegetable oils and is a by-product of the transesterification process for making biodiesel. Glycerin is often used in the manufacture of soap and pharmaceuticals.

greenhouse effect. The heating of the atmosphere that results from the absorption of re-radiated solar radiation by certain gases, especially carbon dioxide and methane.

heterogeneous catalyst. A catalyst that does not dissolve in a reaction solution that can easily be removed and reused.

homogeneous catalyst. A catalyst that dissolves in a reaction solution.

hydrofined diesel. A "second-generation" biodiesel produced by hydrogenating the vegetable-oil feedstock.

indirect-injection engine. A (typically older) diesel engine in which the fuel is injected into a prechamber, where it is partly combusted, before it enters the cylinder.

methanol. A volatile, colorless alcohol, originally derived from wood, that often is used as a racing fuel and as a solvent. Also called methyl alcohol.

methyl ester. Biodiesel that is made with the use of methanol.

multifeedstock. Used to describe a biodiesel process technology that is capable of using a wide variety of feedstock inputs.

neat biodiesel. Pure biodiesel or B100.

nitrogen oxides (NO_x). A product of combustion and a contributing factor in the formation of smog and ozone.

oleochemicals. Chemicals derived from biological oils or fats.

particulate emissions. Substances discharged into the air during combustion. Typically they are fine particles such as carbonaceous soot and various organic molecules.

petrodiesel. Petroleum-based diesel fuel, usually referred to simply as diesel.

photobioreactor. A special type of bioreactor designed to provide optimal illumination, mixing, CO_2 mass transfer, and nutrients to a phototrophic liquid suspension, especially algae as a feedstock for biodiesel.

photosynthesis. A process by which plants and other organisms use light to convert carbon dioxide and water into a simple sugar. Photosynthesis provides the basic energy source for almost all organisms.

pour point. The temperature below which a fuel will not pour. The pour point for biodiesel is higher than that for petrodiesel.

products of combustion. Substances formed during combustion. The products of complete fuel combustion are carbon dioxide and water. Products of incomplete combustion can include carbon monoxide, hydrocarbons, soot, tars, and other substances.

relocalization. An international, grassroots movement that emphasizes local economic development, renewable-energy projects, organic farming, and other sustainable community-based and owned activities in response to the twin dangers of peak oil and global warming.

renewable diesel. A term sometimes used as a generic category for biodiesel and other diesel-like fuels based on renewable feedstocks. More recently in the U.S. used to describe a diesel-like fuel produced from biological material using a process called "thermal depolymerization."

renewable energy. An energy source that renews itself or that can be used today without diminishing future supply.

soydiesel. A term used in the United States for biodiesel made from soybean oil.

sustainable. Used to describe material or energy sources that, if carefully managed, will provide at current levels indefinitely.

thermal depolymerization. A process that involves turning virtually any carbon-based waste material, including vegetable oil, animal fats, plant material, etc., into usable hydrocarbon fuels.

transesterification. A chemical process that uses an alcohol to react with the triglycerides contained in vegetable oils and animal fats to produce biodiesel and glycerin.

triglycerides. Fats composed of three fatty-acid chains linked to a glycerol molecule.

viscosity. The ability of a liquid to flow. A high-viscosity liquid flows slowly, while a low-viscosity liquid flows quickly.

yellow grease. A term used in the United States to refer to recycled cooking oils.

NOTES

Chapter 1

1. W. Robert Nitske and Charles Morrow Wilson, *Rudolf Diesel: Pioneer of the Age of Power* (Norman: University of Oklahoma Press, 1965), 15–17. Unless noted otherwise, most of the information on Rudolf Diesel that follows is from the same source.
2. In a later account of this first engine test given in a speech by Diesel on April 13, 1912, in Saint Louis, Missouri, the first engine was said to be fueled by powdered coal dust and the explosion nearly fatal. This later account is at odds with earlier records of the experiment.
3. Walter Kaiser, "Rudolf Diesel and the Second Law of Thermodynamics," *German News Magazine*, June/July 1997. www.germanembassy-india.org/news/ june97/76 gn16.htm. (This monthly magazine is published by the German embassy of India in New Delhi.)
4. According to most sources, Diesel did not begin to experiment with coal dust as a fuel for his engine until 1897.
5. Gerhard Knothe, "Historical Perspectives on Vegetable Oil-Based Diesel Fuels," *Inform: International News on Fats, Oils and Related Materials* 12 (November 2001): 1104. (*Inform* is a monthly publication of the American Oil Chemists' Society in Champaign, Illinois.)

Chapter 2

1. W. Robert Nitske and Charles Morrow Wilson, *Rudolf Diesel: Pioneer of the Age of Power* (Norman: University of Oklahoma Press, 1965), p. 214.
2. Ibid., p. 213.
3. Ibid., p. 208.
4. Ibid., p. 259.
5. Gerhard Knothe, "Historical Perspectives on Vegetable Oil-Based Diesel Fuels," *Inform: International News on Fats, Oils and Related Materials* 12 (November 2001): 1106.
6. Darryl Melrose, telephone interview by the author, March 23, 2004.
7. Knothe, "Historical Perspectives," 1105.
8. Ibid.
9. Manfred Wörgetter, e-mail interview by the author, January 14, 2004.
10. Knothe, 1107.
11. Martin Mittelbach, telephone interview by the author, January 15, 2004.
12. Manfred Wörgetter, e-mail to the author, May 27, 2004.
13. Werner Körbitz, "The Biodiesel Market Today and Its Future Potential," in *Proceedings of the Plant Oils as Fuels—Present State of Science and Future Developments Symposium* (held in Potsdam, Germany, February 16–18, 1997), (Berlin: Springer-Verlag, 1998), p. 4.
14. Werner Körbitz, telephone interview by the author, January 12, 2004.
15. Lourens du Plessis, telephone interview by the author, February 2, 2004.
16. Charles Peterson, telephone interview by the author, January 12, 2004.

17. Ibid.

18. Ibid.

19. H. E. Haines and J. Evanoff, "Environmental and Regulatory Benefits Derived from the Truck in the Park Biodiesel Emissions Testing and Demonstration in Yellowstone National Park" (paper presented at the Bioenergy '98 Conference of the U.S. Department of Energy Regional Bioenergy Program in Madison, Wisconsin, October 1998; paper revised December 8, 1998).

Chapter 3

1. Werner Körbitz, telephone interview by the author, January 12, 2004.

2. Jessica Ebert, "Nano-Style Biodiesel Production," *Biodiesel Magazine*, October 2007, biodieselmagazine.com/article.jsp?article_id=1859&q=&page=2.

3. All yield figures below are from the chart "Vegetable Oil Yields" located at journeytoforever.org/biodiesel_yield.html.

4. James A. Duke, *Handbook of Energy Crops (1983)*, an electronic publication on NewCROP, the New Crop Resource Online Program hosted by the Purdue University Center for New Crops & Plant Products Web site, located at www.hort.purdue.edu/newcrop/duke_energy/dukeindex.html. Some of the oil-crop information that follows is based on the same source.

5. Werner Körbitz, *New Trends in Developing Biodiesel World-wide* (Vienna: Austrian Biofuels Institute, 1999). All subsequent percentages for global biodiesel raw material sources are from a chart contained in this report.

6. John Sheehan et al., *A Look Back at the U.S. Department of Energy's Aquatic Species Program: Biodiesel from Algae* (Golden, Colo.: National Renewable Energy Laboratory, July 1998), p. ii.

7. Ibid., p. iii.

8. J. Connemann and J. Fischer, "Biodiesel in Europe 2000: Biodiesel Processing Technologies and Future Market Development" (paper presented at the symposium Biodiesel—Fuel from Vegetable Oils for Compression-Ignition Engines at the Technische Akademie Esslingen, May 17, 1999, Ostfildern/Stuttgart, Germany).

9. John Sheehan et al., *Life Cycle Inventory of Biodiesel and Petroleum Diesel for Use in an Urban Bus* (Golden, Colo.: National Renewable Energy Laboratory, May 1998).

10. Ibid.

11. Hosein Shapouri, James A. Duffield, and Michael Wang, *The Energy Balance of Corn Ethanol: An Update, Agricultural Economic Report No. 814* (Washington, D.C.: U.S. Department of Agriculture, Office of the Chief Economist, Office of Energy Policy and New Uses, July 2002).

12. Werner Körbitz, *The Technical, Energy and Environmental Properties of BioDiesel* (Vienna: Körbitz Consulting, 1993).

Chapter 4

1. Onno Syassen, "Diesel Engine Technologies for Raw and Transesterified Plant Oils as Fuels: Desired Future Qualities of the Fuels," in *Proceedings of the Plant Oils as*

Fuels—Present State of Science and Future Developments Symposium (held in Potsdam, Germany, February 16–18, 1997), (Berlin: Springer-Verlag, 1998), p. 52.

2. Bosch, "At High Pressure: 75 years of Bosch Diesel Injection," archive .bosch.com/en/archive/theme_11_2002.htm.

3. National Biodiesel Board, "Lambert International Airport," www.biodiesel .org/resources/users/stories/lambert.shtm.

4. All-biodiesel bus fleet in Gratz. www.trendsetter-europe.org/index.php?ID=714.

5. National Biodiesel Board, "School Buses," www.nbb.org/markets/sch/default.asp.

6. National Biodiesel Board, "Medford, New Jersey School District," www .biodiesel.org/resources/users/stories/medfordnj.shtm.

7. National Biodiesel Board, "Farmer Use," www.nbb.org/pdfjiles/farmer _use.pdf.

8. National Biodiesel Board, "Easier on Marine Environment," www.nbb.org/ markets/mar/default.asp.

9. Mario Osava, "Biodiesel Trains on the Right Track," Inter Press Service News Agency Web site, ipsnews.net/africa/interna.asp?idnews=21707.

10. Denver Lopp and Dave Stanley, *Soy-Diesel Blends Use in Aviation Turbine Engines* (West Lafayette, Ind.: Purdue University, Aviation Technology Department, 1995).

11. World's First Jet Flight Powered Entirely on Renewable Biodiesel Fuel; Green Flight International and Biodiesel Solutions partner to set a new precedent in the use of renewable fuels in transportation, press release dated October 5, 2007, www.green-flightinternational.com/pr.htm.

12. "CytoSol, Exxon Demonstrate Solvent Capabilities in Texas," *Feedstocks: News about Industrial Products Made from Soy* 4, no. 3 (1999): 1, 2.

13. John Van de Vaarst, telephone interviews by the author, May 6, 2003, and June 27, 2003.

14. Jenna Higgins, telephone interview by the author, July 3, 2003. Updated, October 19, 2007.

15. Bob Cerio, telephone interview by the author, June 27, 2003.

16. Ralph Mills, telephone interview by the author, July 9, 2003.

17. Joel Glatz, telephone interview by the author, July 4, 2003.

18. The Vermont Biofuels Association merged with Renewable Energy Vermont in January 2008. The author was a founding member of the VBA in 2003.

19. Stephan Chase, telephone interview by the author, July 7, 2003.

20. Amelot's Opportunity in Massachusetts Increases with Governor's Order to Use Biodiesel for Winter 2007 and Beyond, press release, dated July 16, 2007, www.primenewswire.com/newsroom/news.html?d=122970.

21. National Biodiesel Board press release, February 9, 2004.

Chapter 5

1. USDA, "EU: Biodiesel Industry Expanding Use of Oilseeds," September 20, 2003, Production Estimates and Crop Assessment Division, Foreign Agricultural Service, U.S. Department of Agriculture, 1, 2. www.fas.usda.gov/pecad2/ highlights/2003/ 09/biodiese13/index.htm.

2. The EU-15 members (Austria, Belgium, Denmark, Finland, France, Germany, Greece, Ireland, Italy, Luxembourg, Portugal, Spain, Sweden, the Netherlands, and the United Kingdom) plus new (in 2004) members Cyprus, the Czech Republic, Estonia, Hungary, Latvia, Lithuania, Malta, Poland, Slovakia, and Slovenia. At the start of 2007, the number of EU member states grew to 27 with the addition of Romania and Bulgaria.

3. Anne Prieur-Vernat, Stéphane His, "Biofuels in Europe," December 20, 2006, ifp, France, www.ifp.fr/IFP/en/events/panorama/IFP-Panorama07_06-Biocarburants_Europe_VA.pdf.

4. Bruno Waterfield and Charles Clover, "Set aside suspended by European Union," Telegraph.co.uk, September 26, 2007, www.telegraph.co.uk/earth/main.jhtml?xml=/earth/2007/09/26/easet126.xml.

5. Used frying oil is generally called waste vegetable oil (WVO) in the United States, but the different terms refer to the same commodity.

6. Raffaello Garofalo, telephone interview by the author, February 24, 2004.

7. The percentage represented by feedstock cost can vary from about 60 percent to as much as 85 percent, depending on the feedstock and the production technology used. In 2007 and 2008, the dramatic rise in global vegetable oil prices pushed the percentage up to approximately 85 percent.

8. Biofuels: Emerging Developments and Existing Opportunities (New York: Technical Insights Inc., 2002). For a summary, go to the Web site, www.theinfoshop.com/study/ti12890_biofuels.html.

9. Werner Körbitz, telephone interview by the author, January 12, 2004.

10. Werner Körbitz and Jens Kossmann, "Production and Use of Biodiesel," in New and Emerging Bioenergy Technologies, Risø Energy Report 2, ed. Hans Larsen, Jens Kossmann, and Leif Sønderberg (Roskilde, Denmark: Risø National Laboratory, November 2003), p. 31.

11. Garofalo, interview.

12. USDA, EU: Biodiesel Industry Expanding Use of Oilseeds, 5.

13. Op. cit., Anne Prieur-Vernat, Stéphane His, "Biofuels in Europe," 6.

14. European Union Increases Biofuels Targets, Seema Patel, BCS, Incorporated, September 2007, www.brdisolutions.com/default.aspx.

15. Garofalo, interview.

16. Biofuels in Europe, McDermott Newsletter, October 17, 2006, www.mwe.com/index.cfm/fuseaction/publications.nldetail/object_id/cd74f77c-ece5-4732-a5ac-d03264ff81fb.cfm.

17. Op. cit., Garofalo.

18. Energy and Transport Directorate-General of the European Commission, "Energy Taxation: Commission Proposes Transitional Periods for Accession Countries," europa.eu.int/comm/energy_transport/mm_dg/newsletter/n180–2004–0 1–30_en.html#EN%2001.

19. Elizabeth Rosenthal, "Europe, Cutting Biofuel Subsidies, Redirects Aid to Stress Greenest Options," New York Times, January 22, 2008, www.nytimes.com/2008/01/22/business/worldbusiness/22biofuels.html?_r=1&ref=business&oref=slogin.

20. WWF & the EU Biofuels Communication, February 2006, assets.panda.org/downloads/wwf_on_biofuels_comm_q_a_2006___final_080206.pdf.

21. The Roundtable on Sustainable Biofuels: Ensuring Biofuels Deliver on Their Promise of Sustainability, cgse.epfl.ch/page65660.html.

22. Roundtable on Sustainable Palm Oil Website, History of RSPO, www.rspo.org/History_of_RSPO.aspx.

23. Michael Hogan, "EU biodiesel firms blame politicians as demand falls," Reuters, *International Herald Tribune*, March 22, 2007, www.iht.com/articles/2007/03/22/business/diesel.php.

24. "The EU biodiesel industry unanimously agrees to initiate legal action against US "B99" unfair biodiesel exports," EBB press release, December 3, 2007, www.ebb-eu.org/EBBpressreleases/EBB%20press%20B99%20outcome%20of%20General%20Assembly%2030%20November%20FINAL.pdf.

25. Sarah Smith, "EU pursues legal action against U.S.-subsidized biodiesel," *Biodiesel Magazine*, February 2008, www.biodieselmagazine.com/article.jsp?article_id=2030.

26. Austrian Biofuels Institute, Biodiesel—A Success Story: The Development of Biodiesel in Germany, a report for the International Energy Agency, Bioenergy Task 27, Liquid Biofuels (Vienna: Austrian Biofuels Institute, June 2001, update February 2002), 21.

27. Most of the recent German biodiesel plant expansion information is based on the chart "Biodieselproduktionskapazitaten in Deutschland," which is located at the Web site www.iwr.de/biodiesel/kapazitaeten.html.

28. ADM Expands Hamburg Oil Facility with Palm Oil Refinery, May 15, 2006, Green Car Congress, www.greencarcongress.com/2006/05/adm_expands_ham.html.

29. UFOP, *Biodiesel in Bus Fleets: The Kreiswerke Heinsberg's GmbH and Stadtwerke Neuwied's Experience* (Bonn: UFOP), pp. 1–34.

30. Hans Plaettner-Hochwarth and Klaus Schreiner, *Biodiesel und Sportschiffart in der Eurigio Bodensee* (Bonn: UFOP, 2001), pp. 1–33.

31. Dieter Bockey, *Biodiesel Production and Marketing in Germany: The Situation and Perspective* (Berlin: UFOP, 2002), p. 8.

32. Jane Burgermeister, "German Biodiesel Industry Peaks, Trouble Ahead," RenewableEnergy Access.com, August 24, 2007, www.renewableenergyaccess.com/rea/news/story?id=49745.

33. Dieter Bockey, *Potentials for raw materials for the production of biodiesel: An Analysis* (Berlin: UFOP, 2006), p. 9.

34. "Germany Phasing Out Price Protection for Biodiesel," August 2, 2006, www.dw-world.de/dw/article/0,2144,2116260,00.html.

35. Op. cit., Burgermeister.

36. "German biodiesel industry faces collapse over taxes, US subsidies, competition from the South," June 3, 2007, biopact.com/2007/06/german-biodiesel-industry-faces.html.

37. Michael Hogan, "German biodiesel output collapses," Reuters, January 15, 2008, www.reuters.com/article/GlobalAgricultureandBiofuels08/idUSL1589672020080115.

38. Ibid.

39. "German firm Petrotec stops biodiesel production," October 23, 2007, uk.reuters .com/article/oilRpt/idUKL239959920071023.

40. Op. cit., "German biodiesel output collapses."

41. Op. cit., Anne Prieur-Vernat, Stéphane His, "Biofuels in Europe," 4.

42. Garofalo, interview.

43. France is second only to Spain in EU ethanol production, but ethanol is generally not as important as biodiesel in the EU due to low corn production and a higher proportion of diesel engines compared to the United States.

44. *Biodiesel: Documentation of the World-Wide Status 1997, a report for the International Energy Agency (IEA), commissioned by the BLT-Federal Institute for Agricultural Engineering* (Wieselburg, Austria: Austrian Biofuels Institute, 1997), 20.

45. EU cuts back on biofuel crop subsidies, October 18, 2007, www.euractiv.com/en/ sustainability/eu-cuts-back-biofuel-crop-subsidies/article-167713.

46. Marie-Cécile Hénard and Xavier Audran, *France, Agricultural Situation, French Biofuel Situation, Global Agricultural Information Network report FR3044* (Washington, D.C.: U.S. Department of Agriculture, Foreign Agricultural Service, 2003), p. 3.

47. Marie-Cécile Hénard and Xavier Audran, France, *Bio-Fuels, French Biofuel Production Plans, 2007, Global Agricultural Information Network Report FR7001* (Washington, D.C.: U.S. Department of Agriculture, Foreign Agricultural Service, May 1, 2007), p. 3.

48. Biodiesel: Boom or bust, ICIS, February 5, 2007, www.icis.com/Articles/2007/02/12/4500682/biodiesel-boom-or-bust.html.

49. France May Cut Biofuel Tax Advantage, *World Energy*, November 27, 2007, www.worldenergy.net/public_information/show_news.php?nid=117.

50. Marie-Cécile Hénard, *France, Oilseeds and Products, French Biofuel Production Booms, 2005, Global Agricultural Information Network Report FR6005* (Washington, D.C.: U.S. Department of Agriculture, Foreign Agricultural Service, January 20, 2006), p. 6.

51. Op. cit., Marie-Cécile Hénard and Xavier Audran, *Agricultural Situation*, 2003, p. 6.

52. French railways continue 'Zero Oil' program with biodiesel, August 12, 2007, biopact.com/2007/08/french-railways-continue-zero-oil.html.

53. Op. cit., 4.

Chapter 6

1. *Biodiesel: Documentation of the World-Wide Status 1997, a report prepared for the International Energy Agency (IEA), commissioned by the BLT-Federal Institute for Agricultural Engineering* (Wieselburg, Austria: Austrian Biofuels Institute, 1997), p. 23.

2. Liquid Biofuels Network, *Liquid Biofuels Activity Report* (France: EUBIONET, Liquid Biofuels Network, April 2003), p. 32.

3. Sandro Perini, *Italy, Trade Policy Monitoring, Biofuels, Global Agricultural Information Network Report IT7009* (Washington, D.C.: U.S. Department of Agriculture, Foreign Agricultural Service, April 3, 2007), p. 2. .

4. Italy signs biofuel deal to help save energy, January 11, 2007, www.tmcnet.com/altpowermag/articles/4493-italy-signs-biofuel-deal-help-save-energy.htm.

5. Anne Prieur-Vernat, Stéphane His, "Biofuels in Europe," December 20, 2006, ifp, France, 5, www.ifp.fr/IFP/en/events/panorama/IFP-Panorama07_06-Biocarburants_Europe_VA.pdf.

6. Svetlana Kovalyova, "Italy '07 biodiesel output seen down 40 pct," Reuters, October 30, 2007, uk.reuters.com/article/environmentNews/idUKL2958155220071030.

7. Reuters UK, "Tesco to Sell Rapeseed Biodiesel," May 3, 2004, www.reuters .co.uk/newsPackageArticle.jhtml?type=topNews&storyID=502389&section=news. (Web page has expired.)

8. Bio Diesel, "WhatDiesel," December 21, 2007, www.whatdiesel.co.uk/viewarticle.aspx?articleid=1020.

9. Green Shop, "The First UK Garage to Sell Biodiesel," www.greenshop.co .uk/news/Biodiesel-Opening-2002.htm.

10. Michael Shirek, "McDonald's to convert used cooking oil to biodiesel," *Biodiesel Magazine*, July 2007, www.biodieselmagazine.com/article.jsp?article_id=1716.

11. Penetration of biofuels continues to grow through legal mandates and availability, Energy & Enviro Finland, July 20, 2007, www.energy-enviro.fi/index.php?PAGE=912&NODE_ID=912&LANG=1.

12. Karen McLauchlan, "Petroplus Fueling Biodiesel Drive," *The Evening Gazette*, January 22, 2004, icteesside.icnetwork.co.uk/0400business/0004tod/page .cfm?objectid= 13845660&method=full&siteid=50080. (Web page has expired.)

13. Greenergy press releases, April 14, 2004, December 6, 2005, and June 15, 2006.

14. Terry Macalister, "UK biodiesel producer slows refinery expansion," Guardian Unlimited, September 26, 2007, business.guardian.co.uk/story/0,,2177611,00.html.

15. D1 Oils press release, September 28, 2007, www.d1plc.com/news.php?article=164.

16. Manfred Wörgetter, e-mail to the author, May 27, 2004.

17. BDV press release, September 21, 2006, www.biodiesel-vienna.at/pdf/presseaussendungen/PA_21092006_120044.pdf.

18. Fortune Management Inc. press release, April 16, 2007, www.gate-energy.com/cms/upload/PDF-Investors-Relation_EN/070416_PI_Enns_e_final.pdf.

19. All-biodiesel bus fleet in Gratz. www.trendsetter-europe.org/index.php?ID=714.

20. "Polish Biodiesel," Biodiesel.pl, www.biodiesel.pl. (Web page has expired.)

21. Interfax Poland Business Daily, February 26, 2007, www.accessmylibrary.com/coms2/summary_0286–29768772_ITM.

22. Ewa Krukowska, "Polish Biodiesel Output Seen Surging on Law Change," Reuters, March 14, 2006, www.planetark.com/dailynewsstory.cfm/newsid/35623/story.htm.

23. Katarzyna Marcinkowska, "Poland tanking up on biodiesel," *Warsaw Business Journal*, September 18, 2006, www.wbj.pl/?command=article&id=33867.

24. WMF/EP, *Poland, Bio-Fuels, New Tax Incentives Are Not Enough for Polish Biofuel Producers*, Global Agricultural Information Network Report PL7028 (Washington, D.C.: U.S. Department of Agriculture, Foreign Agricultural Service, May 30, 2007), p. 2.

25. Spanish Ministry of Economics, www2.mineco.es/Mineco/Comunicacion/ Noticias/RATO+BIODIESEL.htm. (Page now removed from Web site.)

26. Arantxa Medina, *Spain, Bio-Fuels, Update, Global Agricultural Information Network report SP7024* (Washington, D.C.: U.S. Department of Agriculture, Foreign Agricultural Service, July 26, 2007), p. 2.

27. Austrian Biofuels Institute, *Annual Report 2002* (Vienna: Austrian Biofuels Institute, 2002), p. 4.

28. Sybille de La Hamaide, "French producer sees price threat to EU biodiesel," Reuters, January 14, 2008, www.reuters.com/article/RussiaInvestment08/ idUSL1447899020080114.

29. Austrian Biofuels Institute, *Biodiesel: Documentation of the World-Wide Status 1997*, pp. 25, 26.

30. Martin Cvengros and Ján Cvengros, "Review on Development and Legislation of Biodiesel Production and Utilization in Slovakia" (a paper presented at the Techagro Fair in Brno, the Czech Republic, April 2002), 3.

31. Ibid., 7, 8.

32. "Slovak refinery Slovnaft to start selling biodiesel oil on Slovak market Friday," Interfax Czech Republic Business Daily, September 29, 2006, www.accessmylibrary.com/coms2/summary_0286–18721527_ITM.

33. P. H. Mensier, "L'emploi des Huiles Végétales Comme Combustible dans les Moteurs," Oléagineux, Février (1952), 69.

34. European Energy Crops InterNetwork, "Evolution of Rape in Belgium and Its Utilization as Biofuel," document ID B10082, November 7, 1997, btgs1.ct.utwente.nl/eeci/archive/biobase/B10082.html. (Web site now discontinued.)

35. "Cargill, Belgian Firms to Build Biodiesel Plant," Reuters, January 30, 2006, www.planetark.com/dailynewsstory.cfm/newsid/34738/story.htm.

36. Neochim to produce biodiesel at BASF site in Feluy, September 11, 2006, www.prdomain.com/companies/B/BASF/newsreleases/200691135520.htm.

37. Proviron press release, www.proviron.be/Proviron_GB/news/news.php.

38. Arantxa Medina, Andy Jessen, *Iberian Peninsula, Trade Policy Monitoring, Portugal's Biofuels Policy, Global Agricultural Information Network Report P07001* (Washington, D.C.: U.S. Department of Agriculture, Foreign Agricultural Service, January 25, 2007), p. 2.

39. "Biodiesel is a boon for farmers," The Copenhagen Post Online, August 24, 2007, www.cphpost.dk/get/103227.html.

40. Daka Biodiesel Web site, www.dakabiodiesel.com/page575.asp.

41. Anna Mudeva, "Dutch Rush to Produce Biofuel as Oil Prices Surge," Reuters, August 15, 2005, www.planetark.com/dailynewsstory.cfm/newsid/32047/newsDate/ 15-Aug-2005/story.htm.

42. GAVE News, Sunoil Biodiesel B.V., October 30, 2007, www.gave.novem.nl/gave/index.asp?id=25&detail=1881.

43. Netherlands Biodiesel Industry Association forms, Biofuels International, www.biofuels-news.com/news/netherlands_assocforms.html.

44. Asa Lexmon, Sweden, Bio-Fuels, Bio-Fuels Annual, 2006, Global Agricultural Information Network Report SW6013 (Washington, D.C.: U.S. Department of Agriculture, Foreign Agricultural Service, June 1, 2006), 5.

45. Dr. Sigitas Lazauskas, Lithuanian Institute of Agriculture, "Non-food Crop Activity in the Baltic Sea Region," published in the Interactive European Network for Industrial Crops and Their Applications, Newsletter Number 21, December 2003, www.ienica.net/newsletters/newsletter21.pdf.

46. "What Is Biodiesel?" www.betancalibration.com/pdf/BioDiesel.pdf.

47. Mestilla press release, September 17, 2007, Statoil signs deal on acquiring 42.5-pct stake in biodiesel plant Mestilla, www.mestilla.lt/en/news.

48. Kate Snipes, Lithuania, Agricultural Situation, Biofuel and Biomass, Production and Plans, 2007, Global Agricultural Information Network Report LH7001 (Washington, D.C.: U.S. Department of Agriculture, Foreign Agricultural Service, June 1, 2006), p. 2.

49. DeltaRiga Web site, www.deltariga.com/eng/index.html.

50. Rico Cruz, phone interview by the author, April 13, 2004.

51. Mila Boshnakova, Bulgaria, Grain and Feed, Biofuels Market in Bulgaria, 2006, Global Agricultural Information Network Report BU6006 (Washington, D.C.: U.S. Department of Agriculture, Foreign Agricultural Service, June 1, 2006), p. 9.

52. Stamatis Sekliziotis, Greece, Bio-Fuels, Biofuel Activity in Greece, 2007, Global Agricultural Information Network Report GR7003 (Washington, D.C.: U.S. Department of Agriculture, Foreign Agricultural Service, February 20, 2007), pp. 2, 3.

53. Radovan Tavzes, Ministry for the Environment and Spatial Planning, Dr Mirko Bizjak, Slovenian Environment Agency, "The use of biofuels in transport in the Republic of Slovenia in 2006," Ministry for the Environment and Spatial Planning, Ljubljana, June 2007, 6. www.ebb-eu.org/legis/SLOVENIA_4th%20report%20 Dir2003_30_report_EN.pdf.

54. Nafta Lendava to build Biodiesel refinery in Slovenia, New Europe, January 31, 2007, www.neurope.eu/view_news.php?id=69742.

55. Green Biofuels Ireland Biodiesel Plant Under Construction, Irish Bioenergy Association, April 30, 2007, www.irbea.org/index.php?option=com_content&task=view&id=288&Itemid=44.

56. "Biodiesel projects gain finance boost," The Diplomat, December, 2006, www.the diplomat.ro/econ_news_1206.htm.

57. "Serbia Set to Start its First Biodiesel Plant in Spring," Reuters, Energia.gr, January 12, 2007, www.energia.gr/indexengr.php?newsid=12761&lang=en.

58. Hungary's largest biodiesel plant to kick off production in May, Portfolio.hu, January 29, 2007, www.portfolio.hu/en/cikkek.tdp?cCheck=1&k=2&i=10893.

59. The use of bioenergy in Transportation, Nordic Bioenergy Project, www.nordicenergy.net/bioenergy/text.cfm?path=96&id=527.

60. Giles Clark, "New biodiesel plant inaugurated by Neste at Porvoo," Biofuel review, May 31, 2007, www.biofuelreview.com/content/view/1007/.

61. RUSBIODIESEL to Build Biofuel Plant in Russia, Russian Biofuels Association Web site, March 29, 2007, www.biofuels.ru/biodiesel/news/760/.

62. Y.Vassilieva, K.Svec, C.Brown, Russian Federation, Bio-Fuels, Annual, 2007, Global Agricultural Information Network Report RS7044 (Washington, D.C.: U.S. Department of Agriculture, Foreign Agricultural Service, June 4, 2007), 9.

63. Ibid.

64. Russia, Eastern Europe to supply European biodiesel feedstock demand, *Biodiesel Magazine*, May, 2007, biodieselmagazine.com/article.jsp?article_id=1579.

65. Ukraine's Ternopol oblast proposes Belarus to start joint biodiesel venture, July 12, 2007, law.by/work/EnglPortal.nsf/0/6F3B66CD951FBB2CC225731600423E27? OpenDocument.

66. Paul R. Kleindorfer and Ülkü G. Öktem, "Economic and Business Challenges For Biodiesel Production in Turkey," The Wharton School of the University of Pennsylvania, September 2007, opim.wharton.upenn.edu/risk/library/2007_ PRK-UGO_BiodieselTurkey.pdf.

Chapter 7

1. *Investigation into the Role of Biodiesel in South Africa, a report to the Department of Science and Technology* (Pretoria: CSIR Transportek, March 2003).

2. UNIDO, *CDM Investor Guide, South Africa* (Vienna: United Nations Industrial Development Organization, February 2003), p. 29.

3. Earthlife Africa/WWF, *Employment Potential of Renewable Energy in South Africa* (Johannesburg: Earthlife Africa / Denmark: WWF, November 2003), p. 45.

4. "A Pioneering Bio-Fuel Project Holds Out Unprecedented Empowerment Opportunities," Echo, supplement to *The Natal Witness*, Thursday, July 29, 2003.

5. Darryl Melrose, telephone interview by the author, March 23, 2004.

6. 2007—The year for biofuels, *Food & Beverage*, www.developtechnology.com/web/content/view/19210/31/.

7. SA biofuel strategy 'positive,' News24.com, January 8, 2007, www.news24.com/News24/Technology/News/0,,2-13-1443_2052240,00.html.

8. Mike Cohen and Carli Lourens, "South Africa Cuts Biofuels Target, Excludes Corn," Bloomberg.com, December 6, 2007, www.bloomberg.com/apps/news?pid= 20601116&sid=aEHcB8saxLH0&refer=africa.

9. Mbuyisi Mgibisa, "Lack of legislation holds back biofuels," Fin24.com, September 16, 2007, www.fin24.co.za/articles/default/display_article.aspx?ArticleId=1518– 25_2184755.

10. D1 Africa Web site, www.d1africa.com/index.php.

11. Op. cit., 2007—The year for biofuels.

12. Dominique Patton, "Mali leads the pack in biodiesel production," *Business Daily*, October 30, 2007, www.bdafrica.com/index.php?option=com_content&task= view&id=3975&Itemid=5822.

13. Alana Herro, "Eye on Mali: Jatropha Oil Lights Up Villages," Worldwatch Institute, June 1, 2007, www.worldwatch.org/node/5101.

14. Giles Clark, "Energem acquires jatropha biodiesel project in Mozambique," *Biofuel Review*, August 2, 2007, www.biofuelreview.com/content/view/1123/.

15. Elnette Oelofse, "Four Little Seeds Helping to Empower Africa," September 2002, the International Oracle Syndicate, www.oraclesyndicate.org/pub_e/e .oelofse/pub_ 9–02_2.htm#top. (Web page has expired.)

16. Biodiesel production begins in Ghana, Checkbiotech, August 31, 2007, www.check biotech.org/green_News_Biofuels.aspx?infoId=15519.

17. Mark-Anthony Vinorkor, "Ghana to save $240M through Biodiesel," *The Ghanaian Times*, August 8, 2003, www.reeep.org/index.cfm?articleid=760. (Web page has expired.)

18. Trees for Clean Energy project: Kenyan farmers to benefit from biofuels in semi-arid zones, Biopact, November 7, 2007, biopact.com/2007/11/trees-for-clean-energy-project-kenyan.html.

19. Alari Alare, "Japanese firm gives Sh1.3b for bio-fuels," *The Standard*, November 19, 2007, www.eastandard.net/hm_news/news.php?articleid=1143977648.

20. Godwin Agaba, "Rwanda: Bio-Diesel Project," *New Times* (Kigali), September 10, 2007, allafrica.com/stories/200709110339.html.

21. Ericsson mobile service stations use biodiesel in Nigeria, *Biodiesel Magazine*, December 2006, biodieselmagazine.com/article.jsp?article_id=1284.

22. Mugabe opens Zim's first biodiesel plant, Mail & Guardian online, November 15, 2007, www.mg.co.za/articlepage.aspx?area=/breaking_news/breaking_news__ business/&articleid=325023&referrer=RSS.

23. Terry Macalister, "Forget the tiger in your tank—Saudis harness seed power," Guardian Unlimited, February 16, 2005, www.guardian.co.uk/business/2005/feb/ 16/saudiarabia.oilandpetrol.

24. Good News India, "Rising Bio-diesel Tide," September 18, 2003, www .goodnews india.com/Pages/content/updates/story/117_0_4_0_C/.

25. Good News India, "Honge Oil Proves to Be a Good Biodiesel," www .goodnews india.com/Pages/content/discovery/honge.html.

26. Ibid.

27. S. Srinivasan, "Isolated Hamlet in Indian Forest Gets Electricity from Seed-Powered Generator," Associated Press, October 15, 2003.

28. RenewingIndia.org, "Biodiesel: First Trial Run on Train," www .renewingindia.org/news1jan_biodiesel.html. (Web page has expired.)

29. India Railways to Run on Biodiesel, Green Car Congress, May 18, 2007, www.green carcongress.com/2007/05/indian_railways.html.

30. DaimlerChrysler biodiesel project enters final phase, December 5, 2006, www.the hindubusinessline.com/2006/12/06/stories/2006120605260300.htm.

31. The Hindu Business Line, "IOC to Start Field Trials of Biodiesel," December 10, 2003, www.thehindubusinessline.com/businessline/blnus/14101710.htm. (Web page has expired.)

32. "Five Biodiesel Plants Sanctioned in State," The Central Chronicle, December 31, 2003.

33. Biofuel: The little shrub that could—maybe, Nature News, October 10, 2007, www.nature.com/news/2007/071010/full/449652a.html.

34. Southern Online starts sale of biodiesel from SBT biodiesel plant, Iris, July 12, 2007, www.myiris.com/newsCentre/newsPopup.php?fileR=20070712221857121& dir=2007/07/12&secID=livenews.

35. Naturol to start India's first biodiesel plant, October 10, 2007, newKerala.com. www.newkerala.com/oct.php?action=fullnews&id=10224.

36. India announces biodiesel policy, UPI, August 16, 2007, www.upi.com/International_Security/Energy/Briefing/2007/08/16/india_announces_ biodiesel_policy/5383/.

37. Biodiesel can contribute to India's energy security: Report, Livemint.com, May 16, 2007, www.livemint.com/2007/05/16152536/Biodiesel-can-contribute-to-In.html.

38. Biodiesel Project in the Royal Thai Navy, R&D Division, Royal Thai Naval dock-yard, 2005, www.navy.mi.th/dockyard/doced/Homepage/Botkwuam/Navy_ Biodiesel.pdf.

39. Govt backs community biodiesel projects, June 10, 2005, www.mcot.org/query.php? nid=39203. (Web page has expired.)

40. Thai oil refiner to produce biodiesel with vegetable oil, People's Daily Online, September 5, 2006, english.people.com.cn/200609/05/eng20060905_299669.html.

41. Thailand to mandate 2% palm oil biodiesel next year, Biopact, June 9, 2007, biopact.com/2007/06/thailand-to-mandate-2-palm oil.html.

42. Thai largest biodiesel producer builds fourth plant, Energy Current, August 23, 2007, www.energycurrent.com/index.php?id=3&storyid=4718.

43. Biodiesel: Documentation of the World-Wide Status 1997, a report prepared for the International Energy Agency (IEA), commissioned by the BLT-Federal Institute for Agricultural Engineering (Wieselburg, Austria: Austrian Biofuels Institute, 1997), 18.

44. Austrian Biofuels Institute, Annual Report 2002 (Vienna: Austrian Biofuels Institute, 2002), 7.

45. Biodiesel firm to list on NASDAQ, China Daily, May 31, 2007, english.people.com.cn/200705/31/eng20070531_379665.html.

46. David Harman, "China's Biodiesel Demand to Exceed Production by 2010," Interfax China Commodities Daily, July 6, 2007, www.resourceinvestor.com/pebble.asp? relid=33620.

47. Zaidi Isham Ismail, "China set to buy additional 1m tons palm oil," Baltimore Times Online, August 4, 2006, www.btimes.com.my/Current_News/BT/Friday/Frontpage/ BT580094.txt/Article/.

48. Jiao Le, "Biodiesel Sweeps China in Controversy," January 23, 2007, www.world watch.org/node/4870.

49. Op. cit., Biofuel: The little shrub that could—maybe.

50. Op. cit., Jiao Le, "Biodiesel Sweeps China in Controversy."

51. Ibid.

52. "About the Company," Pacific Biodiesel, www.biodiesel.com/aboutPac Bio.htm.

53. Yasuji Nagai, "Fields Bloom with New Eco-Friendly Fuel Alternative," The Asahi Shimbun, April 29, 2004.

54. "Government Unveils Biomass-Fuel Project," *The Japan Times*, December 28, 2002, www.japantimes.co.jp/cgi-bin/getarticle.p15?nb20021228a9.htm. (Web page has expired.)

55. Japan for Sustainability, "Making Biodiesel Fuel from Sunflower Oil," www.japanfs.org/db/database.cgi?cmd=dp&num=484&UserNum=&Pass =& AdminPa.

56. Nanohana Project Network, www.eic.or.jp/jfge/english/projects/P13.html (in Japanese).

57. Dane Muldoon, "Toyota in Joint Project to Commercialize Second-Generation Biodiesel Fuel," AutoblogGreen, February 15, 2007, www.autobloggreen.com/2007/02/15/toyota-in-joint-project-to-commercialise-second-generation-biodi/.

58. "S. Korea to impose 3 pct biodiesel rule by 2012," Reuters, September 7, 2007, uk.reuters.com/article/environmentNews/idUKSE028479520070907.

59. Ryu Jin, "Biodiesel Market Already Overcrowded," *The Korea Times*, December 17, 2007, www.koreatimes.co.kr/www/news/biz/2007/12/123_15688.html.

60. Taiwan introduces biodiesel fuel blend for cars, smh.com.au, July 29, 2007, www.smh.com.au/news/Technology/Taiwan-introduces-biodiesel-fuel-blend-for-cars/2007/07/29/1185647724205.html.

61. Bootstrapping Biodiesel in Singapore, Green Car Congress, May 19, 2006, www.greencarcongress.com/2006/05/bootstrapping_b.html.

62. Loh Kim Chin, "Singapore to host two biodiesel plants, investments total over S$80m," Channel NewsAsia, October 26, 2005, www.channelnewsasia.com/stories/singaporebusinessnews/view/175426/1/.html.

63. Department of Environment and Natural Resources, "'Biodiesel' to the Public," denr.gov.ph/article/articleprint/628/-1/152/.

64. Imelda V. Abaño, "The Philippines opts for biodiesel," SciDev.net. May 10, 2007, www.scidev.net/gateways/index.cfm?fuseaction=readitem&rgwid=2&item=News&itemid=3607&language=1.

65. "Philippines to launch world-scale jatropha farm," Reuters, July 2, 2007, uk.reuters.com/article/oilRpt/idUKSIN8406420070702.

66. Philippine Biodiesel Meets International Standards, USAgNet, January 7, 2008, www.wisconsinagconnection.com/story-national.php?Id=36&yr=2008.

67. Indonesia: Palm Oil Production Prospects Continue to Grow, Commodity Intelligence Report (Washington, D.C.: U.S. Department of Agriculture, Foreign Agricultural Service, December 31, 2007), www.pecad.fas.usda.gov/highlights/2007/12/Indonesia_palmoil/.

68. Hanim Adnan, "Palm biodiesel plant to begin ops in Oct," thestar online, September 11, 2007, biz.thestar.com.my/news/story.asp?file=/2007/9/11/business/20070911174605&sec=business.

69. "Malaysian crude palm oil futures largely unchanged," *Daily Times*, August 29, 2007, www.dailytimes.com.pk/default.asp?page=2007%5C08%5C29%5Cstory_29–8-2007_pg5_14.

70. 91 Biodiesel Projects Approved At End-September 2007, Bernama.com, October 30, 2007, www.bernama.com.my/bernama/v3/news_lite.php?id=293113.

71. Naveen Thukral, "Asian biodiesel plants sit idle as costs soar," Reuters, January 14, 2008, www.reuters.com/article/GlobalAgricultureandBiofuels08/idUSKLR59818200 80114?pageNumber=1&virtualBrandChannel=0.

72. Government Considering Planting Jarak Trees For Biodiesel, Bernama.com, November 6, 2007, www.bernama.com.my/bernama/v3/news.php?id=294477.

73. "Government will build four biofuel plants," Jakarta Post.com, April 7, 2006, www.thejakartapost.com/detailbusiness.asp?fileid=20060406.M06&irec=5.

74. Aji Bromokusumo, Indonesia, Bio-Fuels, Biofuels Annual 2007, Global Agricultural Information Network Report ID7019 (Washington, D.C.: U.S. Department of Agriculture, Foreign Agricultural Service, June 13, 2007), 3.

75. Op. cit., Indonesia: Palm Oil Production Prospects Continue to Grow.

76. Ibid.

77. "Indonesia to return to 5 per cent biodiesel blend," Reuters, November 13, 2007, economictimes.indiatimes.com/News/International__Business/Indonesia_to_return_ to_5_per_cent_biodiesel_blend/articleshow/2538259.cms.

78. Andi Haswidi, "Failed policies knock biodiesel production by 85%," The Jakarta Post, January 25, 2008, www.thejakartapost.com/detailbusiness.asp?fileid=20080124.L01&irec=0.

79. "PNG to Develop Coconut Oil As Biodiesel Fuel," TheNational.com, November 14, 2003.

80. Phil Mercer, "Coconut oil powers island's cars," BBC News, May 8, 2007, news.bbc.co.uk/2/hi/asia-pacific/6634221.stm.

81. Michelle Nichols, "PICs See Coconuts As Potential Biofuel Source," Taiwan Times, May 8, 2006, www.pacificmagazine.net/news/2006/05/08/region-pics-see-coconuts-as-potential-biofuel-source.

Chapter 8

1. Biodiesel Industries, Las Vegas, Nevada, press release, March 13, 2003.

2. "Gov't 'must' change the way biodiesel fuel is sold," ABC News, November 5, 2007, www.abc.net.au/news/stories/2007/11/05/2082047.htm?section=business.

3. Suzi Kerr, Brian White, et al., Renewable Energy and the Efficient Implementation of New Zealand's Current and Potential Future Greenhouse Gas Commitments (New Zealand: East Harbour Management Services Limited, August 14, 2002).

4. Errol Kiong, "NZ firm makes bio-diesel from sewage in world first," nzherald.com, May 12, 2006, www.nzherald.co.nz/section/1/story.cfm?c_id=1&ObjectID=10381404.

5. All aboard for 'bioloco' ride, stuff.co.nz, December 8, 2007, www.stuff.co.nz/4316046a13.html.

6. Shell and Argent Energy plan for biodiesel at the pump, Shell press release, February 13, 2007, www.shell.com/home/content/nz-en/news_and_library/2007/argent_biofuels.html.

7. NZ plant will draw biodiesel from tallow, Energy Current, October 12, 2007, www.energycurrent.com/index.php?id=3&storyid=5969.

8. Father of bio-jet fuel launches biofuel cooperatives in Brazil to reduce poverty, Biopact, May 25, 2007, biopact.com/2007/05/father-of-bio-jetfuel-launches-biofuel.html.

9. Morgan Perkins, Brazil, Bio-Fuels, Annual 2006, Global Agricultural Information Network Report BR6008 (Washington, D.C.: U.S. Department of Agriculture, Foreign Agricultural Service, May 26, 2006), 3.

10. Alexander's Gas & Oil Connections, "Latin America Is Turning Clean and Green," www.gasandoil.com/goc/history/welcome.html. (Click on Latin America, then look for the article in the list that appears.)

11. *Renewable Energy: State of the Industry Report 9, April–June 2003* (Morrilton, Ark.: Winrock International, 2003), pp. 7, 9.

12. BR Distribuidora, CVRD ink B20 biodiesel supply deal—Brazil, Business News Americas, May 18, 2007, www.bnamericas.com/story.jsp?sector=9&noticia=393203&idioma=I.

13. Lula Inaugurates Biodiesel Plant and Highlights Potential, *Prensa Latina*, March 25, 2005, www.plenglish.com/Article.asp?ID=%7B439E1E5B-FF6C-4B0A-8A6F-2AC8BE2BD564%7D&language=EN.

14. Brazilian President Kicks Off National Biodiesel Program, Environment News Service, August 5, 2005, www.ens-newswire.com/ens/aug2005/2005-08-05-04.asp.

15. Sebastian Blanco, "Revamped Barralcool, Brazil's first ethanol-biodiesel plant, opens in Sao Paolo," AutoblogGreen, November 24, 2006, www.autobloggreen.com/tag/biodiesel%20ethanol%20plant/.

16. Amazon soy becomes greener, mongabay.com, July 25, 2006, news.mongabay.com/2006/0725-amazon.html.

17. Diesel oil distributor anticipates addition of biodiesel, Agência Brazil, May 8, 2007, www.anba.com.br/ingles/noticia.php?id=14638.

18. Kenneth Rapoza, "Shell adds two new biodiesel centers in Brazil," Market Watch, March 14, 2007, www.marketwatch.com/news/story/shell-adds-two-new-biodiesel/story.aspx?guid=%7BB294A05E-B1D0-4254-A065-5F66CF1C1DCF%7D.

19. Brazil Opens its First Commercial Jatropha Biodiesel Facility, Renewable Energy Access.com, August 10, 2007, www.renewableenergyaccess.com/rea/news/story?id=49611.

20. Brazil to anticipate mixing of 5% biodiesel into diesel, Agência Brazil, October 23, 2007, www.anba.com.br/ingles/noticia.php?id=16303.

21. Carlos Büttner, "Biodiesel: Characterización, Procesos de Elaboración, Y Normas de Control," www.mic.gov.py/combustibles/Presentacin_Bttner.ppt.

22. Paraguay launches plan to become major biofuel exporter, Biopact, March 22, 2007, biopact.com/2007/03/paraguay-launches-plan-to-become-major.html.

23. Paraguay joins the "biofuel me too" boom, Europa Press, May 25, 2007, www.autobloggreen.com/tag/paraguay%20biodiesel/.

24. April Howard, Benjamin Dangl, "The Multinational Beanfield War," *In These Times*, April 12, 2007, www.inthesetimes.com/article/3093/the_multinational_beanfield_war/.

25. "Argentina angling to join biofuels race," Reuters, October 18, 2005, today. reuters.com/news/articlenews.aspx?type=lifeAndLeisureNews&storyID=2005-10-18 T171629Z_01_MAR862077_RTRUKOC_0_US-FOOD-ARGENTINA-BIO-FUEL.xml.

26. Kelly Hearn, "Bio for All," Grist, December 14, 2006, www.grist.org/news/maindish/2006/12/14/hearn/.

27. Repsol YPF Will Invest $30 Million to Build Biodiesel Plant in Argentina, Chron.com, January 11, 2006, www.chron.com/disp/story.mpl/prn/texas/3579598.html.

28. Biodiesel, Argentour,com, www.argentour.com/en/argentina_economy/biodiesel.php.

29. Ton Steever, "U.S. gets most of Argentina's biodiesel exports," Brownfield Network, January 8, 2008, www.brownfieldnetwork.com/gestalt/go.cfm?objectid=5B912B7D-C036-CD9D-AC381CAD3463AA53.

30. Carrie Gibson, "CU Biodiesel Welcome New Production Factory," April 8, 2004, bcn.boulder.co.us/campuspress/messages/1732.html. (Web page has expired.)

31. Hilda Martinez, "COLOMBIA: Biodiesel Push Blamed for Violations of Rights," ipsnews, December 5, 2006, ipsnews.net/news.asp?idnews=35722.

32. Ibid.

33. Oilsource Holding Group Forms Joint Venture With Abundant Biofuels Corp. to Produce Jatropha-Based Biodiesel in Colombia, Grainnet, October22, 2007, www.grainnet.com/articles/Oilsource_Holding_Group_Forms_Joint_Venture_With_Abundant_Biofuels_Corp__to_Produce_Jatropha_Based_Biodiesel_in_Colombia_-49687.html.

34. Michael Kanellos, "Expanding biodiesel in South America," cnet news.com, September 17, 2007, www.news.com/8301-10784_3-9778798-7.html?tag=nefd.blgs.

35. Chile slashes taxes on biofuels to avoid social and health crisis, Biopact, May 18, 2007, biopact.com/2007/05/chile-slashes-taxes-on-biofuels-to.html.

36. Daniel Drosdoff, "Guatemala's Biofuels Success," *Latin Business Chronicle*, March 26, 2007, www.latinbusinesschronicle.com/app/article.aspx?id=1035.

37. Combustibles Ecológicos S.A. Web site, www.guatebiodiesel.com/index_eng.html.

38. Tierramérica, "Costa Rica: Promoting Biodiesel," March 26, 2004, www.tierramerica.net/2004/0105/iecobreves.shtml.

39. Biodiesel Reactor Factory Opens in Costa Rica, Renewable Energy Access.com, April 6, 2006, www.renewableenergyaccess.com/rea/news/story?id=44556.

40. 2,000-acre jatropha plantation announced in Costa Rica, Biofuels Digest, biofuelsdigest.com/blog2/2007/08/29/2000-acre-jatropha-plantation-announced-in-costa-rica/.

41. Proyecto Biomasa, "Biomass Project Nicaragua," www.ibw.com.ni/~biomasa/.

42. Bryan Sims, "Texas Biodiesel Corp. and Insta-Pro set to jumpstart Central American biodiesel project," *Biodiesel Magazine*, March 2007, biodieselmagazine.com/article.jsp?article_id=1516.

43. Mairi Beautyman, "Fish Farm Taps Biodiesel From Fish Guts," Treehugger, July 31, 2007, www.treehugger.com/files/2007/07/honduran_tilapi.php.

44. "Mexico to encourage biodiesel production," Reuters, December 19, 2007,

uk.reuters.com/article/environmentNews/idUKN1941689220071219?pageNumber=
1&virtualBrandChannel=0.

45. City of Brampton, Ontario, "Brampton Transit Powered by Biodiesel," press release, October 24, 2003, www.city.brampton.on.ca/press/03–197.tml. (Web page has expired.)

46. National Biodiesel Board, press release, www.biodiesel.org/resources/memberre leases/20040302_TOPIA_CANADAS_FIRST_BIODIESEL_PUMP.pdf.

47. Canada's Canola Industry Applauds Government's Commitment to Biodiesel, Marketwire, March 19, 2007, www.marketwire.com/mw/release.do?id=641136.

48. Canadian Government Invests $600,000 in Western Biodiesel's New 5-MMGY Biodiesel Plant in Aldersyde, AB, *Biofuels Journal*, November 7, 2007, www.biofuel sjournal.com/articles/Canadian_Government_Invests__600_000_in_Western_Biodi esel_s_New_5_MMGY_Biodiesel_Plant_in_Aldersyde__AB-50276.html.

49. 'World-scale' biofuel plant proposed, Canadian Bioenergy Corporation press release, October 16, 2007, www.canadianbioenergy.com/news.php?nid=14.

50. Chris Anderson, "Count on Canola for Your Biodiesel," *Biodiesel Magazine*, February 2008, www.biodieselmagazine.com/article.jsp?article_id=2063.

51. Scott Learn, "Does canola biodiesel help or hurt climate?" *The Oregonian*, October 21, 2007, www.oregonlive.com/news/oregonian/index.ssf?/base/news/119284713 085770.xml&coll=7&thispage=1.

Chapter 9

1. National Biodiesel Board fact sheet, "Biodiesel Is Part of the Solution to Decrease America's Dependence on Foreign Oil," www.biodiesel.org/pdf_files/fuelfactsheets/ Energy_Security.pdf.

2. Bill Kovarik, "Henry Ford, Charles Kettering, and the 'Fuel of the Future,'" *Automotive History Review*, no. 32 (Spring 1998), 7–27, reproduced on the Web at www.radford.edu/~wkovarik/papers/fuel.html.

3. European Biofuels Group, "A History of Biodiesel/Biofuels," www.eurobg .com/biodiesel_history.html (Web site now discontinued.)

4. Thomas Reed, telephone interview by the author, April 6, 2004.

5. Ibid., as well as information from the biodiesel page on Reed's Web site at www.woodgas.com/biodies.htm.

6. Leon Schumacher, telephone interview by the author, April 2, 2004.

7. Ibid.

8. Bill Ayres, telephone interview by the author, April 9, 2004.

9. Ibid.

10. "Florida Kids and Keys Benefit from Biodiesel," *Biofuels Update: Report on U.S. Department of Energy Biofuels Technology* 4, no. 3 (Fall 1996): 1, 3.

11. Gary Haer, telephone interview by the author, April 14, 2004.

12. "Biodiesel Plants Sprout Up Across the United States," *Biofuels Update: Report on U.S. Department of Energy Biofuels Technology* 5, no. 1 (Winter 1997): 1, 3, 6.

13. Joe Loveshe, e-mails to the author, April 12, 2004, and telephone interview, December 3, 2007.

14. Bob King, telephone interview by the author, May 4, 2004.

15. Pacific Biodiesel Web site, www.biodiesel.com/aboutPacBio.htm.

16. King, interview.

17. Neil Caskey, telephone interview by the author, April 6, 2004.

18. National Biodiesel Board, "Who Are We?" www.nbb.org/aboutnbb/ whoarewe/.

19. Joe Jobe, telephone interview by the author, May 11, 2004.

20. Haer, interview.

21. Nicole Cousino, telephone interview by the author, April 10, 2004.

22. Ibid.

23. Sarah Lewison, telephone interview by the author, April 11, 2004.

24. www.fieldsoffuel.com/

25. Cousino, interview.

Chapter 10

1. Jon Van Gerpen, telephone interview by the author, April 12, 2004.

2. Dennis Griffin, telephone interview by the author, April 19, 2004.

3. Gene Gebolys, telephone interviews by the author, April 14, 2004 and December 3, 2007.

4. Ibid.

5. Ibid.

6. Bob Clark, telephone interview by the author, April 19, 2004.

7. Ibid.

8. Jeff Probst, telephone interview by the author, December 12, 2007.

9. Blue Sun press release, November 7, 2006, www.gobluesun.com/news_article.php?id=42.

10. Op. cit., Probst interview.

11. Dave Nilles, "One Stop Shop," *Biodiesel Magazine*, April/May 2005, biodiesel-magazine.com/article.jsp?article_id=335.

12. Myke Feinman, "Renewable Energy Group Producing Biodiesel at Six of 15 Planned Plants," August 28, 2007, www.grainnet.com/info/articles_2col.html?companyid=3335&type=ci&ID=47705.

13. A Short History of Imperium Renewables, altdotenergy, April 22, 2007, www.altdotenergy.com/2007/04/22/a-short-history-of-imperium-renewables/.

14. John Plaza, telephone interview by the author, December 13, 2007.

15. Seattle Biodiesel Web site, About Us, www.seattlebiodiesel.com/about.html.

16. Giles Clark, "Imperium Renewables launches 100mgpy biodiesel facility," biofuel review, August 22, 2007, www.biofuelreview.com/content/view/1153/.

17. Op. cit., Plaza interview.

18. John Hurley, telephone interview by the author, April 29, 2004.

19. Ibid.

20. Maria "Mark" Alovert, "The Grease Trap: Co-ops part 1," May 21, 2003, www.mail-archive.com/sustainablelorgbiofuel@sustainablelists.org/msg49725.html.

21. Piedmont Biofuels Cooperative, Services, www.biofuels.coop/services.shtml.

22. Lyle Estill, telephone interview by the author, March 19, 2006.

23. Ibid.

24. Jerrel Branson, telephone interview by the author, November 13, 2003.

25. Joe Jobe, telephone interview by the author, May 11, 2003.

Chapter 11

1. Neil Caskey, telephone interview by the author, April 6, 2004.

2. Charles Hatcher, telephone interview by the author, April 23, 2004.

3. National Biodiesel Board, "Biodiesel Poised to Be a Significant Contributor to the U.S. Alternative Fuels Market," January 2000, www.biodiesel.org/ resources/reports database/reports/gen/20000102_gen-212.pdf.

4. H. Josef Hebert, "EPA Targets Off-Road Vehicles, Marine Vessels in New Pollution Controls," Associated Press, May 12, 2004.

5. National Biodiesel Board, press release, October 22, 2004, www.biodiesel.org/ resources/pressreleases/gen/20041022_tax_incentive_passage.pdf.

6. National Biodiesel Board press release, December 17, 2007, nbb.grassroots.com/ 08Releases/HouseSenateEnergyBillApproved/.

7. National Biodiesel Board press release, April 16, 2007, www.biodiesel.org/resources/ pressreleases/gen/20070416_renewabledieselnrfinal.pdf.

8. Bill Ayres, telephone interview by the author, April 9, 2004.

9. National Biodiesel Board, press release, July 24, 2007, www.biodiesel.org/resources/ pressreleases/far/20070724_farm%20bill%20news%20release%207–24–07.pdf

10. Ian Swanson, "EU officials face off with U.S. over biodiesel credit," TheHill.com, November 13, 2007, thehill.com/business—lobby/eu-officials-face-off-with-u.s.-over-biodiesel-credit-2007–11–13.html.

11. Ibid.

12. Gene Gebolys, interview by the author, December 3, 2007.

13. Ibid.

14. E-mail correspondence between the author and Manning Feraci, January 9, 2008.

15. E-mail correspondence between the author and Scott Hughes, January 9, 2008.

16. Gov. Pawlenty calls on state to increase use of biodiesel-blended fuel to "B20" by 2015, HometownSource.com, August 9, 2007, hometownsource.com/index. php?option=com_content&task=view&id=2111&Itemid=29.

17. Andrea Johnson, "Minnesota gets extension of waiver on B2 mandate," Tri-State Neighbor, January 18, 2006, www.tristateneighbor.com/articles/2006/01/18/tri_ state_news/top_stories/news26.txt.

18. Op. cit., Gov. Pawlenty.

19. National Biodiesel Board, press release, June 12, 2003.

20. Illinois Government News Network, press release, August 22, 2006, www.illinois. gov/PressReleases/ShowPressRelease.cfm?SubjectID=17&RecNum=5200.

21. Energy Law Alert: Washington State Adopts Minimum Renewable Fuel Content Requirements, stoel.com, April 4, 2006, www.stoel.com/showalert.aspx?Show=2316.

22. Mark Baard, "Backroom Tussling Over Biodiesel," Wired.com, February 2, 2005, www.wired.com/science/discoveries/news/2005/02/66455?currentPage=all.

23. Bob King, telephone interview by the author, May 4, 2004.

24. Ibid.

25. Rico Cruz, telephone interview by the author, April 13, 2004.

26. Jobe, interview.

27. Daryl Reece, telephone interview by the author, November 24, 2003.

Chapter 12

1. National Biodiesel Board, "Bonnie Raitt Fuels Up with Cleaner Burning Biodiesel on Tour," press release, September 17, 2002.

2. National Biodiesel Board, "Bonnie Raitt to Host Benefit for Biodiesel Education in Tennessee," press release, December 2, 2005, www.biodiesel.org/resources/press releases/gen/20051201_bonnieraittknoxville.pdf.

3. National Biodiesel Board, "Biodiesel Awards Recognize Daryl Hannah, Industry Leaders," press release, February 2, 2004.

4. Sustainable Biodiesel Alliance Web site, www.sustainablebiodieselalliance.com/.

5. Biodiesel Industries Inc., "Biodiesel Industries Opens First Biodiesel Filling Station in US," www.pipeline.to/biodiesel/. (Web page has expired.)

6. San Joaquin Valley Clean Cities Coalition, "First Public Biodiesel Fueling Station to Open in San Francisco on May 23, 2001," May 23, 2001, www.valleycleancities. org/Articles/05162001D.html. (Web page has expired.)

7. SolarAccess, "Biodiesel Power for Colorado School Buses," November 7, 2003, www.solaraccess.com/news/story?storyid=5489. (Web page has expired.)

8. School Bus Pollution Report Card 2006, Union of Concerned Scientists, May 2006, www.ucsusa.org/clean_vehicles/big_rig_cleanup/clean-school-bus-pollution.html.

9. Gail R. Frahm, "St. Johns School Buses Rolling 1 Million Miles on Biodiesel," Michigan Soybean Promotion Committee press release, April 27, 2004.

10. It should be noted that some earlier fleet tests have noted a slight decrease in fuel efficiency with the use of biodiesel, so the jury is still out on this issue.

11. National Biodiesel Board, "Back to School with Biodiesel," press release, October 6, 2003.

12. American Public Transportation Association, Bus and Trolleybus Power Sources, www.apta.com/research/stats/bus/power.cfm.

13. Clean Cities Program (U.S. Department of Energy), "Alternative Fuel Success Stories, Five Seasons Transportation and Parking," www.ccities.doe.gov/success/five_seasons.shtml.

14. Clean Cities Program (U.S. Department of Energy), "Alternative Fuel Success Stories, Bi-State Transit Agency," www.eere.energy.gov/cleancities/progs/ new_success_ddown.cgi?25.

15. "Demand for 'green shippers' driving demand for biodiesel," Farm Week, September 12, 2007, farmweek.ilfb.org/viewdocument.asp?did=10728&drvid=114&r=0. 5858423.

16. National Biodiesel Board, "Truckers Can Now Find Biodiesel Locations Using Mapping Software," press release, August 23, 2007, www.biodiesel.org/resources/pressreleases/gen/20070824_promiles_press_releasefinal.pdf.

17. National Biodiesel Board, "National Trucking Company's Biodiesel Study Shows Positive Results," press release, March 21, 2007, www.biodiesel.org/resources/press releases/fle/20070321_decker.pdf.

18. Donald A. Heck, "The 2 Million Mile Haul: Year One Summary," Iowa Central Community College, www.2millionmilehaul.com/news/2mmhaul.doc.

19. Michael Shirek, "Locomotive fleet switches to biodiesel," *Biodiesel Magazine*, August 7, 2007, www.biodieselmagazine.com/article.jsp?article_id=1760.

20. Mary Hopkin, "Richland company on track with biodiesel," *Tri-City Herald*, January 6, 2008, www.tri-cityherald.com/tch/business/story/9557419p-9469839c.html.

21. National Biodiesel Board, "U.S. Military Facilities Increasingly Fill Up With Biodiesel," press release, June 16, 2003, www.biodiesel.org/resources/pressreleases/fle/20030616_military_users.pdf.

22. Ibid.

23. National Biodiesel Board, "U.S. Navy to Produce Its Own Biodiesel," press release, October 30, 2003, www.nbb.org/resources/pressreleases/gen/ 2003 103 0_navy_to_produce_biodiesel.pdf.

24. Ibid.

25. Jessica Williams, "Embarking on Biodiesel," *Biodiesel Magazine*, June 2005, biodiesel magazine.com/article.jsp?article_id=428.

26. Andrew Foley, telephone interview by the author, January 25, 2008.

27. The Soy Daily, "Biodiesel Pump Opens to the Public at Yellowstone," October 15, 2002, www.thesoydailyclub.com/BiodieselBiobased/ yellowstone10182002.asp. (Web page has expired.)

28. Kentucky Soybean News, "Mammoth Cave Becoming an Environmental Leader," www.kysoy.org/news/mammothcave.htm.

29. National Biodiesel Board, "Biodiesel Users: Channel Islands National Park," www.biodiesel.org/resources/users/stories/channelisle.shtm.

30. "Yosemite to Produce Its Own Biodiesel," *The Biodiesel Bulletin*, December 1, 2003.

31. Ron Kotrba, "National Park Power," *Biodiesel Magazine*, October 2006, biodiesel-magazine.com/article.jsp?article_id=1184&q=&page=1.

32. City of San Francisco, "Mayor Newsom Announces San Francisco's 1,500 Diesel Vehicle Fleet Now 100% Biodiesel," press release, November 29, 2007, sfgov.org/site/mayor_index.asp?id=71707.

33. Rebekah Kebede, "Garbage trucks spruce up with biodiesel," Reuters, June 14, 2007, www.reuters.com/article/environmentNews/idUSN1449494220070614.

34. National Biodiesel Board, "Harvard Makes Smart Move to Biodiesel," press release, February 20, 2004.

35. "Purdue Switches to Biodiesel," *The Biodiesel Bulletin*, April 4, 2004.

36. SnowJournal.com, "Cranmore Resort First in East to Use Biodiesel to Power Groomers," February 25, 2004, www.snowjournal.com/article766.html. (Web page has expired.)

37. Mount Sunapee Resort, "Mount Sunapee Resort Adds Biodiesel Fuel to Its Integrated Energy Management Program, press release, October 15, 2007, www.mtsunapee.com/mtsunapeewinter/aboutsunapee/pressroom/releases/biodiesel.asp.

38. National Biodiesel Board, "Biodiesel-Fueled Dragster Sets Record," press release, September 18, 2002.

39. National Biodiesel Board, "Biodiesel Earns High Marks at 2003 Michelin Challenge Bibendum," press release, September 29, 2003.

40. Bio-diesel bike sets world speed record, motoring.co.za, September 16, 2007, www.motoring.co.za/index.php?fArticleId=4034872&fSectionId=&fSetId=381.

41. US biodiesel capacity faces massive shutdown, ICIS, October 23, 2007, www.icis.com/Articles/2007/10/23/9072437/us-biodiesel-capacity-faces-massive-shutdown.html.

42. John Plaza, telephone interview by the author, December 13, 2007.

43. Martin LaMonica, "Biodiesel maker Imperium Renewables slims down workforce," c/net news.com, January 11, 2008, www.news.com/8301–11128_3–9848719–54.html.

44. Mark Steil, "Biodiesel industry sputtering," Minnesota Public Radio, November 30, 2007, minnesota.publicradio.org/display/web/2007/11/29/biodiesel/.

45. Tom Bryan, telephone interview by the author, December 19, 2007.

46. Op. cit.

Chapter 13

1. Naveen Thukral, "Asian biodiesel plants sit idle as costs soar," Reuters, January 14, 2008, www.reuters.com/article/GlobalAgricultureandBiofuels08/idUSKLR5981820080114?pageNumber=1&virtualBrandChannel=0.

2. Ibid.

3. Jed Seybold, "Taking the Pulse of the Biodiesel Industry," Biodiesel Magazine, February 2008, www.biodieselmagazine.com/article.jsp?article_id=2060&q=&page=1.

4. Justin Moresco, "Mixed Signals for Biofuels," Red Herring, January 7, 2008, www.redherring.com/Home/23410.

5. Raffaello Garofalo, telephone interview by the author, February 24, 2004.

6. Ibid.

7. Bill Ayres, telephone interview by the author, April 9, 2004.

8. Tom Leue, telephone interview by the author, April 30, 2004.

9. Joe Jobe, telephone interview by the author, January 4, 2008.

10. Daryl Reece, telephone interview by the author, November 24, 2003.

11. Mark Zappi, Rafael Hernandez, et al., A Review of the Engineering Aspects of the Biodiesel Industry, a Report for the Mississippi Biomass Council (Jackson, Miss.: Mississippi University Consortium for the Utilization of Biomass, Mississippi State University, August 2003), p. 15.

12. Libby Tucker, "Oregon's biofuels industry split over local producers' role,"

BNET.com, September 27, 2007, findarticles.com/p/articles/mi_qn4184/is_2007 0927/ai_n21025279.

13. Leue, interview.

14. Sustainable Biodiesel summit, www.sustainable-biodiesel.org/.

15. Dieter Bockey, Situation and Development Potential for the Production of Biodiesel—An International Study (Berlin: Union zur Förderung von Oel- und Proteinpflanzen e.V., 2002).

16. Werner Körbitz, telephone interview by the author, January 12, 2004.

17. Anne Prieur-Vernat, Stéphane His, "Biofuels in Europe," December 20, 2006, ifp, France, www.ifp.fr/IFP/en/events/panorama/IFP-Panorama07_06-Biocarburants_ Europe_VA.pdf.

18. Jerrel Branson, telephone interview by the author, November 13, 2003.

19. Jon Van Gerpen, telephone interview by the author, April 12, 2004.

20. Branson, interview.

21. Dennis Griffin, telephone interview by the author, April 19, 2004.

22. Jobe, interview.

23. Jeff Probst, telephone interview by the author, December 12, 2007.

24. Leon Schumacher, telephone interview by the author, April 2, 2004.

25. "Wild Jatropha Stirs Hope of Biodiesel Bounty in India," Reuters, November 2, 2006, enviro.org.au/enews-description.asp?id=668.

26. Gene Gebolys, telephone interview by the author, December 3, 2007.

27. Scott Harper, "ODU experiment on turning sewage into algae-based biodiesel is flourishing," The Virginian-Pilot, January 19, 2008, hamptonroads.com/2008/01/ odu-experiment-turning-sewage-algaebased-biodiesel-flourishing.

28. "Algae Emerges as a Potential Fuel Source," New York Times, December 2, 2007, www.nytimes.com/2007/12/02/us/02algae.html?_r=1&ex=1197176400&en=9024c 864d1c9eba6&ei=5070&emc=eta1&oref=slogin.

29. Ibid.

30. Latest Development on Cultivating Algae for Biodiesel Production, Mass Media Distribution Newswire, October 25, 2007, www.mmdnewswire.com/ltest-develop ment-2465.html.

31. Karen Woodmansee, "Oil from Algae," Nevada Appeal, May 13, 2007, www.nevadaappeal.com/article/20070513/BUSINESS/105130100.

32. Valcent Products, Inc. press release, December 12, 2007, www.valcent.net/s/News Releases.asp?ReportID=277235&_Type=News-Releases&_Title=Initial-Data-From- the-Vertigro-Field-Test-Bed-Plant-Reports-Average-Product. . . .

33. Jobe interview.

34. Rico Cruz, telephone interview by the author, April 13, 2004.

35. John Schmitz, "Biodiesel Has Oregon Veggie Seed Growers Edgy," Capital Press, February 18, 2005, www.harvestcleanenergy.org/enews/enews_0305/enews_0305_ Seed_Growers.htm.

36. Charles Shaw, "Special Report: Fueling Controversy," GNN, April 27, 2006, gnn.tv/articles/2247/Special_Report_Fueling_Controversy.

37. Vegetable Oils: Situation and Outlook, Agriculture and Agri-Food Canada, Market Analysis Division, October 26, 2007, www.agr.gc.ca/mad-dam/print_e.php?s1=pubs &s2=bi&s3=php&page=bulletin_20_07_2007–05–04&PHPSESSID=0df934f28686 6b31c199351aba8193ee.

38. Ibid.

39. Keith Bradsher, "An Oil Quandary: Costly Fuel Means Costly Calories," *New York Times*, January 19, 2008, www.nytimes.com/2008/01/19/business/worldbusiness/19palmoil.html?ei=5070&en= 11d076a2c5b5841e&ex=1201410000&emc=etal&pagewanted=all.

40. Tun-Hsiang (Edward) Yu, telephone interview by the author, January 9, 2008.

41. Sybille de La Hamaide, "Biofuel impact on farm prices overplayed," Reuters, January 14, 2008, www.reuters.com/article/environmentNews/idUSL1419899820080114 ?feedType=RSS&feedName=environmentNews&rpc=22&sp=true.

42. Op. cit. Yu.

43. Gillian Murdoch, "Snacks light fuse of Indonesian 'forest bomb,'" Reuters, November 8, 2007, www.alertnet.org/thenews/newsdesk/SP65072.htm.

44. Michael Sullivan, Martin Kaste and Emily Harris, "Biodiesel Demand Grows Across Continents," NPR, September 2, 2007, www.npr.org/templates/story/story.php?storyId=14063007.

45. Bites of the roundtable, Friends of the Earth, www.foe.co.uk/campaigns/corporates/news/roundtable.html.

46. John Plaza, telephone interview by the author, December 13, 2007.

47. Jobe interview.

48. James Kanter, "Europe May Ban Imports of Some Biofuel Crops," *New York Times*, January 15, 2008, www.nytimes.com/2008/01/15/business/worldbusiness/15biofuel. html?_r=1&ex=1358053200&en=310b230962f0adfe&ei=5088&partner=rssnyt& emc=rss&oref=slogin.

49. Op. cit. Jobe.

50. Ibid.

51. Schumacher, interview.

52. Jobe, interview.

53. Garofalo, interview.

54. Probst, interview.

55. Gebolys, interview.

56. Bob King, telephone interview by the author, May 4, 2004.

57. Gary Haer, telephone interview by the author, April 14, 2004.

58. Darryl Melrose, telephone interview by the author, March 23, 2004.

59. Gebolys, interview.

60. Reece, interview.

INDEX

A

academic institutions, 278–279
ADM (Archer Daniels Midland), 92, 239
Africa, 137–146
 see also specific countries
Africa Eco Foundation, 143
Ag Environmental Products (AEP), 202–203
Ag Processing Inc. (AGP), 202–203
Agency for Environment and Energy
  Management (ADEME), 84
agribusiness concerns, 306–308
agricultural equipment
 potential for biodiesel use, 63–64
 tractors, 23–25
Air Force, 273–276
algae
 overview, 45–47
 research programs, 303–306
 sewage treatment plants, 171
alkyl esters, 34
Allgemeine Gesellschaft für Dieselmotoren
  (General Society for Diesel Engines), 11
Allied Biodiesel Industries, 323
Alovert, Maria "Mark," 37
alternative fuels, biodiesel as, 315–317
American Jobs Creation Act (2004), 246
American Soybean Association (ASA), 207
Angola, 145
animal fats, 44–45, 171
Aquatic Species Program: Biodiesel from Algae
  (NREL), 45–47
Archer Daniels Midland (ADM), 92, 239
Argentina, 179–181
Army, 273–276
Asia, 146–168
ASTM D 975 standard, 58
ASTM standards, 242
Atwood, Matt, 307
Australia, 169–170, 321
Austria, 111–114, 321
 main producers, 112–113
 McDonald's, 114
 research programs, 111–112
 standards, 114
 tax incentives, 113
Austrian Biofuels Institute, 26, 321
Austrian Federal Institute of Agricultural

Engineering, 19–21
Austurias, Ricardo, 185
automotive industry
 potential for biodiesel use, 59–60
 use of diesel engines, 15–16
aviation industry, potential for biodiesel use, 66
Ayres, Bill, 202–203, 293

B

B100 blend, issues with, 57
B100 community, 210, 258
B20 blend, 53
bacterial slime, 57
Bai, Lakshmi, 147
Barnickel, Christoph, 6
batch process, 36–37, 215
beef tallow. see animal fats
Belgium, 120–121
Berkeley Biodiesel Collective, 324
Berkeley Ecology Center, 278
Biluck, Jr., Joe, 62
biodegradability, 51–52
biodiesel
 as an alternative fuel, 315–317
 biodegradability, 51–52
 as a bridge technology, 317–319
 comparison with petrodiesel, 49–53
 cost, xv
 definition, xxvi–xxvii
 early adoption of, 100–101
 flash point, 52–53
 heating with, 67–74
 life cycle studies, 51
 lubricity, 67
 production potential, 47–49
 solvent properties, 67
 testing, 23–25
 toxicity, 51–52
 transportation and storage, 52–53
 U.S. demand, 215
Biodiesel America, 326
Biodiesel Basics and Beyond (Kemp), 37
Biodiesel Conference & Expo, 293–294
Biodiesel Discussion Forum, 212, 327
Biodiesel Homebrew Guide 2004 (Alovert), 37
biodiesel hubs, 296

biodiesel industry
    challenges, 87–90, 283–285, *284*
    crisis in Germany, 96–99
    impact of ethanol industry, 259–260
    key issues, 292–296
    petroleum industry involvement, 238–239
    state of, xvi
biodiesel process technology, 36–37, 214–218
biodiesel pumps, 265–266
Biodiesel*Now*.com, 212, 327
BioFuel Africa, 140
biofuels
    overview, xxvii
    early adoptions, 196–197
    research programs, 19–21
    second-generation biofuels, 85, 89, 97–98,
        156
Biofuels Association of Australia, 321
Biofuels Corporation, 110–111
Biofuels Research Cooperative (Straight
    Vegetable Oil), 324
bioheat, 67–74
biomass, xxvi–xxvii
Bio-Willie, 264
Blagojevich, Rod, 255
blends, 53, 57
BLT, research programs, 19–21
Blue Sun Biodiesel, 226–228
boats, potential for biodiesel use, 64–65
Bolivia, 184–185
Boonekamp, Loek, 310
Bosch, Robert, 55
Boulder Biodiesel Cooperative, 324
BQ-9000, 295–296
Branson, Jerrel, 238–239, 300, 301
Brazil, 173–177
Brevard BioDiesel, 324
bridge technology, biodiesel as, 317–319
Bryan, Tom, 285
Buehler, Kurt, 275
Bulgaria, 127–128
bulk-buy co-ops, 234
Bullard, Kent, 277
Bundesanstalt für Landtechnik, 19–21
Burke, Tom, 266
Burlington Biodiesel Coop, 324
Busch, Adolphus, 10
buses
    Brazil, 174
    diesel engines in, 61
    India, 148–149
    school buses, 62, 220, 266–269

    using biodiesel, 61–62
    see also public transportation; school buses
Bush administration, climate change study, xxiii

C

California
    Berkeley Ecology Center, 278
    Channel Islands National Park, 277
    cooperatives, 324–325
    Imperial Western Products, 224–225
    San Francisco, 278
    Yosemite National Park, 277
camelina oil, 40–41
Campbell, Colin, xxii
Campbell, Tim, 273
Canada, 189–192, 321–322
Canadian Renewable Fuels Association,
    321–322
canola oil. *see* rapeseed
CAP (Common Agricultural Policy), 78–81
Carels, George, 3
Caribbean, 188
Carlstein, Ricardo, 180, 186
Carter, Jimmy, 265
Caskey, Neil, 207–208, 241
catalysts, transesterification process, 34–36
caustic soda, 35
CCC (Commodity Credit Corporation), 249
celebrity endorsements, 263–264
Cerio, Bob, 70–71
Channel Islands National Park, 277
Chase, Stephan, 73–74
Chavanne, G., 16
Cheney, Dick, 245
Chile, 184
China, 152–154, 309–310
Clark, Bob, 224–225
Clean School Bus USA, 267
climate change, effects of, xxiii
Clinton, Bill, 245
clouding, 38, 56–57
CNG (compressed natural gas), as an oil alter-
    native, xxv
coconut oil, 39, 160–161, 167
coking test, 30
cold weather issues, 56–57
Collaborative Biodiesel Tutorial, 326
Colombia, 181–183
Colorado
    Blue Sun Biodiesel, 226–228
    cooperatives, 324–325

ski industry, 279–280
transesterification experiments, 198–199
Columbus Foods, 204–205
commercial trucking, 62–63, 270–272
Commodity Credit Corporation (CCC), 249
Common Agricultural Policy (CAP), 78–81
commuter trains. *see* railroads
compressed natural gas (CNG), as an oil alternative, xxv
Connecticut Waste Vegetable Oil List, 324
continuous-flow process, 36–37, 215
cooperatives, 234–238, 324–325
co-processed renewable diesel, 247–249
corn oil, 43–44
corn production, 309
corporate agribusiness, 306–308
Corvallis Biodiesel Cooperative, 324
Costa Rica, 186
Council for Scientific and Industrial Research (CSIR), 26–27
Cousino, Nicole, 210–211, 213
Cruz, Rico, 259, 306
CSIR (Council for Scientific and Industrial Research), 26–27
CytoSol, 67
Czech Republic, 115–116

D

D 6751 standard, 58
D1 Oils, 110, 140, 145
da Silva, Luiz Inacio Lula, 175
de Lavigne, Chris, 291
Decker Truck Line, Inc., 271–272
Defferrari, Edmundo, 180
Deffeyes, Kenneth, xxii
deforestation, 88, 176, 311–313
Denmark, 122–123
Die Moto, 282, *282*
Diesel, Rudolph
  disappearance of, 3–5
  early years, 5–6
  heat engine, 7–8
  Munich Power and Machinery Exhibition, 10–11
  Paris Exposition, 12–13
  vision of, 13–14
diesel engines
  alternate biofuels research, 19–21
  applications, 59–66
  basic features, 54–56
  prototype, 8–10, *10*

diesel fuel. *see* petrodiesel
diester, 101
Diester Industrie, 102
DIN E 51.606 standard, 95
direct injection (DI), 56
Distribution Drive, 326
Dog River Alternative Fuels, 233–234
Dominican Republic, 188
DonauWind, 112–113
Douglas, George, 305
drag racing, 281
du Plessis, Lourens, 27–28
Dunow, Tobias, 97

E

electrical generators, potential for biodiesel use, 66
Energy Action Plan, 84–85
Energy Bill (2002), 245–246
Energy Independence and Security Act (2007), 247
Energy Policy Act (1992), 242–243
energy-efficiency ratios, 51
engine warranties, 26, 58–59
environmental concerns, 88, 176, 311–313
EPA approval, 243–244
EPA registration, 257
EPAct (1992), 242–243
esters, 33–34
Estill, Lyle, 237–238
ethanol
  Brazil, 173
  criticisms, 88
  Ford's support of, 196–197
  gasohol, 18
  impact on biodiesel industry, 259–260
  production in France, 100
  transesterification process, 34
ethyl esters, 34
EU 14214 standard, 85
Europe, 322
  challenges of biodiesel industry, 87–90
  Common Agricultural Policy (CAP), 78–81
  Directive for the Promotion of Biofuels, 85
  Directive on Energy Taxation, 86
  Directive on Fuel Quality, 85
  Energy Action Plan, 84–85
  legislation, 83–87
  production plants, 77–78, 79
  regulations, 83–87
  role of UFO, 80

European Biodiesel Board (EBB), 77, 83, 322
Evanoff, Jim, 276
events, 280–282
Evergreen Biofuels, 140
executive orders, 245
exhaust, odor of, 31

F

false flax, 40–41
Farm Bill (2007), 250, 252
farm equipment
    potential for biodiesel use, 63–64
    tractors, 23–24
fast food companies
    Kentucky Fried Chicken, 154
    McDonald's, 103–104, 109, 114, 174, 199
Fat and Protein Research Foundation (FPRF),
    209
Fat of the Land, 210–211, 326
feedstock flexibility, 224
feedstock neutrality, 221, 252
feedstocks
    algae, 45–47
    animal fats, 44–45, 171
    camelina oil, 40–41
    coconut oil, 39
    corn oil, 43–44
    costs, 290–292, 300–303
    feedstock flexibility, 224
    feedstock neutrality, 221
    Germany, 96–97
    hemp oil, 43
    Japan, 155
    jatropha, 40
    mustard oil, 42, 49
    oilseed-bearing trees (OSBT), 146
    outsourcing, 88
    palm oil, 38
    peanut oil, 41
    rapeseed oil. *see* rapeseed
    rising prices, 283
    safflower oil, 42
    soybean oil, 42–43
    sunflower oil, 41
    supply decrease, 310
    used vegetable oil, 44
Feraci, Manning, 253
Finland, 132
fish oil. *see* animal fats
Five Seasons Transportation & Parking
    (FST&P), 269

flash point, 52–53
fleets, potential for biodiesel use, 60–61
Florida, 324
Foley, Andrew, 275–276
food-versus-fuel issue, 308–310
Ford, Henry, 196–197
fossil fuels. *see* petroleum supply
France, 100–104
free fatty acids, 35–36
French National Railways (SNCF), 103
*From the Fryer to the Fuel Tank: The Complete
    Guide to Using Vegetable Oil as an
    Alternative Fuel* (Tickell), 37
fuel economy standards, 300
fuel-injection technology, 55
Fuji Keizai USA, 291
Funk, Mike, 279

G

Gagliardo, Michael, 205
Garofalo, Raffaello
    EBB, 83
    EU directives, 86–87
    future of biodiesel industry, 314
    key issues for biodiesel industry, 292–296
    market for biodiesel, 89
    straight vegetable oil, 99
    vegetable oil prices, 81
gasohol, 18
Gebolys, Gene
    biodiesel as a bridge technology, 317–319
    feedstock costs, 303
    feedstock neutrality, 253
    future of biodiesel industry, 315–316
    tax incentives, 251
    World Energy Alternatives LLC, 221–224
Gehrke, Russel, 281
gel point, 56
General Society for Diesel Engines (Allgemeine
    Gesellschaft für Dieselmotoren), 11
generators, potential for biodiesel use, 66
genetically engineered crops, 306–308
Germany, 93–99, 322
Ghana, 143–144
Glatz, Joel, 72–73
global economy
    effects of oil prices, xxii–xxiii
    oil prices, 17–19
global warming, xxiii
glow plugs, 55
glycerin, 34

GMO issue, 306–308
GoBiodiesel Cooperative, 325
gold-of-pleasure, 40–41
government policies
    France, 101–102
    India, 149–150
    Italy, 106–107
    Japan, 155
    Poland, 115
    Spain, 117
    United Kingdom, 108
grain alcohol. *see* ethanol
Grassley, Chuck, 246
Grazer Verkehrsbetriebe (GVB), 61–62
Grease Works! Biodiesel Cooperative, 325
Greece, 128–129
Green Energy Parks Program, 277
Green Highway Tour, 263
Green*ergy*, 109
Gregoire, Christine, 256
Griffin, Douglas, 219–221, 301
Griffin Industries, 219–221
Guatemala, 184–186

H
Haer, Gary, 209, 316
Hageman, Roger, 269–270
Haiti, 188
Hamilton, Tony, 271
Hannah, Daryl, 264
Harris Jr., David, 279
Harvard University, 278–279
Hatcher, Charles, 242, 252
Hatcher, Patrick, 304
Hearst, William Randolph, 197
heat engine, 7–8
heating with biodiesel, 67–74
heavy industrial equipment, potential for
    biodiesel use, 62–63
Hegland, Ed, 255, 285
hemp oil, 43
heterogeneous catalyst, 35
Hettler, Wayne, 268
Higgins, Jenna, 70
highway taxes, 210, 235, 237
Holmgren, Jennifer, 304
homebrew, 209–213
homogeneous catalysts, 35
Honduras, 188
honge oil, 147
Horn, Mathias, 167

Hughes, Scott, 254
Hungary, 131–132
Hurley, John, 233
hybrid rapeseed, 153
hydrogen-powered fuel cells, xxv

I
Idaho, 28–32
ignition temperature, 52–53
Illinois, 255–256
Imperial Western Products, 224–225
Imperium Renewables, 230–232, *231*
    see also Plaza, John
India, 146–150, 309–310
Indian Railways, 147–149
Indiana, 279
Indonesia, 165–167, 250–251
industrial hemp, 43
injection pump technology, 55
injector coking test, 30
internal combustion engine, 196
Iowa
    Decker Truck Line, Inc., 271–272
    Five Seasons Transportation & Parking
        (FST&P), 269
    Renewable Energy Group, Inc. (REG),
        228–230
Ireland, 130
Isman, Barb, 191
Italy, 106–107

J
Jamaica, 188
Japan, 154–156, 322–323
jatropha
    overview, 40
    Brazil, 176–177
    Caribbean, 188
    Chile, 184
    China, 153
    Colombia, 182–183
    Costa Rica, 186
    D1 Oils, 110
    and future of biodiesel industry, 317–318
    Guatemala, 185
    India, 148–149
    Malaysia, 164, 166
    Mali, 141
    Middle East, 145
    Mozambique, 142

Nicaragua, 187
Philippines, 161–162
Rwanda, 144
South Africa, 138–139
test planting, 302–303
Jobe, Joe
    algae, 306
    B100 community, 258
    biodiesel performance, 281
    ethanol industry, 260
    farmers, 208–209, 301
    future of biodiesel industry, 313–314
    palm oil, 312
    petroleum industry, 239
    quality, 295
    tax incentives, 251
Journey to Forever, 326
Junek, Hans, 20–21

K

Kemp, Bill, 37
Kentucky, 219–221, 276
Kentucky Fried Chicken, 154
Kenya, 144
King, Bob, 205–206, 258–259, 316
Körbitz, Werner, 26, 33, 299
Kyoto Protocol, 84, 105, 139, xxiv

L

Latvia, 127
Laurance, William, 313
legislation
    American Jobs Creation Act (2004), 246
    Energy Bill (2002), 245–246
    Energy Independence and Security Act
        (2007), 247
    Energy Policy Act (1992), 242–243
    executive orders, 245
    Farm Bill (2007), 250, 252
    Responsible Renewable Energy Tax Credit
        Act, 248–249
    state initiatives, 253–257, 297
Leue, Tom, 295, 298
Lincoln, Blanche, 246
Lithuania, 126
Local B100 Forum, 327
locomotive. see railroads
lubricity, 67
Luckmann, Alfred, 3
lye, 35

M

Mack Truck Company, 15
Madagascar, 145
maize, 43–44
Malaysia, 162–165, 163, 250–251
Mali, 141–142
Mammoth Cave National Park, 276–277
MAN AG
    "salad oil engine" design, 54
    use of diesel engines, 15
mapping software, 271
marine industry
    NOPEC Corporation, 203–204
    potential for biodiesel use, 64–65
Marines, 273–276
market pull approach, 226
mass transit. see public transportation
Massachusetts, 278–279
McDiesel, 174, 199
McDonald's, 103–104, 109, 114, 174, 199
Meier, Edward D., 10
Mellon, Andrew, 197
Melrose, Darryl, 138, 317
Merkel, Angela, 97–99
Mersereau, Jim, 280
methanol, 34, xxvii
methyl esters, 34
Metz, Bob, 246
Mexico, 188
Middle East, 145–146
middle-distillate fuels, U.S. consumption of, 47
Midwest Biofuels, 202
Mielke, Thomas, 303
Minnaar, Johan, 139
Minnesota, 254–255
Missouri, 200–201, 296
Mittelbach, Martin, 20–26
Morales, Evo, 184–185
Morey, Samuel, 196
moringa, 143
motorcycle racing, 280–282, 282
Mozambique, 142
Munich Power and Machinery Exhibition,
    10–11
mustard oil, 42, 49
Myers, Jim, 307

N

Nanohana Project Network, 322–323
National Biodiesel Accreditation Commission
    (NBAC), 58

National Biodiesel Board (NBB)
    biodiesel production estimates, 48
    contact information, 323
    formation of, 208–209
    membership dues, 257
    NBAC, 58
    public biodiesel pumps guide, 266
    see also Jobe, Joe
national parks, 31, 276–277
National Renderers Association (NRA), 209
National Renewable Energy Laboratory
        (NREL), 209, 305, 323
    Aquatic Species Program: Biodiesel from
        Algae, 45–47
Navy, 273–276
NBAC, 58
NBB. see National Biodiesel Board (NBB)
neem seeds, 143
Nelson, Willie, 263–264
The Netherlands, 123–124
New Hampshire, ski industry, 280
New Zealand, 171–172
Newsom, Gavin, 278
Nicaragua, 187
Nigeria, 144–145
nitrous oxide (NOx) emissions, 225, 244
Nixon, Richard M., 18, 198
NOPEC Corporation, 203–204
North Carolina
    Burlington Biodiesel Coop, 324
    Piedmont Biofuels, 235–238, 236, 298, 325
Norway, 132

O

OBI, 26
OceanAir Environmental, 222–223
off-spec biodiesel, 294–296
oil
    demand for, xxi
    prices, 17–19, 289–292, xxii–xxiii
    rate of production, xxi
    reserve figures, xxii
oil embargo, 17–19
oil palm. see palm oil
oil transformation. see transesterification
oilseed-bearing trees (OSBT), 146
oleochemicals, 22
Oliva, Patrick, 281
Olympia Biodiesel, 325
OMV AG, funding for biodiesel testing, 24
ON C 1190 standard, 25, 114

ON C 1191 standard, 114
OPEC
    countries, xxii
    oil crisis, 197–198
    oil embargo, 17–19
    oil reserves, xxiii
Oregon, 297
    cooperatives, 324–325
Organization of the Petroleum Exporting
        Countries (OPEC). see OPEC
Osterreichische Normungsinstitut (ON), 25
Osterreichisches Biotreibstoff Institut, 26

P

Pacific Biodiesel, 205–206, 258–259
    see also King, Bob
Pacific Islands, 167–168
Pakistan, 146
palm kernel oil, 38
palm oil
    overview, 38
    Colombia, 182
    environmental concerns, 311–313
    Malaysia, 162–165, 163
    problem of, 312
    world production, 163
Panama, 187
Papua New Guinea, 167
Paraguay, 177–179
Parente, Expedito, 173
Paris Exposition, 12–13
passenger trains. see railroads
Passmore, Gary, 274
Pawlenty, Tim, 255
peanut oil, 41
Peru, 183–184
Peterson, Charles, 28–30, 198
petrodiesel
    comparison with biodiesel, 49–53
    used in diesel engines, 16
petroleum supply, xix
Petroplus, 109
Philippines, 160–162
Pickering, Doug, 202
Piedmont Biofuels, 235–238, 236, 298, 325
piñon tree, 185
Plaza, John, 230–232, 283, 311–312
Poland, 114–115
Portugal, 122
potassium hydroxide (KOH), 35
poultry fat. see animal fats

pour point, 56
Probst, Jeff, 226–228, 302
ProMiles XF software, 271
public biodiesel pumps, 265–266
public transportation, 61–62, 94, 113–114,
     269–270
     see also mass transit
Purdue University, 279

Q

quality, as a key issue for biodiesel industry,
     294–296

R

racing events, 280–282, 282
railroads
     biodiesel use, 272–273
     Diesel's research on, 15
     French National Railways (SNCF), 103
     Indian Railways, 147–149
     New Zealand, 172
     potential for biodiesel use, 65
Raitt, Bonnie, 263
rapeseed
     overview, 40
     Germany, 96–97, 99
     hybrid, 153
Ratzlaff, Karin, 97
reactors, 235–236
recycled cooking oils. see used vegetable oil
Reece, Daryl, 31, 205–206, 260, 296, 318
Reed, Thomas, 198–199
relocalization movement, 238
rendered animal fats. see animal fats
Renewable Diesel, 247–249
Renewable Energy Group, Inc. (REG), 228–230
Renewable Energy Road Map, 84–85
Renewable Fuels Standard, 247
Republic of Congo, 145
research programs
     algae, 303–306
     Austria, 111–112
     biofuels, 19–21
     BLT, 19–21
     China, 152
     Idaho, 28–32
     South Africa, 26–28
Responsible Renewable Energy Tax Credit Act,
     248–249
Ricci, James, 221–224

road taxes, 210, 235, 237
Roaring Fork Biodiesel Coop, 325
Rockerfeller family, 197
Roman, Kaia, 211–212
Romania, 130–131
rotation crops, 302
Royal Dutch/Shell, accounting scandal, xxii
Russia, 133–134
Rwanda, 144

S

safflower oil, 42
Samples, Dave, 273
San Francisco, 278
Sapp, Claude, 305
Sasol, 140
Saudi Arabia, 145
school buses, 62, 220, 266–269
Schrum, Peter, 98
Schumacher, Leon, 200–201, 302, 314
Scotland, 111
Seattle Biodiesel, LLC. see Imperium
     Renewables
second-generation biofuels, 85, 89, 97–98, 156
Senegal, 145
Serbia, 131
sewage treatment plants, and algae, 171
Simon, William E., 198
Simplot Transportation, over-the-road test,
     30–31
Singapore, 159–160
Singh, Sarju, 303
ski industry, 279–280
slime, 57
Slovakia, 118–120
Slovenia, 129–130
sludge, 57
small producers, political issues, 257–259
soapmaking process, 197
SoCo Biodiesel Co-op, 325
sodium hydroxide (NaOH), 35
solid catalyst, 35
solvent properties, 67
South Africa, 26–28, 137–140
South Korea, 157–158
soybean oil, 42–43, 178, 207–209
Spain, 116–118
splash and dash, 89–90, 250–252
Srinivasa, Udupi, 146–147
Standard Oil Company, 197
standards

ASTM D 975, 58
ASTM Standard, 242
ON C 1190, 25, 114
ON C 1191, 114
Czech Republic, 116
D 6751, 58
DIN E 51.606, 95
EU 14214, 85
fuel economy, 300
global initiative for, 89
Japan, 156
Renewable Fuels Standard, 247
Slovakia, 119
South Africa, 139
state initiatives, 253–257
STN 656530 standard, 119
straight vegetable oil (SVO)
    Biofuels Research Cooperative, 324
    vehicle conversions, 261–262
    see also vegetable oil
Stroburg, Jeff, 284
subsides, 249–250
sunflower oil, 41
Sustainable Biodiesel Alliance (SBA), 264
Sustainable Biodiesel Summit (SBS), 258, 298
Sweden, 124–126

T

Tacoma Biodiesel, 325
Taiwan, 158–159
tallow. see animal fats
tax incentives
    Austria, 113
    Belgium, 121
    Czech Republic, 116
    Energy Bill (2002), 246–247
    France, 102
    Germany, 93, 96–99
    Greece, 129
    Ireland, 130
    Italy, 107
    The Netherlands, 123–124
    Poland, 115
    Spain, 117
    splash and dash, 89–90
    Sweden, 125–126
    trends, 283
    United Kingdom, 108
taxis, 93–94
Teall, Russell, 266, 274
Tevis, Jack, 74

Thailand, 150–151
thermal depolymerization (TDP), 248
Tickell, Joshua, 37, 211–212
Tobias, Martin, 284
Toll Rail, 172
toxicity, 51–52
tractors, biodiesel testing in, 23–25
trains. see railroads
transesterification
    by-products, 82–83
    process of, 21–22, 33–37
    soapmaking process, 197
transportation of biodiesel, 52–53
triglycerides. see vegetable oil
Truck in the Park Project, 31
trucking industry, 62–63, 270–272
Turkey, 134–135
Twin Rivers Technology, 203

U

Union for the Promotion of Oil and Protein
    Plants (Union zur Förderung von Oel-
    und Proteinpflanzen, or UFOP), 322
United Kingdom
    Allied Biodiesel Industries, 323
    government policies, 108
    main producers, 109–111
    tax incentives, 108
United Soybean Board (USB), 207
United States
    biodiesel process technology, 214–218
    cooperatives, 234–238
    demand for biodiesel, 215
    energy consumption, xxiv
    and Kyoto Protocol, xxiv
    large producers, 218–232
    production costs, 217
    production plants, 218
    small producers, 232–238, 257–259
    see also National Biodiesel Board (NBB)
U.S. Bioenergy Program, 249
U.S. military, 273–276
USDA, 249
used vegetable oil, 25–26, 44, 80, 217
    see also vegetable oil

V

Van de Vaarst, John, 69
Van Gerpen, Jon, 215–216, 300–301
vegetable oil

composition of, 34
early experiments, 16–17
prices, 81
straight vegetable oil (SVO), 261–262
use in Germany, 99
see also used vegetable oil
Veggie Avenger, 326
Veggie Van, 211–212
Verkuijl, Hugo, 141
Vermont, 233–234, 280
Verry, Tom, 271
Vogel & Noot, 112
von Linde, Carl, 7–8

W
Washington, 256, 325
waste cooking oils. *see* used vegetable oil
waste exchange, 46
water-wash procedure, 37
West Central Cooperative, 204

West Indies, 188
Williams, Frank, 60–61
wood alcohol. *see* methanol
Wörgetter, Manfred, 19–21, 22, 111–112
World Energy Alternatives LLC, 221–224
World Wildlife Fund (WWF), support of
    biodiesel, 88–89
Worldwatch Institute, xxv

Y
"yellow grease". *see* used vegetable oil
Yellowstone National Park, 31, 276–277
Yosemite National Park, 277
Yoshida, Soichiro "Sol," 154
Yu, Tun-Hsiang (Edward), 310

Z
Zimbabwe, 145

"This logo identifies paper that meets the standards of the Forest Stewardship Council. FSC is widely regarded as the best practice in forest management, ensuring the highest protections for forests and indigenous peoples."

The Chelsea Green Publishing Company is committed to preserving ancient forests and natural resources. We elected to print this title on 30% postconsumer recycled paper, processed chlorine-free. As a result, for this printing, we have saved:

24 Trees (40' tall and 6-8" diameter)
8,685 Gallons of Wastewater
17 million BTU's Total Energy
1,115 Pounds of Solid Waste
2,092 Pounds of Greenhouse Gases

Chelsea Green Publishing made this paper choice because we and our printer, Thomson-Shore, Inc., are members of the Green Press Initiative, a nonprofit program dedicated to supporting authors, publishers, and suppliers in their efforts to reduce their use of fiber obtained from endangered forests. For more information, visit: www.greenpressinitiative.org.

Environmental impact estimates were made using the Environmental Defense Paper Calculator.
For more information visit: www.papercalculator.org.

the politics and practice of sustainable living

CHELSEA GREEN PUBLISHING

Chelsea Green Publishing sees books as tools for effecting cultural change, and seeks to empower citizens to participate in reclaiming our global commons and become its impassioned stewards. If you enjoyed *Biodiesel* please consider these other great books related to energy and the environment.

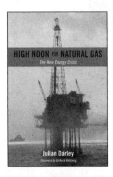

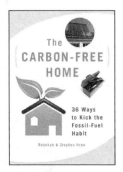

High Noon for Natural Gas	*The Citizen-Powered Energy Handbook*	*The Carbon-Free Home*
The New Energy Crisis	*Community Solutions to a Global Crisis*	*36 Remodeling Projects to Help*
JULIAN DARLEY	GREG PAHL	*Kick the Fossil-Fuel Habit*
ISBN 978-1-931498-53-1	ISBN 978-1-933392-12-7	STEPHEN AND
Paperback • $18	Paperback • $21.95	REBEKAH HREN
		ISBN 978-1-933392-62-2
		Paperback • $35

Green Guides
Energy; Water; Composting; Reduce, Reuse, Recycle; Biking to Work; Greening Your Office
Each, $7.95 • Set of six, $44.95

CHELSEA
GREEN
PUBLISHING

the politics and practice of sustainable living

For more information or to request a catalog,
visit **www.chelseagreen.com** or
call toll-free **(800) 639-4099**.